			3A (13)	4A (14)	5A (15)	6A (16)	7A (17)	8A (18)
							Hydrogen 1 **H** 1.0079	Helium 2 **He** 4.0026
			Boron 5 **B** 10.811	Carbon 6 **C** 12.011	Nitrogen 7 **N** 14.0067	Oxygen 8 **O** 15.9994	Fluorine 9 **F** 18.9984	Neon 10 **Ne** 20.1797
	1B (11)	2B (12)	Aluminum 13 **Al** 26.9815	Silicon 14 **Si** 28.0855	Phosphorus 15 **P** 30.9738	Sulfur 16 **S** 32.066	Chlorine 17 **Cl** 35.4527	Argon 18 **Ar** 39.948
Nickel 28 **Ni** 58.693	Copper 29 **Cu** 63.546	Zinc 30 **Zn** 65.39	Gallium 31 **Ga** 69.723	Germanium 32 **Ge** 72.61	Arsenic 33 **As** 74.9216	Selenium 34 **Se** 78.96	Bromine 35 **Br** 79.904	Krypton 36 **Kr** 83.80
Palladium 46 **Pd** 106.42	Silver 47 **Ag** 107.8682	Cadmium 48 **Cd** 112.411	Indium 49 **In** 114.82	Tin 50 **Sn** 118.710	Antimony 51 **Sb** 121.757	Tellurium 52 **Te** 127.60	Iodine 53 **I** 126.9045	Xenon 54 **Xe** 131.29
Platinum 78 **Pt** 195.08	Gold 79 **Au** 196.9665	Mercury 80 **Hg** 200.59	Thallium 81 **Tl** 204.3833	Lead 82 **Pb** 207.2	Bismuth 83 **Bi** 208.9804	Polonium 84 **Po** (209)	Astatine 85 **At** (210)	Radon 86 **Rn** (222)
110 Discovered Nov. 1994	111 Discovered Dec. 1994							

(10)

Europium 63 **Eu** 151.965	Gadolinium 64 **Gd** 157.25	Terbium 65 **Tb** 158.9253	Dysprosium 66 **Dy** 162.50	Holmium 67 **Ho** 164.9303	Erbium 68 **Er** 167.26	Thulium 69 **Tm** 168.9342	Ytterbium 70 **Yb** 173.04	Lutetium 71 **Lu** 174.967

Americium 95 **Am** (243)	Curium 96 **Cm** (247)	Berkelium 97 **Bk** (247)	Californium 98 **Cf** (251)	Einsteinium 99 **Es** (252)	Fermium 100 **Fm** (257)	Mendelevium 101 **Md** (258)	Nobelium 102 **No** (259)	Lawrencium 103 **Lr** (260)

Student Solutions Manual to accompany

Chemistry
& Chemical Reactivity
THIRD EDITION
Kotz & Treichel

Alton J. Banks
North Carolina State University

Saunders Golden Sunburst Series
Saunders College Publishing
Harcourt Brace College Publishers

Fort Worth   Philadelphia   San Diego   New York   Orlando   Austin
San Antonio   Toronto   Montreal   London   Sydney   Tokyo

Banks; Student Solutions Manul to accompany Chemistry and
Chemical Reactivity, 3E. Kotz and Treichel.

ISBN 0-03-001309-7

67 021 76543

To the student:

The skills involved in solving chemistry problems are acquired only by discovering the paths which connect the available data to the desired piece(s) of information. These paths are found by repeated efforts. Therefore, this **Solutions Manual** will benefit you most if used as a reference--after you have attempted to solve a problem.

The selected Study Questions have been chosen by the author of your text to allow you to discover the range and depth of your understanding of chemical concepts. The importance of mastering the "basics" cannot be overemphasized. You will find that the text, **Chemistry & Chemical Reactivity,** has a wealth of study questions to assist you in your study of the science we call Chemistry.

Many of the questions contained in your book—and this manual—have multiple parts. In many cases, comments have been added to aid you in following the "path" from gathering available data and applicable unity-factors to arriving at the desired "answer". In these "multiple-step" questions, you may get an answer which differs slightly from those given here. This may be a result of "rounding" intermediate answers. The procedure followed in this manual was to report each intermediate answer to the appropriate number of significant figures, and to calculate the "final" answer without intermediate rounding. In cases involving atomic and molecular weights, those quantities were expressed with one digit more than the number needed for the data provided.

A word of appreciation is due to several people. Thanks go to Dr. John C. Kotz for the many conversations held during the development of this manual. Additionally, I would like to thank my wife, Dr. Catherine Hamrick Banks for her invaluable assistance in typing and proofreading this manuscript. Rarely is a person twice blessed with a patient wife and chemical colleague. Thanks also go to Jonathan and Jennifer for assisting in many ways while their father was in the throes of completing a manuscript.

We have worked diligently to remove all errors from the manual, but remain certain that some have escaped the many inspections. I accept responsibility for those errors.

This manuscript was typed using Microsoft® Word on a Macintosh® computer. Microsoft® Excel was used for the spreadsheet solutions. Atomic and molecular masses were determined using Softshell's Molecular Mass Calculator. Graphics were prepared with SuperPaint™ and ChemIntosh™, and graphs were drawn with Cricket Graph™.

Alton J. Banks
Department of Chemistry
North Carolina State University
Raleigh, North Carolina 27695

Table of Contents

Chapter 1:
Matter and Measurement

Density

14. What mass of ethylene glycol possesses a volume of 500. mL of the liquid?

$$\frac{500.\ mL}{1} \cdot \frac{1\ cm^3}{1\ mL} \cdot \frac{1.1135\ g}{cm^3} = 557\ g$$

16. Calculate the mass (in g) of 500. mL of water, given D = 0.997 g/cm^3 at 25 °C.

$$Mass = D \cdot V = \frac{0.997\ g}{1\ cm^3} \cdot 500.\ mL = 498\ g$$

Express this mass in kilograms: $498\ g \cdot \frac{1\ kg}{1000\ g} = 0.498\ kg$

18. Calculate the density of olive oil if 1 cup (225 mL) has a mass of 205 g:

$$Since\ Density = \frac{Mass}{Volume}\ then\ \frac{205\ g\ olive\ oil}{225\ mL\ olive\ oil} = 0.911\ g/mL\ or\ 0.911\ g/cm^3$$

20. The metal will displace a volume of water that is equal to the volume of the metal.

Hence the difference in volumes of water (20.7-6.7) corresponds to the volume of metal.

Since 1 mL = 1 cm^3, the density of the metal is then:

$$\frac{Mass}{Volume} = \frac{37.5\ g}{14.0\ cm^3}\ or\ 2.68\ \frac{g}{cm^3}\ .$$

From the list of metals provided, the metal with a density closest to this is **Aluminum**.

Temperature

22. Express 25 °C in kelvins:

$$K = (25\ °C + 273)\ or\ 298\ kelvins$$

24. Make the following temperature conversions:

	°C	K
(a)	16	16 + 273.15 = 289
(b)	370 - 273	370
	3.7 x 10^2 - 2.73 x 10^2 = 1.0 x 10^2	

	°C	K
(c)	- 40	- 40 + 273.15 = 230
		(note no decimal point after -40)

26. The accepted value for a normal human temperature is 98.6 °F.

    On the Celsius scale this corresponds to:

    $$°C = \frac{5}{9} (98.6 - 32) = 37 \ °C$$

    So the gallium should melt in your hand.

Elements and Atoms

28. Names for the following elements:
     a. C - carbon   c. Cl - chlorine   e. Mg - magnesium
     b. Na - sodium   d. P - phosphorus   f. Ca - calcium

30. Symbols for the following elements:
     a. lithium - Li   c. iron - Fe   e. cobalt - Co
     b. titanium - Ti   d. silicon - Si   f. zinc - Zn

Units and Unit Conversions

32. Express 19 cm in millimeters, and in meters:

$$\frac{19 \ cm}{1} \cdot \frac{10 \ mm}{1 \ cm} = 190 \ mm \qquad and \qquad \frac{19 \ cm}{1} \cdot \frac{1 \ m}{100 \ cm} = 0.19 \ m$$

34. Express the area of a 2.5 cm x 2.1 cm stamp in cm^2; in m^2:

$$2.5 \ cm \cdot 2.1 \ cm = 5.3 \ cm^2$$

$$5.3 \ cm^2 \cdot \left(\frac{1 \ m}{100 \ cm}\right)^2 = 5.3 \times 10^{-4} \ m^2$$

2

36. Express 800. mL in cm^3; in liters(L); in m^3 :

$$\frac{800.\ mL}{1\ beaker} \cdot \frac{1\ cm^3}{1\ mL} = \frac{800.\ cm^3}{1\ beaker}$$

$$\frac{800.\ cm^3}{1\ beaker} \cdot \frac{1\ L}{1000\ cm^3} = \frac{0.800\ L}{1\ beaker}$$

$$\frac{800.\ cm^3}{1\ beaker} \cdot \frac{1\ m^3}{1 \times 10^6\ cm^3} = \frac{8.00 \times 10^{-4}\ m^3}{1\ beaker}$$

38. Express the mass of a U. S. quarter (5.63 g) in units of kilograms; in milligrams:

$$\frac{5.63\ g}{1\ quarter} \cdot \frac{1\ kg}{1 \times 10^3\ g} = \frac{5.63 \times 10^{-3}\ kg}{quarter}$$

$$\frac{5.63\ g}{1\ quarter} \cdot \frac{1 \times 10^3\ mg}{1\ g} = \frac{5.63 \times 10^3\ mg}{quarter}$$

40. Complete the following table of masses.

MILLIGRAMS	GRAMS	KILOGRAMS
693	0.693	6.93×10^{-4}
156	0.156	1.56×10^{-4}
2.23×10^6	2.23×10^3	2.23

42. Express the dimensions 8 1/2 x 11 inches in centimeters:

$$\frac{8.5\ in}{1} \cdot \frac{2.54\ cm}{1\ in} = 21.59\ cm \quad \text{or 22 cm (to 2 significant figures)}$$

$$\frac{11\ in}{1} \cdot \frac{2.54\ cm}{1\ in} = 27.94\ cm \quad \text{or 28 cm (to 2 significant figures)}$$

The area in square centimeters would be:

$$21.59\ cm \cdot 27.94\ cm = 603.2\ cm^2 \quad \text{or } 6.0 \times 10^2\ cm^2 \text{ (to 2 significant figures)}$$

Note that multiplying 22 cm x 28 cm will provide an answer of 620 cm^2. One good habit to develop is to **round once** *after* all your calculations are done. This habit will eliminate the cumulative roundoff errors that can occur.

Significant Figures

44. Calculate the average of 10.3 g, 9.234 g, and 8.35 g:

$$\frac{10.3\ g\ +\ 9.234\ g\ +\ 8.35\ g}{3} = 9.29\ g\ (3\ sf\ are\ allowed)$$

46. Express the product of three numbers to the proper number of significant figures:

$$(0.000523) \cdot (0.0263) \cdot (263.28) = 0.00362$$

The answer is limited to three significant figures by the first and second terms.

48. Express the result of the calculation to the proper number of significant figures:

$$(0.0345) \cdot \left(\frac{25.35 - 2.4}{1.678\ x\ 10^3}\right) = 4.72\ x\ 10^{-4}$$

The difference (25.35-2.4) leaves a number with 3 sf's (one digit past the decimal). 0.0345 has 3 sf's. 1.678 x 10^3 has 4 sf's. The final answer can then have 3.

50. Solve for n and report the answer to the correct number of significant figures:

$$\underset{\underset{4\ sf}{\uparrow}}{\overset{\overset{3\ sf}{\downarrow}}{\frac{36.3}{760.0}}} \cdot \underset{\underset{3\ sf}{\uparrow}}{75.0} = n \cdot \underset{\underset{3\ sf}{\uparrow}}{0.0821} \cdot \underset{\underset{4\ sf}{\uparrow}}{298.3}$$

$$\frac{\frac{36.3}{760.0}\ x\ 75.0}{0.0821\ x\ 298.3} = 0.1463\ or\ 0.146\ \text{to the correct number of significant figures.}$$

4

Percent

52. Total mass of bracelet = 17.6 g; Mass of Silver = 14.1g

$$\% \text{ Ag} = \frac{\text{mass of Ag}}{\text{mass of bracelet}} = \frac{14.1 \text{ g}}{17.6 \text{ g}} \cdot 100 = 80.1 \%$$

$\% \text{ Cu} = 100.0 - 80.1 = 19.9 \%$

54. Mass of sulfuric acid in 500. mL of battery acid solution which is 38.08 % H_2SO_4:

Mass = Density • Volume

$$= \frac{1.285 \text{ g}}{1 \text{ cm}^3} \cdot \frac{1 \text{ cm}^3}{1 \text{ mL}} \cdot 500 \text{ mL} = 643 \text{ g of acid solution}$$

However, only 38.08 % is sulfuric acid. So the mass of sulfuric acid is:

$$643 \text{ g of acid solution} \cdot \frac{38.08 \text{ g sulfuric acid}}{100 \text{ g acid solution}} = 245 \text{ g sulfuric acid}$$

General Questions

56. Express 1.97 Å in nm; in pm:

$$1.97 \text{ Å} \cdot \frac{1 \times 10^{-10} \text{ m}}{1 \text{ Å}} \cdot \frac{1 \text{ nm}}{1 \times 10^{-9} \text{ m}} = 0.197 \text{ nm}$$

$$1.97 \text{ Å} \cdot \frac{1 \times 10^{-10} \text{ m}}{1 \text{ Å}} \cdot \frac{1 \text{ pm}}{1 \times 10^{-12} \text{ m}} = 197 \text{ pm}$$

58. Express the volume of a cube with edge length 0.563 nm in nm^3 ; in cm^3 :

Since v = 1 • w • h then $(0.563 \text{ nm})^3 = 0.178 \text{ nm}^3$

Express the volume in cm^3 :

$$0.178 \text{ nm}^3 \cdot \left(\frac{1 \text{ m}}{1 \times 10^9 \text{ nm}}\right)^3 \cdot \left(\frac{1 \times 10^2 \text{ cm}}{1 \text{ m}}\right)^3 = 1.78 \times 10^{-22} \text{ cm}^3$$

60. Compare a troy ounce to an avoirdupois ounce:

   A troy ounce = 31.103 g

   Calculate the mass of an avoirdupois ounce in grams:

$$\frac{453.59237 \text{ g}}{1 \text{ lb}} \bullet \frac{1 \text{ lb}}{16 \text{ oz avoir}} = \frac{28.350 \text{ g}}{1 \text{ oz avoir}}$$

   A troy ounce is larger than an avoirdupois ounce.

62. Calculate the mass of platinum in 1.53 g of a compound that is 65.0% platinum:

$$\frac{1.53 \text{ g compound}}{1} \bullet \frac{65.0 \text{ g Pt}}{100.0 \text{ g compound}} = 0.995 \text{ g Pt}$$

64. Calculate the mass lost on popping one kernel of corn:

   0.125 g before popping

   <u>0.106</u> g after popping

   0.019 g lost during popping

   So, the percent mass lost during popping is: $\frac{0.019 \text{ g}}{0.125 \text{ g}} \bullet 100 = 15\%$

   Calculate the number of kernels of popcorn in a pound of popcorn:

   Given that there are 454 g per pound,

$$\frac{1 \text{ kernel}}{0.125 \text{ g}} \bullet \frac{454 \text{ g}}{1 \text{ lb}} = 3632 \text{ kernels} \quad \text{or } 3630 \text{ kernels (to 3 sf's)}$$

66. According to Archimedes' Principle, the object will displace a volume of water equal to its own volume. So, let's calculate the volume of the brass piece.

$$\frac{154 \text{ g brass}}{1} \bullet \frac{1 \text{ cm}^3}{8.56 \text{ g}} = 18.0 \text{ cm}^3$$

   The displacement of the equivalent amount of water (18.0 mL) will provide a reading in the graduated cylinder of (50.0+18.0) or 68.0 mL.

68. This calculation can best be thought of in two steps. First let's calculate the number of tons of water used in the city per year. Then we'll calculate the mass of NaF needed to provide a 1 ppm solution of F.

Tons of water per year:

$$\frac{1.5 \times 10^5 \text{ people}}{1} \cdot \frac{175 \text{ gal}}{1 \text{ people} \cdot \text{day}} \cdot \frac{365 \text{ days}}{1 \text{ yr}} \cdot \frac{8.34 \text{ lb}}{1 \text{ gal}} \cdot \frac{1 \text{ T}}{2000 \text{ lb}} = \frac{4.00 \times 10^7 \text{ T water}}{\text{yr}}$$

A 1 ppm solution of fluoride requires $\frac{1 \text{ T fluoride}}{1.00 \times 10^6 \text{ T water}}$.

NaF is only 45.3 % fluoride, so we can calculate the amount of NaF by use of the fraction: $\frac{100 \text{ T NaF}}{45.3 \text{ T fluoride}}$.

Then the amount of NaF needed to provide 1 ppm solution for 4.00×10^7 T water is:

$$\frac{4.00 \times 10^7 \text{ T water}}{\text{yr}} \cdot \frac{1 \text{ T fluoride}}{1.00 \times 10^6 \text{ T water}} \cdot \frac{100 \text{ T NaF}}{45.3 \text{ T fluoride}} = 90 \text{ T NaF}$$

70. Calculate the volume of the "cylinder" of copper wire with a density of 8.94 g/cm^3 and a mass of 125 lb:

$$125 \text{ lb} \cdot \frac{454 \text{ g}}{1 \text{ lb}} \cdot \frac{1 \text{ cm}^3}{8.94 \text{ g}} = 6350 \text{ cm}^3$$

Since the volume of a cylinder is equal to $\pi r^2 h$, we can calculate the "length" of the wire (the height of the cylinder) since we are told that the diameter of the wire is 9.50 mm. Expressing the diameter in centimeters and converting the diameter into a radius (1/2 the diameter) we obtain:

$$\text{Volume} = \pi r^2 h$$

$$6350 \text{ cm}^3 = 3.1415 \cdot (0.475 \text{ cm})^2 \cdot h$$

then $h = \dfrac{6350 \text{ cm}^3}{3.1415 \cdot (0.475 \text{ cm})^2} = 8.96 \times 10^3 \text{ cm}$

Expressing this number in feet:

$$8.96 \times 10^3 \text{ cm} \cdot \frac{1 \text{ in}}{2.54 \text{ cm}} \cdot \frac{1 \text{ ft}}{12 \text{ in}} = 294 \text{ ft}$$

72. To calculate the density of the metal, first let's calculate the volume of the piece of metal:

$$2.35 \text{ cm} \bullet 1.34 \text{ cm} \bullet 1.05 \text{ cm} = 3.31 \text{ cm}^3$$

Then the density can be calculated by dividing the mass (29.454g) by the volume:

$$D = \frac{29.454 \text{ g}}{3.31 \text{ cm}^3} = 8.91 \frac{\text{g}}{\text{cm}^3}$$

Note that this answer is obtained by dividing the mass by the **unrounded** volume (3.30645 cm^3).

Given this calculated density, the metal in question has to be **nickel**.

74. First, let's determine the mass of asteroid dust deposited on the earth's surface:

$$\frac{5.1 \times 10^{14} \text{ m}^3}{1} \bullet \frac{1 \times 10^4 \text{ cm}^2}{1 \text{ m}^2} \bullet \frac{0.02 \text{ g asteroid}}{1 \text{ cm}^2 \text{ earth}}$$

Given that approximately 20% of the asteroid's mass settled as dust, then the total mass of the asteroid is 5 times this mass or 5×10^{17}g asteroid(to 2 sf's).

Calculate the volume of the asteroid:

Assuming the density was $2\frac{\text{g}}{\text{cm}^3}$, the volume of the asteroid is then:

$$V = \frac{M}{D} = \frac{1 \text{ cm}^3}{2 \text{ g}} \bullet \frac{5.1 \times 10^{17} \text{ g}}{1} = 2.55 \times 10^{17} \text{ cm}^3 \text{ or } 3 \times 10^{17} \text{ cm}^3 \text{ (to 1 sf)}$$

Assuming that the asteroid is a sphere, that is has a volume $= 4/3 \, \pi \, r^3$,

The radius may be calculated: $4/3 \, \pi \, r^3 = 2.55 \times 10^{17} \text{ cm}^3$

$$r = \sqrt[3]{\frac{3}{4} \bullet \frac{2.55 x 10^{17} cm^3}{\Pi}} = 4 x 10^5 cm \text{ and the diameter will be 8} \times 10^5 \text{ cm (to 1 sf)}$$

76. The diameter of the "quantum corral" of iron atoms is 143 Å. The diameter in nm is:

$$143 \text{ Å} \bullet \frac{1 \times 10^{-10} \text{ m}}{1 \text{ Å}} \bullet \frac{1 \times 10^9 \text{ n m}}{1 \text{ m}} = 14.3 \text{ nm}$$

Conceptual Questions

78. Determine the average diameter of lead shot:

We can use the formula $V = 4/3\pi r^3$ if we first determine the volume of the lead shot.

The formula for Density allows us to solve for V: $V = \dfrac{M}{D}$

Let's incorporate the average mass in our calculation:

$$V = \dfrac{M}{D} = \dfrac{2.31\text{ g}}{25\text{ beads}} \cdot \dfrac{1\text{ cm}^3}{11.3\text{ g}} = 8.18 \times 10^{-3}\ \dfrac{\text{cm}^3}{\text{bead}}$$

Now we can use the volume formula to calculate the radius of an average lead shot bead:

$$V = \dfrac{4}{3}\pi r^3 = 8.18 \times 10^{-3}\text{ cm}^3 \quad \text{and rearranging the equation to solve for } r^3:$$

$$r^3 = 3 \cdot 8.18 \times 10^{-3}\text{ cm}^3 \cdot \dfrac{1}{4} \cdot \dfrac{1}{3.1416} = 1.95 \times 10^{-3}\text{ cm}^3$$

$r = 0.125$ cm or 1.25 mm or a diameter of 2.50 mm

An alternative method of separating shot on the basis of its diameter would be to pour the lead shot into several sieves in which the holes of the sieves are of varying diameters. A third method would be simply using calipers to measure the diameter of several samples of lead shot.

81. One clue to an element's identity is its density. Determine the density of the metal sample and compare it to the literature value for silver. This datum would provide one clue as to the identity of the metal sample.

83. Experimental method to determine the density of an irregularly shaped piece of metal: Immerse the sample of metal in an accurately measured volume of water. The **increase** in volume of the water would correspond to the volume of the irregularly shaped piece of metal. (Archimedes' method). Weighing the metal sample would provide a mass. With these data, one can calculate the density of the metal.

85. Experimental method to determine the density of a liquid: There are several methods. One of which would be to weigh an accurately known volume of the liquid. An empty dry weighed 10.0 mL graduated cylinder could be filled to the 10.0 mL

mark with the liquid, and reweighed. The mass of liquid divided by the volume would provide the density of the liquid. The accuracy of this measurement would be determined by the balance used to determine the masses and the measured volume of liquid. The accuracy of volume is probably good to ± 0.1 mL. The calculated density would have at most three significant figures—limited by the 10.0 mL volume.

87. Describe experiments to:

(a) Separate salt (NaCl) from water:

Heating a solution of salt dissolved in water so that the solution boiled, capturing, and condensing the water vapor would separate salt from water. This process is known as distillation.

(b) Separate iron filings from small pieces of lead:

Iron is diamagnetic, i.e. it is attracted to a magnetic field. Placing the mixture of metal pieces on a stiff sheet of paper, and drawing a bar magnet underneath the paper will cause the iron filings to follow the magnet, separating the two metals. Like Part a of this problem, this is an example of a physical technique

(c) Separate elemental sulfur from sugar:

Pour the two solids into a beaker of water, stir and allow the sugar to dissolve, while the sulfur remains in the solid state. Filter the mixture though a conical filter paper to separate the solid sulfur from the solution. Carefully distilling the water off the sugar (see part a) would separate the sugar from the water.

Chapter 2
Atoms and Elements

The Composition of Atoms

21. Mass number for

    a. Be (at. no. 4) with 5 neutrons :   9

    b. Ti (at. no. 22) with 26 neutrons : 48

    c. Ga (at. no. 31) with 39 neutrons : 70

23. Mass number (A)   = no. of protons + no. of neutrons;

    Atomic number (Z) = no. of protons

    a. $^{23}_{11}\text{Na}$ b. $^{39}_{18}\text{Ar}$ c. $^{70}_{31}\text{Ga}$

25.

substance	protons	neutrons	electrons
a. calcium-40	20	20	20
b. tin- 119	50	69	50
c. plutonium-244	94	150	94

Note that the number of protons and electrons are equal for any neutral atom. The number of protons is always equal to the atomic number. The mass number equals the sum of the numbers of protons and neutrons.

27.

Symbol	^{45}Sc	^{33}S	^{17}O	^{56}Mn
Number of protons	21	16	8	25
Number of neutrons	24	17	9	31
Number of electrons in the neutral atom	21	16	8	25

Isotopes

29. For americium - 241 (at. no. 95) :   # protons : 95

                                      # neutrons : (241 - 95) = 146

                                      # electrons : 95

31. Isotopes of element X (at. no. 9) will have an atomic number of 9.

Hence $^{19}_{9}X$ $^{20}_{9}X$, and $^{21}_{9}X$ are isotopes of element X.

Atomic Mass

33. The atomic mass of lithium is:
$$(0.0750)(6.015121) + (0.9250)(7.016003) = 6.94 \text{ amu}$$

35.   The average atomic weight of gallium is 69.723. The percentage abundances of the two isotopes are determined by :
$$(x)(68.9257) + (1 - x)(70.9249) = 69.723$$
Simplifying gives:

$$68.9257 \text{ amu } x + 70.9249 \text{ amu } - 70.9249 \text{ amu } x = 69.723 \text{ amu}$$
$$-1.9992 \text{ amu } x = (69.723 \text{ amu } - 70.9249)$$
$$-1.9992 \text{ amu } x = -1.202 \text{ amu}$$
$$x = 0.6012$$

So the relative abundance of isotope 69 is 60.12 % and that of isotope 71 is 39.88 %

The Periodic Table

37. The elements in Group 4A are:

carbon (C - nonmetal), silicon (Si - metalloid), germanium (Ge - metalloid), tin (Sn - metal), lead (Pb - metal)

39. The seventh period is incomplete presently. The majority of these elements are known as the **Actinides** (transition metals), and most **undergo** spontaneous **radioactive decay**.

The Mole

41. The number of grams in:

a. $2.5 \text{ mol B} \cdot \dfrac{10.811 \text{ g B}}{1 \text{ mol B}} = 27 \text{ g B}$ (2 sf)

b. $0.015 \text{ mol } O_2 \cdot \dfrac{32.00 \text{ g } O_2}{1 \text{ mol } O_2} = 0.48 \text{ g } O_2 \qquad (2 \text{ sf})$

c. $1.25 \times 10^{-3} \text{ mol Fe} \cdot \dfrac{55.847 \text{ g Fe}}{1 \text{ mol Fe}} = 0.0698 \text{ g Fe} \qquad (3 \text{ sf})$

d. $653 \text{ mol He} \cdot \dfrac{4.0026 \text{ g He}}{1 \text{ mol He}} = 2.61 \times 10^3 \text{ g He} \qquad (3 \text{ sf})$

43. Moles represented by:

a. $127.08 \text{ g Cu} \cdot \dfrac{1 \text{ mol Cu}}{63.546 \text{ g Cu}} = 1.9998 \text{ mol Cu}$

b. $20.0 \text{ g Ca} \cdot \dfrac{1 \text{ mol Ca}}{40.08 \text{ g Ca}} = 0.499 \text{ mol Ca}$

c. $16.75 \text{ g Al} \cdot \dfrac{1 \text{ mol Al}}{26.982 \text{ g Al}} = 0.6208 \text{ mol Al}$

d. $0.012 \text{ g K} \cdot \dfrac{1 \text{ mol K}}{39.1 \text{ g K}} = 3.1 \times 10^{-4} \text{ mol K}$

e. $5.0 \text{ mg Am} \cdot \dfrac{1 \text{ g Am}}{1.0 \times 10^3 \text{ mg}} \cdot \dfrac{1 \text{ mol Am}}{243 \text{ g Am}} = 2.1 \times 10^{-5} \text{ mol Am}$

45. One mole of sodium has a mass of 22.99 g (its atomic weight to four significant figures). So 50.4 g of sodium represents

$$50.4 \text{ g Na} \cdot \dfrac{1 \text{ mol Na}}{22.99 \text{ g Na}} = 2.19 \text{ moles Na}$$

The number of Na atoms would be:

$$2.19 \text{ moles Na} \cdot \dfrac{6.0221 \times 10^{23} \text{ Na atoms}}{1 \text{ mole Na}} = 1.32 \times 10^{24} \text{ atoms of Na}$$

47. How many moles of Pt in 15.0 troy ounces ?

$$15.0 \text{ troy oz} \cdot \dfrac{31.1 \text{ g Pt}}{1 \text{ troy oz Pt}} \cdot \dfrac{1 \text{ mol Pt}}{195.1 \text{ g Pt}} = 2.39 \text{ mol Pt}$$

The density of Pt is 21.45 g/cm^3. What is the size of the block ?

$$15.0 \text{ troy oz} \cdot \dfrac{31.1 \text{ g Pt}}{1 \text{ troy oz Pt}} \cdot \dfrac{1 \text{ cm}^3}{21.45 \text{ g Pt}} = 21.7 \text{ cm}^3$$

49. The number of atoms of chromium in 35.67 g Cr :

$$35.67 \text{ g Cr} \cdot \dfrac{1 \text{ mol Cr}}{51.996 \text{ g Cr}} \cdot \dfrac{6.0221 \times 10^{23} \text{ atoms Cr}}{1 \text{ mol Cr}} = 4.131 \times 10^{23} \text{ atoms Cr}$$

51. The average mass of one atom of copper:

$$\frac{63.546 \text{ g Cu}}{1 \text{ mol Cu}} \cdot \frac{1 \text{ mol Cu}}{6.0221 \times 10^{23} \text{ atoms Cu}} = 1.0552 \times 10^{-22} \text{ g Cu/atom}$$

General Questions

53. Given that the average atomic mass for potassium is 39.0983 amu, the lighter isotope, ^{39}K would be the more abundant of the remaining isotopes. Remember that atomic masses are **weighted**, that is the average atomic mass is closer to the most abundant isotope.

55. The mass spectrum shows the isotopes : $^{121}_{51}Sb$ and $^{123}_{51}Sb$. The most abundant isotope is $^{121}_{51}Sb$, with 51 protons, 70 neutrons, and 51 electrons. The approximate atomic mass for Sb is 122.

57.   a. Symbol: Ti ; Atomic number: 22 ; Atomic mass: 47.88

        b. Ti is in Period 4, Group 4B (or 4)

            Other elements in the group include zirconium (Zr), hafnium (Hf), and Rutherfordium (Rf)

        c. Lack of chemical reactivity and density make Ti a good choice for surgical applications.

        d. Properties: melting point, 1660 °C; boiling point, 3287 °C; low density; good strength; easily fabricated; excellent corrosion resistance
            Uses: Alloyed with Al, Mo, Mn, Fe and others. Titanium alloys are used for aircraft and missiles, propeller shafts—and other ship parts exposed to sea water. TiO_2 is also used extensively as both a house paint and artists pigment.

59.   a. Three elements in the series with the greatest density: At. no. 27 (Cobalt), At. no. 28 (Nickel), and At. no. 29 (Copper). The density of all three **metals** is approximately 9 g/cm^3.
        b. The element in the second period with the largest density is Boron (atomic number 5) while the element in the third period with the largest density is Aluminum (atomic number 13). Both of these elements belong to group 3A.

c. Elements from the first 36 elements with very **low densities** are **gases**. These include Hydrogen, Helium, Nitrogen, Oxygen, Fluorine, Neon, Chlorine, Argon and Krypton.

61. How many moles of Pt in 1 troy ounces ?

a. 1 troy oz $\cdot \dfrac{31.1 \text{ g Pt}}{1 \text{ troy oz Pt}} \cdot \dfrac{1 \text{ mol Pt}}{195.1 \text{ g Pt}} =$ 0.159 mol Pt (assuming 1 troy ounce

is an exact conversion)

b. How many grams of Pt can you buy for $5000?

$5000 $\cdot \dfrac{1 \text{ troy ounce}}{\$420} \cdot \dfrac{31.1 \text{ g Pt}}{1 \text{ troy oz Pt}} =$ 370. g Pt

370. g Pt $\cdot \dfrac{1 \text{ mol Pt}}{195.1 \text{ g Pt}} =$ 1.90 mol Pt

63. The "volume of the cylinder" is $\pi r^2 \ell$ where ℓ is the length in cm:

25 ft $\cdot \dfrac{12 \text{ in}}{1 \text{ ft}} \cdot \dfrac{2.54 \text{ cm}}{1 \text{ in}} =$ 760 cm (to 2 sf)

Volume $= \pi (0.10 \text{ cm})^2 (760 \text{ cm}) =$ 23.9 cm^3 or 24 cm^3

Mass of Cu : $\dfrac{8.92 \text{ g}}{1 \text{ cm}^3} \cdot 24 \text{ cm}^3 =$ 210 g Cu

210 g Cu $\cdot \dfrac{1 \text{ mol Cu}}{63.546 \text{ g Cu}} =$ 3.4 mol Cu

and 3.4 mol Cu $\cdot \dfrac{6.0221 \times 10^{23} \text{ atom Cu}}{1 \text{ mol Cu}} =$ 2.0 x 10^{24} atom Cu

65.   Crossword puzzle:
Clues:
1-2 A metal used in ancient times: tin (Sn)
3-4 A metal that burns in air and is found in Group 5A: bismuth (Bi)

1-3 A metalloid: antimony (Sb)
2-4 A metal used in U.S. coins: nickel (Ni)

1. A colorful nonmetal: sulfur (S)

2. A colorless gaseous nonmetal: nitrogen (N)

3. An element that makes fireworks green: boron (B)

4. An element that has medicinal uses: iodine (I)

1-4 An element used in electronics: silicon (Si)

2-3 A metal used with Zr to make wires for superconducting magnets: niobium (Nb)

Using these solutions, the following letters fit in the boxes:

Box 1: S   Box 2: N   Box 3: B   Box 4: I

Conceptual Questions

68. One would predict that calcium's analogues in Group 2A would also form 1 : 1 oxides, MO.

70. In general the relative abundance of elements 1-36 decreases as the atomic number increases. One exception to this trend occurs in the "even" elements, with iron (atomic number 26) being a bit more abundant that surrounding elements.
    A difference in the abundances of the "even" and "odd" elements in this range is that the "odd" elements are less abundant than the "even" elements with atomic number ±1.

72. a. Submicroscopic particles in a sample of solid: **a, b, f** (regularly distributed particles, closely packed)

   b. Submicroscopic particles in a sample of liquid: **h, i**

   c. Submicroscopic particles in a sample of gas: **c, d, g** (particles widely separated)

   d. Submicroscopic particles in a sample of element : **a, d, f** (particles of one kind)

   e. Submicroscopic particles in a sample of compound: **c**

   (stoichiometric combination of elements)

   f. Submicroscopic particles in a sample of solution: **b, g, i**

   (solute particles distributed uniformly in solvent particles)

   g. Submicroscopic particles in a sample of heterogeneous mixture: **e, h**

   (particles of more than one type, mixed nonuniformly)

74.   a. Using our present atomic weights (based on carbon-12) the relative masses of O : H are:

$$\frac{\text{at. mass O}}{\text{at. mass H}} = \frac{15.9994}{1.00797} = 15.8729$$

If H $\equiv$ 1.0000 amu, the atomic mass of O would be

$$15.8729 \cdot 1.0000 = 15.873 \text{ amu}$$

Similarly for Carbon:

$$\frac{\text{at. mass C}}{\text{at. mass H}} = \frac{12.0000}{1.00797} = 11.9051$$

If H is 1.0000 amu, the atomic mass of C would be

$$11.9051 \cdot 1.0000 = 11.9051 \text{ amu}$$

The number of particles associated with one mole is:

$$\frac{11.9051}{12.0000} = \frac{X}{6.0221367 \times 10^{23}} \quad \text{and } X = 5.9745 \times 10^{23} \text{ particles}$$

b. Using the ratio from part a

$$\frac{\text{at. mass H}}{\text{at. mass O}} = \frac{1.00797}{15.9994} = 0.0630005$$

If O $\equiv$ 16.0000 amu, the atomic mass of H would be

$$0.0630005 \cdot 16.0000 = 1.00801 \text{ amu}$$

Similarly for Carbon, the ratios of the atomic masses of C to O is:

$$\frac{\text{at. mass C}}{\text{at. mass O}} = \frac{12.0000}{15.9994} = 0.750028 \text{ , and the atomic mass of C is}$$

$$0.750028 \cdot 16.0000 = 12.00045 \text{ amu}$$

The number of particles associated with one mole is:

$$\frac{12.00045}{12.0000} = \frac{X}{6.0221367 \times 10^{23}} \quad \text{and } X = 6.02236 \times 10^{23} \text{ particles}$$

76. The ratio of the atomic masses of P/O :

    0.744 g of P combined with 0.960 g of O.   (1.704 - 0.744)

    Given that the compound formed is P_4O_{10}, the masses of phosphorus and oxygen correspond to the ratio of 4 phosphorus atoms to 10 oxygen atoms. Dividing the masses by the appropriate number of atoms

$$\frac{0.744 \text{ g P}}{4} = 0.186 \text{ g P} \qquad \text{and} \qquad \frac{0.960 \text{ g O}}{10} = 0.0960 \text{ g O}$$

So the mass of $\frac{P}{O}$ is $\frac{0.186 \text{ g P atoms}}{0.0960 \text{ g O atoms}} = 1.94$

If the atomic mass of oxygen is 16.000 amu, the atomic mass of phosphorus is

$$16.000 \text{ amu} \cdot \frac{1.94 \text{ g P atoms}}{1.0 \text{ g O atoms}} = 31.0 \text{ amu}$$

Chapter 3
Compounds and Molecules

Molecular Formulas

13. Molecules with more O atoms per molecule:

$$C_{12}H_{22}O_{11} \qquad\qquad C_{10}H_{17}N_3O_6S$$

   Sucrose contains **11 oxygen atoms per molecule** while glutathione contains 6 per molecule.

   Atoms of all kinds:

   Summing the subscripts indicates that **sucrose has 45** atoms per molecule while **glutathione has 37**.

15. Total atoms of each element in a formula unit:

	element	# atoms
a. CaC_2O_4	Ca	1
	C	2
	O	4
b. $C_6H_5CHCH_2$	C	8
	H	8
c. $Cu_2CO_3(OH)_2$	Cu	2
	C	1
	O	5
	H	2
d. $Pt(NH_3)_2Cl_2$	Pt	1
	N	2
	H	6
	Cl	2
e. $K_4Fe(CN)_6$	K	4
	Fe	1
	C	6
	N	6

17. Molecular formula for lactic and citric acids:

   The molecular formula may be written by adding all atoms of a particular element:

   lactic acid : $\quad C_3H_6O_3$ $\qquad\qquad$ citric acid : $\quad C_6H_8O_7$

Ions and Ion Charges

19. Charges on ions of Aluminum and Selenium.

    Metallic aluminum will lose 3 electrons forming a +3 ion.

    Selenium (an analogue of O and S) will form a -2 ion.

21. Most commonly observed ion for:

    a. Magnesium—like all the alkaline earth metals : +2

    b. Zinc : +2

    c. Iron : +3 (although iron—as most transition metals—can form more than one ion—in this case also +2)

    d. Gallium : +3 (an analogue of Aluminum)

23. The symbol and charge for the following ions:

    a. Strontium ion Sr^{+2}
    b. Aluminum ion Al^{+3}
    c. Sulfide ion S^{-2}
    d. Cobalt(II) ion Co^{+2}
    e. Titanium(IV) ion Ti^{+4}
    f. Hydrogen carbonate ion HCO_3^{-1}
    g. Perchlorate ion ClO_4^{-1}
    h. Ammonium ion NH_4^{+1}

Ionic Compounds

25. Formula, Charge, and Number of ions in:

	cation	# of	anion	# of
a. K_2S	K^{+1}	2	S^{-2}	1
b. $NiSO_4$	Ni^{+2}	1	SO_4^{-2}	1
c. $(NH_4)_3PO_4$	NH_4^{+1}	3	PO_4^{-3}	1
d. $Ca(ClO)_2$	Ca^{+2}	1	ClO^{-1}	2
e. $KMnO_4$	K^{+1}	1	MnO_4^{-1}	1

27. Cobalt oxide

      Cobalt(II) oxide CoO cobalt ion : Co^{+2}

      Cobalt(III) oxide Co_2O_3 Co^{+3}

29. Provide correct formulas for compounds:

    a. $AlCl_3$ The tripositive aluminum ion requires three chloride ions.

    b. NaF Sodium is a monopositive cation. Fluoride is a mononegative anion.

    c. and d. are correct formulas

31. The formula for magnesium oxide is MgO. Magnesium oxide has a much higher melting point than NaCl owing to the +2,-2 charges for the magnesium and oxide ions, compared to the monopositive and mononegative ions in NaCl. The increased electrostatic attraction in MgO results in a greater energy (higher temperature) needed to disrupt the solid lattice.

Naming Compounds

33. Names for the ionic compounds

    a. K_2S Potassium sulfide

    b. $NiSO_4$ Nickel(II) sulfate

    c. $(NH_4)_3PO_4$ Ammonium phosphate

    d. $Ca(ClO)_2$ Calcium hypochlorite

35. Formulas for the ionic compounds

    a. Ammonium carbonate $(NH_4)_2CO_3$

    b. Calcium iodide CaI_2

    c. Copper(II) bromide $CuBr_2$

    d. Aluminum phosphate $AlPO_4$

    e. Silver(I) acetate $AgCH_3CO_2$

37. Names and formulas for ionic compounds

cation	anion CO_3^{2-}	anion Br^-	anion NO_3^-
K^+	K_2CO_3 Potassium carbonate	KBr Potassium bromide	KNO_3 Potassium nitrate
Ba^{2+}	$BaCO_3$ Barium carbonate	$BaBr_2$ Barium bromide	$Ba(NO_3)_2$ Barium nitrate
NH_4^+	$(NH_4)_2CO_3$ Ammonium carbonate	NH_4Br Ammonium bromide	NH_4NO_3 Ammonium nitrate

39. Names of binary nonmetal compounds

    a. Nitrogen trifluoride NF_3

    b. Hydrogen iodide   HI

    c. Boron tribromide BBr_3

    d. Hexane C_6H_{14}

41. Formulas for:

    a. Butane C_4H_{10}

    b. Dinitrogen pentaoxide N_2O_5

    c. Nonane C_9H_{20}

    d. Silicon tetrachloride $SiCl_4$

    e. Diboron trioxide B_2O_3

Molar Mass and Moles

43. Molar mass of the following: (with atomic weights expressed to 4 significant figures)

    a. Fe_2O_3 $(2)(55.85) + (3)(16.00) = 159.7$

    b. BF_3 $(1)(10.81) + (3)(19.00) = 67.81$

    c. N_2O $(2)(14.01) + (1)(16.00) = 44.02$

    d. $MnCl_2 \cdot 4\,H_2O$ $(1)(54.94) + (2)(35.45) + (8)(1.008) + (4)(16.00) = 197.9$

    e. $C_6H_8O_6$ $(6)(12.01) + (8)(1.008) + (6)(16.00) = 176.1$

45. Moles represented by 1.00 g of the compounds:

Compound	Molar mass	Moles in 1.00 g
a. CH_3OH	32.04	0.0312
b. Cl_2O	98.92	0.0101
c. NH_4NO_3	80.04	0.0125
d. $MgSO_4 \cdot 7\ H_2O$	246.48	4.06×10^{-3}

47. Moles of acrylonitrile in 2.50 kg:

  1. Molar mass of C_3H_3N:

$$(3)(12.01) + (3)(1.008) + (1)(14.01) = 53.06 \text{ g/mol}$$

  2. Moles:

$$2.50 \times 10^{+3} \text{g} \cdot \frac{1 \text{ mol } C_3H_3N}{53.06 \text{ g}} = 47.1 \text{ mol } C_3H_3N$$

49. To begin, express the mass of aspirin, sodium hydrogen carbonate, and aspirin in grams. Convert those masses to moles, using the molar mass of each substance.

  a. Moles of aspirin:
$$0.324 \text{ g } C_9H_8O_4 \cdot \frac{1 \text{ mol } C_9H_8O_4}{180.16 \text{ g } C_9H_8O_4} = 1.80 \times 10^{-3} \text{ mol } C_9H_8O_4$$

  Moles of sodium hydrogen carbonate:

$$1.904 \text{ g } NaHCO_3 \cdot \frac{1 \text{ mol } NaHCO_3}{84.007 \text{ g } NaHCO_3} = 2.266 \times 10^{-2} \text{ mol } NaHCO_3$$

  Moles of citric acid:
$$1.000 \text{ g } C_6H_8O_7 \cdot \frac{1 \text{ mol } C_6H_8O_7}{192.13 \text{ g } C_6H_8O_7} = 5.205 \times 10^{-3} \text{ mol } C_6H_8O_7$$

  b. Molecules of aspirin per tablet:
$$1.80 \times 10^{-3} \text{ mol } C_9H_8O_4 \cdot \frac{6.022 \times 10^{23} \text{ molecules aspirin}}{1 \text{ mol aspirin}}$$

$$= 1.08 \times 10^{21} \text{ molecules aspirin}$$

51. The number of moles of SO_3 in 1.00 lb of the compound:

$$1.00 \times 10^3 \text{ g SO}_3 \cdot \frac{454 \text{ g SO}_3}{1.00 \text{ lb SO}_3} \cdot \frac{1 \text{ mol SO}_3}{80.06 \text{ g SO}_3} = 12.5 \text{ mol SO}_3$$

The number of molecules of SO_3 :

$$12.5 \text{ mol SO}_3 \cdot \frac{6.022 \times 10^{23} \text{ molecules SO}_3}{1 \text{ mol SO}_3} = 7.52 \times 10^{24} \text{ molecules SO}_3$$

The number of sulfur atoms is easily calculable by noting that in each molecule of sulfur trioxide, there is one (1) atom of sulfur.

$$7.52 \times 10^{24} \text{ molecules SO}_3 \cdot \frac{1 \text{ atom S}}{1 \text{ molecule SO}_3} = 7.52 \times 10^{24} \text{ atoms S}$$

Similarly in each molecule of sulfur trioxide, there are three (3) atoms of oxygen.

$$7.52 \times 10^{24} \text{ molecules SO}_3 \cdot \frac{3 \text{ atoms O}}{1 \text{ molecule SO}_3} = 2.26 \times 10^{25} \text{ atoms O}$$

Percent Composition

53. Mass percent for: [4 significant figures]

a. PbS: $(1)(207.2) + (1)(32.06) = 239.3$ g/mol

$$\%\text{Pb} = \frac{207.2 \text{ g Pb}}{239.3 \text{ g PbS}} \times 100 = 86.59 \%$$

$$\%\text{S} = 100.00 - 86.60 = 13.41 \%$$

b. C_3H_8: $(3)(12.01)+(8)(1.008) = 44.09$ g/mol

$$\%\text{C} = \frac{36.03 \text{ g C}}{44.09 \text{ g C}_3\text{H}_8} \times 100 = 81.71\%$$

$$\%\text{H} = 100.00 - 81.71 = 18.29\%$$

c. $CoCl_2 \cdot 6\,H_2O$: $(1)(58.93)+(2)(35.45)+(12)(1.008)+(6)(16.00) = 237.93$ g/mol

$$\%\text{Co} = \frac{58.93 \text{ g Co}}{237.93 \text{ g CoCl}_2 \cdot 6\,\text{H}_2\text{O}} \times 100 = 24.77\%$$

$$\%\text{Cl} = \frac{70.90 \text{ g Cl}}{237.93 \text{ g CoCl}_2 \cdot 6\,\text{H}_2\text{O}} \times 100 = 29.80\%$$

$$\%H \ = \ \frac{12.09 \ g \ H}{237.93 \ g \ CoCl_2 \cdot 6 \ H_2O} \ x \ 100 \ = 5.08\%$$

$$\%O \ = \ \frac{96.00 \ g \ O}{237.93 \ g \ CoCl_2 \cdot 6 \ H_2O} \ x \ 100 \ = \ 40.35\%$$

d. NH_4NO_3: $(2)(14.01) + (4)(1.008) + (3)(16.00) = 80.05$ g/mol

$$\%H = \frac{4.032 \ g \ H}{80.05 \ g \ NH_4NO_3} \ x \ 100 \ = 5.04 \ \%H$$

$$\%O = \frac{48.00 \ g \ O}{80.05 \ g \ NH_4NO_3} \ x \ 100 \ = 59.96 \ \%O$$

$$\%N = 100.00 - (5.04 + 59.96) \ = \ 35.00 \ \%N$$

55. For vinyl chloride, C_2H_3Cl

a. Molar mass:

$$(2) \ (12.01) \ + \ (3)(1.008) \ + \ (1)(35.45) \ = \ 62.50 \ g/mol$$

b. Mass percent of each element:

$$\% \ C \ = \ \frac{24.02 \ g \ C}{62.49 \ g \ C_2H_3Cl} \ x \ 100 \ = \ 38.44 \ \% \ C$$

$$\% \ H \ = \ \frac{3.024 \ g \ H}{62.49 \ g \ C_2H_3Cl} \ x \ 100 \ = \ 4.84 \ \% \ H$$

$$\% \ Cl \ = \ \frac{35.45 \ g \ Cl}{62.49 \ g \ C_2H_3Cl} \ x \ 100 \ = \ 56.72 \ \% \ Cl$$

c. Mass of C in 454 g of C_2H_3Cl

$$454 \ g \ C_2H_3Cl \cdot \frac{38.44 \ g \ C}{100.00 \ g \ C_2H_3Cl} \ = \ 175 \ g \ C$$

Empirical and Molecular Formulas

57.   The empirical formula ($C_2H_3O_2$) would have a mass of 59.04 g.

Since the molar mass is 118.1 g/mol we can write

$$\frac{1 \ empirical \ formula}{59.04 \ g \ succinic \ acid} \cdot \frac{118.1 \ g \ succinic \ acid}{1 \ mol \ succinic \ acid} \ = \ \frac{2.0 \ empirical \ formulas}{1 \ mol \ succinic \ acid}$$

So the molecular formula contains 2 empirical formulas (2 x $C_2H_3O_2$) or $C_4H_6O_4$.

59. Calculate the empirical formula of acetylene by calculating the atomic ratios of carbon and hydrogen in 100 g of the compound.

$$92.26 \text{ g C} \cdot \frac{1 \text{ mol C}}{12.011 \text{ g C}} = 7.681 \text{ mol C}$$

$$7.74 \text{ g H} \cdot \frac{1 \text{ mol H}}{1.008 \text{ g H}} = 7.678 \text{ mol H}$$

Calculate the atomic ratio: $\dfrac{7.68 \text{ mol C}}{7.68 \text{ mol H}} = \dfrac{1 \text{ mol C}}{1 \text{ mol H}}$

The atomic ratio indicates that there is 1 C atom for 1 H atom (1:1). The empirical formula is then CH. The formula mass is 13.01. Given that the molar mass of the compound is 26.02 g/mol, there are two formula units per molecular unit, hence the molecular formula for acetylene is C_2H_2 .

61. Calculate the empirical formula of a nitrogen oxide which is 36.84 % N.

This compound must contain (100.00-36.84) 63.16 % O. With this knowledge we can calculate the molar ratio of atoms in 100 g of the compound.

$$63.16 \text{ g O} \cdot \frac{1 \text{ mol O}}{16.00 \text{ g O}} = 3.948 \text{ mol O}$$

$$36.84 \text{ g N} \cdot \frac{1 \text{ mol N}}{14.01 \text{ g N}} = 2.630 \text{ mol N}$$

Calculating the ratio of atoms:

$$\frac{3.948 \text{ mol O}}{2.630 \text{ mol N}} = \frac{1.5 \text{ mol O}}{1 \text{ mol N}}$$

The empirical formula of this oxide indicates that there are 1.5 O atoms for each N atom. Since we cannot have a fractional part of an atom we write the empirical formula to express this ratio-- N_2O_3 .

63. Empirical and Molecular formula for Mandelic Acid:

$$63.15 \text{ g C} \cdot \frac{1 \text{ mol C}}{12.0115 \text{ g C}} = 5.258 \text{ mol C}$$

$$5.30 \text{ g H} \cdot \frac{1 \text{ mol}}{1.0079 \text{ g H}} = 5.28 \text{ mol H}$$

$$31.55 \text{ g O} \cdot \frac{1 \text{ mol O}}{15.9994 \text{ g O}} = 1.972 \text{ mol O}$$

Using the smallest number of atoms, we calculate the ratio of atoms:

$$\frac{5.258 \text{ mol C}}{1.972 \text{ mol O}} = \frac{2.666 \text{ mol C}}{1 \text{ mol O}} \quad \text{or} \quad \frac{2\ 2/3 \text{ mol C}}{1 \text{ mol O}} \quad \text{or} \quad \frac{8/3 \text{ mol C}}{1 \text{ mol O}}$$

So 3 mol O combine with 8 mol C and 8 mol H so the empirical formula is $C_8H_8O_3$.
The formula mass of $C_8H_8O_3$ is 152.15. Given the data that the molar mass is 152.15
g/mL, the molecular formula for mandelic acid is $C_8H_8O_3$.

65. Empirical and molecular formula for cacodyl :

$$22.88 \text{ g C} \quad \bullet \quad \frac{1 \text{ mol C}}{12.011 \text{ g C}} = 1.905 \text{ mol C}$$

$$5.76 \text{ g H} \quad \bullet \quad \frac{1 \text{ mol H}}{1.008 \text{ g H}} = 5.71 \text{ mol H}$$

$$71.36 \text{ g As} \bullet \quad \frac{1 \text{ mol As}}{74.922 \text{ g As}} = 0.9525 \text{ mol As}$$

Expressing these values as ratios of each element to one element we obtain:

$$\frac{1.905 \text{ mol C}}{0.9525 \text{ mol As}} = \frac{2.000 \text{ mol C}}{1 \text{ mol As}}$$

$$\frac{5.71 \text{ mol H}}{0.9525 \text{ mol As}} = \frac{5.99 \text{ mol H}}{1 \text{ mol As}}$$

From these ratios we know that for each As atom in the molecule there are 2 C atoms and
6 H atoms, for an empirical formula of AsC_2H_6.
Adding the atomic weights of the empirical formula gives a mass of :

empirical formula =   1(74.922) + 2(12.011) + 6(1.008) = 104.99

$$\frac{210 \text{ g cacodyl}}{1 \text{ mol cacodyl}} \bullet \frac{1 \text{ empirical formula}}{104.99 \text{ g cacodyl}} = \frac{2 \text{ empirical formulas}}{1 \text{ mol cacodyl}}$$

Multiplying each element in the empirical formula by two yields a molecular formula of
$As_2C_4H_{12}$.

67.   The amount of water present in the 1.687 g sample of epsom salt is:

$$(1.687 \text{ g hydrate} - 0.824 \text{ g MgSO}_4) = 0.863 \text{ g water}$$

$$0.863 \text{ g water} \cdot \frac{1 \text{ mol water}}{18.02 \text{ g water}} = 0.0479 \text{ mol water}$$

Calculate the number of moles of the anhydrous salt:

$$0.824 \text{ g MgSO}_4 \cdot \frac{1 \text{ mol MgSO}_4}{120.4 \text{ g MgSO}_4} = 0.00684 \text{ mol MgSO}_4$$

Calculating the ratios of water to anhydrous salt gives:

$$\frac{0.0479 \text{ mol water}}{0.00684 \text{ mol MgSO}_4} = \frac{7 \text{ mol water}}{1 \text{ mol MgSO}_4}$$

so there are **7 molecules of water** for each formula unit of MgSO$_4$.

69. Given the masses of sulfur and fluorine involved, we can calculate the molar composition of the compound:

$$1.256 \text{ g S} \cdot \frac{1 \text{ mol S}}{32.06 \text{ g S}} = 0.03918 \text{ mol S}$$

The mass of fluorine present is: 5.722 g compound - 1.256 g S = 4.466 g F

$$4.466 \text{ g F} \cdot \frac{1 \text{ mol F}}{19.00 \text{ g F}} = 0.2351 \text{ mol F}$$

Calculating atomic ratios:

$$\frac{0.2351 \text{ mol F}}{0.03918 \text{ mol S}} = \frac{6 \text{ mol F}}{1 \text{ mol S}} \quad \text{indicating that the value of } \mathbf{x} \text{ is 6.}$$

71. Galena has the formula PbS. The percentage of galena that is lead is :

$$\% \text{ Pb} = \frac{207.2 \text{ g Pb}}{239.27 \text{ g PbS}} \times 100 = 86.60 \% \text{ Pb}$$

To produce 2.00 kg Pb:

$$2.00 \text{ kg Pb} \cdot \frac{100.0 \text{ g PbS}}{86.60 \text{ g Pb}} = 2.31 \text{ kg PbS}$$

73. $Ca_3(PO_4)_2 + C + SiO_2 \longrightarrow P_4$ (not a complete equation)

Calculate the % P in calcium phosphate:

$$\% P = \frac{61.95 \text{ g P}}{310.18 \text{ g Ca}_3(PO_4)_2} \times 100 = 19.97 \% P$$

To produce 15.0 kg of phosphorus:

$$15.0 \text{ kg P} \cdot \frac{100.0 \text{ g Ca}_3(PO_4)_2}{19.97 \text{ g P}} = 75.1 \text{ kg Ca}_3(PO_4)_2$$

General Questions

75. $NaCl \xrightarrow{\text{electricity}} Na + Cl_2$

Caculate the % Na and % Cl in NaCl:

$$\% Na = \frac{22.99 \text{ g Na}}{58.44 \text{ g NaCl}} \times 100 = 39.34 \% Na$$

$$\% Cl = 100.00 - 39.34 = 60.66 \% Cl$$

$$2.00 \text{ MT NaCl} \cdot \frac{39.34 \text{ g Na}}{100.00 \text{ g NaCl}} = 0.787 \text{ MT Na}$$

$$2.00 \text{ MT NaCl} \cdot \frac{60.66 \text{ g Cl}}{100.00 \text{ g NaCl}} = 1.21 \text{ MT Cl}$$

77. Molecular formulas for TNT and alanine:

$C_7H_5N_3O_6$ $C_3H_7N_1O_3$

   TNT   serine

Molar mass:

TNT : $7(12.011) + 5(1.0079) + 3(14.0067) + 6(15.9994) = 227.1$ g/mol

serine : $3(12.011) + 7(1.0079) + 1(14.0067) + 3(15.9994) = 105.1$ g/mol

TNT: % N $\frac{42.02}{227.13} \times 100 = 18.50 \%$ % C $\frac{84.08}{227.13} \times 100 = 42.26 \%$

$$\text{serine} \quad \overset{\text{\% N}}{\underset{}{\frac{14.01}{105.09}}} \times 100 = 13.33\,\% \qquad \overset{\text{\% C}}{\underset{}{\frac{36.03}{105.09}}} \times 100 = 34.28\,\%$$

79. Mass of one molecule of oxalic acid, $H_2C_2O_4$:

One mole of molecules of oxalic acid has a mass in grams equal to the molecular weight (molar mass).

Molar Mass $= 2(1.0079) + 2(12.011) + 4(15.9994) = 90.04$ g/ml

Avogadro's number of molecules (6.022×10^{23}) has a mass of 90.04 g, so the mass of one is:

$$\frac{90.04 \text{ g}}{6.022 \times 10^{23} \text{ molecules}} = 1.495 \times 10^{-22} \text{ g/molecule}$$

81. Dimethyl sulfoxide (DMSO):

Molecular formula: C_2H_6SO

Molar mass : $2(12.011) + 6(1.0079) + 1(32.066) + 1(15.9994) = 78.13$ g/mol

Mass percentages:

Carbon : $\dfrac{24.022 \text{ g C}}{78.13 \text{ g DMSO}} \times 100 = 30.75\,\%$ C

Hydrogen : $\dfrac{6.0474 \text{ g H}}{78.13 \text{ g DMSO}} \times 100 = 7.74\,\%$ H

Sulfur : $\dfrac{32.066 \text{ g S}}{78.13 \text{ g DMSO}} \times 100 = 41.03\,\%$ S

Oxygen : $\dfrac{15.9994 \text{ g O}}{78.13 \text{ g DMSO}} \times 100 = 20.48\,\%$ O

Moles of DMSO in 10.0 g:

10.0 g DMSO $\cdot \dfrac{1 \text{ mol DMSO}}{78.13 \text{ g DMSO}} = 0.128$ mol DMSO

Mass of S in 10.0 g DMSO:

$$\overset{\downarrow\,\%\,\text{S}}{10.0 \text{ g DMSO} \cdot \frac{32.066 \text{ g S}}{78.13 \text{ g DMSO}}} = 4.10 \text{ g S}$$

83. a. Molecular formula for ferrocene: $C_{10}H_{10}Fe$

  b. Mass of iron in 0.150 g compound

  $$0.150 \text{ g ferrocene} \cdot \frac{55.85 \text{ g Fe}}{186.0 \text{ g ferrocene}} = 0.0450 \text{ g Fe}$$

  Number of Iron atoms:

  $$0.0450 \text{ g Fe} \cdot \frac{1 \text{ mol Fe}}{55.85 \text{ g Fe}} \cdot \frac{6.022 \times 10^{23} \text{ Fe atoms}}{1 \text{ mol Fe}} = 4.85 \times 10^{20} \text{ Fe atoms}$$

  c. Element with greatest mass percent:

        % H = 5.42   % Fe = 30.02   % C = 64.56

85. Ionic compounds; formulas and names

  b. Li_2Te lithium telluride

  d. MgF_2 magnesium fluoride

  f. In_2S_3 indium sulfide

  Pair a. : consists of two non-metals — a covalent compound is anticipated

  Pair c. : Argon doesn't typically form ionic compounds

  Pair e. and g : consists of two non-metals

87. Formulas for compounds; identify the ionic compounds

a. Sodium hypochlorite	$NaClO$	ionic
b. Aluminum perchlorate	$Al(ClO_4)_3$	ionic
c. Potassium permanganate	$KMnO_4$	ionic
d. Potassium dihydrogen phosphate	KH_2PO_4	ionic
e. Chlorine trifluoride	ClF_3	
f. Boron tribromide	BBr_3	
g. Calcium acetate	$Ca(CH_3CO_2)_2$	ionic
h. Ammonium sulfite	$(NH_4)_2SO_3$	ionic
i. Disulfur dichloride	S_2Cl_2	
j. Phosphorus trifluoride	PF_3	

89. Empirical and Molecular formula of Azulene:

Given the information that azulene is a hydrocarbon, if it is 93.71 % C, it is also (100.00 - 93.71) 6.29 % H.

In a 100.00 g sample of azulene there are

$$93.71 \text{ g C} \cdot \frac{1 \text{ mol C}}{12.011 \text{ g C}} = 7.802 \text{ mol C} \qquad \text{and}$$

$$6.29 \text{ g H} \cdot \frac{1 \text{ mol H}}{1.0079 \text{ g H}} = 6.241 \text{ mol H}$$

The ratio of C to H atoms is: 1.25 mol C : 1 mol H or a ratio of 5 mol C: 4 mol H. (C_5H_4) The mass of such an empirical formula is $\approx$ 64. Given that the molar mass is ~128 g/mol, the molecular formula for azulene is $C_{10}H_8$.

91.
$$I_2 \quad + \quad Cl_2 \longrightarrow \quad I_xCl_y$$
$$0.678 \text{ g} \quad (1.246 - 0.678) \quad 1.246 \text{ g}$$

Calculate the ratio of I : Cl atoms

$$0.678 \text{ g I} \cdot \frac{1 \text{ mol I}}{126.9 \text{ g I}} = 5.34 \times 10^{-3} \text{ mol I atoms}$$

$$0.568 \text{ g Cl} \cdot \frac{1 \text{ mol Cl}}{35.45 \text{ g Cl}} = 1.6 \times 10^{-2} \text{ mol Cl atoms}$$

The ratio of Cl : I is : $\dfrac{1.6 \times 10^{-2} \text{ mol Cl atoms}}{5.34 \times 10^{-3} \text{ mol I atoms}} = 3.00 \dfrac{\text{Cl atoms}}{\text{I atoms}}$

The empirical formula is ICl_3 (FW = 233.3)

Given that the molar mass of I_xCl_y was 467 g/mol, we can calculate the number of empirical formulas per mole:

$$\frac{467 \text{ g/mol}}{233.3 \text{ g/empirical formula}} = 2 \frac{\text{empirical formula}}{\text{mol}}$$

for a molecular formula of I_2Cl_6.

93. Mass of Fe in 15.8 kg of FeS_2 :

$$\% \text{ Fe in FeS}_2 = \frac{55.85 \text{ g Fe}}{119.97 \text{ g FeS}_2} \times 100 = 46.55 \% \text{ Fe}$$

and in 15.8 kg FeS_2

$$15.8 \text{ kg FeS}_2 \cdot \frac{46.55 \text{ kg Fe}}{100.00 \text{ kg FeS}_2} = 7.36 \text{ kg Fe}$$

95. The antimony metal contained in the ore is present in the form of Sb_2S_3. From the molar mass of this compound (339.70 g/mol) and the atomic mass of antimony (121.75 g/mol), we can express this relation as either:

$$\frac{2 \times 121.75 \text{ g Sb}}{339.70 \text{ g Sb}_2\text{S}_3} \quad \text{or} \quad \frac{339.70 \text{ g Sb}_2\text{S}_3}{2 \times 121.75 \text{ g Sb}}$$

The mass of antimony present in 1.00 kg of the ore may then be calculated:

$$1000 \text{ g ore} \cdot \frac{10.6 \text{ g Sb}}{100 \text{ g ore}} = 106 \text{ g Sb.}$$

Using the fraction expressing the % Sb in Sb_2S_3 gives:

$$106 \text{ g Sb} \cdot \frac{339.70 \text{ g Sb}_2\text{S}_3}{243.5 \text{ g Sb}} = 148 \text{ g Sb}_2\text{S}_3 .$$

97. Formula for $Ni(CO)_x$:

    0.364 g $Ni(CO)_x$

    <u>0.125 g Ni</u>

    0.239 g CO

Calculate the moles of Ni atoms and moles of CO molecules

$$0.239 \text{ g CO} \cdot \frac{1 \text{ mol CO}}{28.01 \text{ g CO}} = 8.53 \times 10^{-3} \text{ mol CO molecules}$$

$$0.125 \text{ g Ni} \cdot \frac{1 \text{ mol Ni}}{58.69 \text{ g Ni}} = 2.13 \times 10^{-3} \text{ mol Ni atoms}$$

The ratio of CO : Ni is $\quad \frac{8.53}{2.13} = 4.00 \quad$ The value of x must be 4.

99. Molar mass of ECl_4:

    If 2.50 mol has a mass of 385 g then the molar mass is : $\quad \frac{385 \text{ g}}{2.50 \text{ mol}} = 154 \text{ g/mol}$

The identity of E:

With a molar mass of 154 g/mol, we can obtain information about E by subtracting the mass of 4 mol of Cl atoms from the molar mass for ECl_4 :

154 g ECl_4 /mol - (4 mol x 35.45 g Cl/mol)   or   154 g - 141.8 g Cl/mol = 12 g

One suspects that E is **Carbon**. Carbon tetrachloride is a well-known compound of carbon.

101. Composition of mixture of A (CH_4O) and B (C_2H_6O) :

We are given the data that A and B are mixed in such a ratio that the % C is 49.9 %.
Calculate % C for both A and B:

Compound A :   % C = $\dfrac{12.011 \text{ g C}}{32.042 \text{ g } CH_4O}$ = 37.49 %

        B :   % C = $\dfrac{24.022 \text{ g C}}{46.070 \text{ g } C_2H_6O}$ = 52.14 %

Represent the % A as PA and the % B as PB.
Then PA + PB = 100 g and PB = 100 - PA

Every gram of A contains 0.3749 g C and every gram of B 0.5214 g C.
Expressing that in an equation:

        grams C from A + grams C from B   =   gram C in mixture

        (0.3749 g C)(PA) + (0.5214 g C)(PB)   =   (0.499 g C)(100)
        0.379 g C • PA + (0.5214 g C)(100 - PA) =   49.9 g C
        0.3749 g C • PA + 52.14 g C - 0.5214 g C • PA   =   49.9 g C
                (0.3749 g C - 0.5214 g C) • PA   =   49.9 g C - 52.14 g C
                        -0.1465 g C • PA   =   -2.24 g C
                                PA   =   15.29 and PB   =   84.71

So any mixture should contain 84.71 g of B for every 15.29 g of A. To determine the mass of B per gram of A, we can use this ratio:

        $\dfrac{84.71 \text{ g B}}{15.29 \text{ g A}}$ = $\dfrac{?}{1.00 \text{ g A}}$ or $\dfrac{5.54 \text{ g B}}{1.00 \text{ g A}}$

An alternative approach to this solution provides 5.5 g B (2 sf) as an answer.

103. If 0.15 mol of A_2Z_3 has a mass of 15.9 g and 0.15 mol of AZ_2 has a mass of 9.3 g, we can calculate the molar mass of each.

        A_2Z_3 $\dfrac{15.9 \text{ g } A_2Z_3}{0.15 \text{ mol}}$ = 106 g/mol AZ_2 $\dfrac{9.3 \text{ g } AZ_2}{0.15 \text{ mol}}$ = 62 g/mol

From the compounds compositions we can write:

for A_2Z_3 : (2 mol A atoms)(AW_A) + (3 mol Z atoms)(AW_Z) = 106 g   (1)

for AZ_2 : (1 mol A atoms)(AW_A) + (2 mol Z atoms)(AW_Z) = 62 g   (2)

Solving in the second equation for **mol A atoms** :

$$\text{(1 mol A atoms)}(AW_A) = 62 \text{ g} - \text{(2 mol Z atoms)}(AW_Z)$$

Substituting into the first equation for the first term (multiply by 2) gives:

(2)(62 g - (2 mol Z atoms)(AW_Z)) + (3 mol Z atoms)(AW_Z) = 106 g

124 g - (4 mol Z atoms)(AW_Z) + (3 mol Z atoms)(AW_Z) = 106 g

$$\text{or } 124 \text{ g} - \text{(1 mol Z atoms)}(AW_Z) = 106 \text{ g}$$

$$-\text{(1 mol Z atoms)}(AW_Z) = 106 \text{ g} - 124 \text{ g}$$

$$\text{(1 mol Z atoms)}(AW_Z) = 18 \text{ g}$$

$$AW_Z = \frac{18 \text{ g}}{1 \text{ mol Z atoms}} \qquad AW_Z = 18 \text{ g/mol}$$

If the Atomic weight of Z is 18 g/mol we can solve for the atomic weight of A by using either equation (1) or (2)

(1 mol A atoms)(AW_A) + (2 mol Z atoms)(AW_Z) = 62 g   (2)

$$\text{(1 mol A atoms)}(AW_A) + \text{(2 mol Z atoms)}(\frac{18 \text{ g Z}}{\text{mol Z}}) = 62 \text{ g}$$

$$\text{(1 mol A atoms)}(AW_A) = 62 \text{ g} - 36 \text{ g or} \qquad AW_A = \frac{26 \text{ g}}{1 \text{ mol A atoms}}$$

The atomic weight of A is then 26 g/mol.

105. Weight % of O in compound MO_2 is 15.2 % so % of M in compound is 84.8 %.

From the formula we know that in 1 mol of the compound, there are 2 oxygen atoms per M atom, and 2 mol O atoms would have a mass of (2 x 16.00) 32.00 g.

Since 32.00 g is 15.2 % the compound, the compound must have a mass of:

$$32.00 \text{ g} = 15.2 \text{ % (mass compound)}$$

$$\frac{32.00 \text{ g}}{0.152} = 211 \text{ g}$$

$$\frac{X \text{ g}}{32.00 \text{ g}} = \frac{100 \text{ % mass of compound}}{15.2 \text{ % mass of compound}}$$

211 g is the molar mass of the compound, of which 32 g is oxygen, leaving 179 g for the metal. The formula indicates that this mass (~179 g) is one mol of M atoms ; the metal is either tantalum (AW = 180.8479) or hafnium (AW = 178.49).

Conceptual Questions

107. Li^+ 90 pm (ionic radius)
       F^- 119 pm
       Br   114 pm (atomic radius)

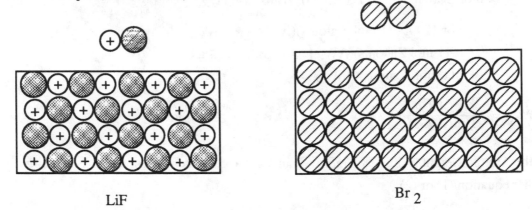

LiF Br_2

The diagram showing LiF indicates a regular arrangment of lithium and fluoride ions— stacked in a cubic array. Note that —with the exception of the ions "on the edges", each ion is surrounded in this 2-D drawing with 4 other "counter-ions".
Compare this to the Br_2 drawing. Note that the "schematic" liquid fills the container without the fixed arrangement associated with the LiF crystal lattice.

110.   Of the ions Cl^-, K^+, Mg^{2+}, Al^{3+}, which has the strongest attraction for water molecules in the vicinity? We can calculate the proportional charge density by dividing the ionic charges by the ionic radii:

Cl^-	K^+	Mg^{2+}	Al^{3+}
$\dfrac{1}{167 \text{ pm}}$	$\dfrac{1}{152 \text{ pm}}$	$\dfrac{2}{86 \text{ pm}}$	$\dfrac{3}{68 \text{ pm}}$

The charge density of Al^{3+} is greatest, and so this cation would have the greatest attraction to the dipole, H_2O.

36

112. The empirical formula may be calculated by determining (1) the moles of each element in a given sample and (2) the ratio of the moles of each element:

mol C $54.0 \text{ g C} \cdot \dfrac{1 \text{ mol C}}{12.01 \text{ g C}} = 4.50 \text{ mol C}$

mol H $6.00 \text{ g H} \cdot \dfrac{1 \text{ mol H}}{1.008 \text{ g H}} = 5.95 \text{ mol H}$

mol O $40.0 \text{ g O} \cdot \dfrac{1 \text{ mol O}}{16.00 \text{ g O}} = 2.50 \text{ mol O}$

The ratio of mol of Carbon and Hydrogen to mol of O:

$\dfrac{4.50 \text{ mol C}}{2.50 \text{ mol O}} = 1.80$ $\dfrac{5.95 \text{ mol H}}{2.50 \text{ mol O}} = 2.38$

The ratio would be $C_{1.80}H_{2.38}O_1$, obviously not a ratio of integral numbers of atoms. A trial and error multiplication of <u>each</u> subscript by 2, 3, 4, and 5 successively shows that when each subscript is multiplied by 5 an **integral ratio of 9 : 12 : 5 results**. The correct formula is then $C_9H_{12}O_5$. Incorrect answers arise in one of the two steps: (1) calculation of moles of each element or (2) calculation of the ratios of atoms present!

115. Calculate:

a. moles of nickel—found by density **once** the volume of foil is calculated.

$V = 1.25 \text{ cm} \times 1.25 \text{ cm} \times 0.0550 \text{ cm} = 8.59 \times 10^{-2} \text{ cm}^3$

$\text{Mass} = \dfrac{8.908 \text{ g}}{1 \text{ cm}^3} \cdot 8.59 \times 10^{-2} \text{ cm}^3 = 0.766 \text{ g Ni}$

$0.766 \text{ g Ni} \cdot \dfrac{1 \text{ mol Ni}}{58.69 \text{ g Ni}} = 1.30 \times 10^{-2} \text{ mol Ni}$

b. Formula for the fluoride salt:

$\text{Mass F} = (1.261 \text{ g salt} - 0.766 \text{ g Ni}) = 0.495 \text{ g F}$

$\text{Moles F} = 0.495 \text{ g F} \cdot \dfrac{1 \text{ mol F}}{19.00} = 2.60 \times 10^{-2} \text{ mol F},$

so 1.30×10^{-2} mol Ni combines with 2.60×10^{-2} mol F , indicating a formula of NiF_2

c. Name: Nickel(II) fluoride

Chapter 4
Principles of Reactivity: Chemical Reactions

Balancing Equations

15.   Balancing equations can be a matter of "running in circles" if a reasonable methodology is not employed. While there isn't one "right place" to begin, generally you will suffer fewer complications if you begin the balancing process using a substance that contains the greatest number of elements or the largest subscript values. Noting that you must have at least that many atoms of each element involved, coefficients can be used to increase the inventory of each atom. In the next few questions, you will see one **emboldened** substance in each equation. This emboldened substance is the one that I judge to be a "good" starting place. One last hint -- modify the coefficients of uncombined elements, i.e. those not in compounds, after you modify the coefficients for compounds containing those elements -- not before!

a. $4 Cr (s) + 3 O_2 (g) \longrightarrow 2 Cr_2O_3 (s)$

   1. Note the need for at least 2 Cr and 3 O atoms.
   2. Oxygen is diatomic -- we'll need an even number of oxygen atoms, so
        try : $2 Cr_2O_3$.
   3. $3 O_2$ would give 6 O atoms on both sides of the equation.
   4. 4 Cr would give 4 Cr atoms on both sides of the equation.

b. $Cu_2S (s) + O_2 (g) \longrightarrow 2 Cu(s) + \mathbf{SO_2} (g)$

   1. A minimum of 2 O in SO_2 is required, and is provided with one molecule of elemental oxygen.
   2. 2 Cu atoms (on the right) indicates 2 Cu (on the left).

c. $\mathbf{C_6H_5CH_3} (\ell) + 9 O_2 (g) \longrightarrow 4 H_2O (g) + 7 CO_2 (g)$

   1. A minimum of 7 C and 8 H is required.
   2. $7 CO_2$ furnishes 7 C and $4 H_2O$ furnishes 8 H atoms.
   3. $4 H_2O$ and $7 CO_2$ furnish a total of 18 O atoms, making the coefficient
        of $O_2 = 9$.

17. a. $3\,MgO\,(s)\ +2\,Fe\,(s)\ \rightarrow\ \mathbf{Fe_2O_3}\,(s)\ +3\,Mg(s)$

   b. $\mathbf{AlCl_3}\,(s)+3\,H_2O\,(\ell)\ \rightarrow\ Al(OH)_3\,(s)+3\,HCl\,(aq)$

   c. $2\,NaNO_3\,(s)\ +\ H_2SO_4\,(\ell)\ \rightarrow\ \mathbf{Na_2SO_4}\,(s)\ +2\,HNO_3\,(g)$

19. Balance:

   a. the synthesis of urea:
   $$CO_2(g)\ +\ 2\,NH_3(g)\ \longrightarrow\ \mathbf{CO(NH_2)_2}(s)\ +\ H_2O(\ell)$$
   1. Note the need for two NH_3 in each molecule of urea, so multiply NH_3 by 2.
   2. $2\,NH_3$ provides the two H atoms for a molecule of H_2O.
   3. Each CO_2 provides the O atom for a molecule of H_2O.

   b. the synthesis of uranium(VI) fluoride
   $$UO_2(s)\ +\ 4\,HF(aq)\ \longrightarrow\ \mathbf{UF_4}(s)\ +\ H_2O(aq)$$
   $$\mathbf{UF_4}(s)\ +\ F_2(g)\ \longrightarrow\ \mathbf{UF_6}(s)$$
   1. The 4 F atoms in UF_4 requires 4 F atoms from HF.   (equation 1)
   2. The H atoms in HF produce 2 molecules of H_2O.   (equation 1)
   3. The 1:1 stoichiometry of $UF_6 : UF_4$ provides a simple balance. (equation 2)

   c. synthesis of titanium metal from TiO_2:
   $$\mathbf{TiO_2}(s)\ +\ 2\,Cl_2(g)\ +\ 2\,C(s)\ \longrightarrow\ TiCl_4(\ell)\ +\ 2\,CO(g)$$
   $$\mathbf{TiCl_4}(\ell)\ +\ 2\,Mg(s)\ \longrightarrow\ Ti(s)\ +\ 2\,MgCl_2(s)$$
   1. The O balance mandates 2 CO for each TiO_2.   (equation 1)
   2. A coefficient of 2 for C provides C balance.   (equation 1)
   3. The Ti balance ($TiO_2 : TiCl_4$) requires 4 Cl atoms, hence $2\,Cl_2$ (equation 1)
   4. The Cl balance requires $2\,MgCl_2$, hence 2 Mg.   (equation 2)

Properties of Aqueous Solutions

21. a. $FeCl_2$ is expected to be soluble.
   b. $AgNO_3$ is soluble. AgI and Ag_3PO_4 are not soluble.
   c. NaCl and $KMnO_4$ are soluble. In general salts of the alkali metals are soluble.

23.   a. any alkali metal, e.g. $NaC_2H_3O_2$

b. any cation except alkali, alkaline earths, or NH_4^+ e.g. Fe_2S_3

c. any alkali metal hydroxide, e.g. KOH

d. Ag^+, Hg^{2+}, or Pb^{2+} form insoluble chlorides although $PbCl_2$ is slightly soluble
   at temperatures above $\sim 30\,°C$.

25.
Compound	Cation	Anion
a. KI	K^+	I^-
b. K_2SO_4	K^+	SO_4^{2-}
c. $KHSO_4$	K^+	HSO_4^-
d. KCN	K^+	CN^-

27.
Compound	Water Soluble	Cation	Anion
a. $BaCl_2$	yes	Ba^{2+}	Cl^-
b. $Cr(NO_3)_2$	yes	Cr^{2+}	NO_3^-
c. $Pb(NO_3)_2$	yes	Pb^{2+}	NO_3^-
d. $BaSO_4$	no		

29. Anions to produce a soluble Cu^{2+} salt: SO_4^{2-}, NO_3^-, $C_2H_3O_2^-$, ClO_3^-, ClO_4^-,
   halides (F^-, Cl^-, Br^-, I^-),

Anions to produce an insoluble Cu^{2+} salt: S^{2-}, O^{2-}, $C_2O_4^{2-}$, SO_3^{2-}, OH^-

Acids and Bases

31. $HNO_3\,(aq) + H_2O\,(\ell) \rightarrow H_3O^+\,(aq) + NO_3^-\,(aq)$
   alternatively $HNO_3\,(aq) \rightarrow H^+\,(aq) + NO_3^-\,(aq)$

33. $H_2C_2O_4(aq) \rightarrow H^+(aq) + HC_2O_4^-(aq)$
   $HC_2O_4^-(aq) \rightarrow H^+(aq) + C_2O_4^{2-}(aq)$

35. $MgO(s) + H_2O(\ell) \rightarrow Mg(OH)_2(aq)$

40

Writing Net Ionic Equations

37. a. $Zn (s) + 2 HCl (aq) \rightarrow H_2 (g) + ZnCl_2 (aq)$

   1. Write species as they exist in aqueous solution.

   $Zn (s) + 2 H^+ (aq) + 2 Cl^- (aq) \rightarrow H_2 (g) + Zn^{2+} (aq) + 2 Cl^- (aq)$

   2. Remove any species which appear <u>in exactly the same form</u> on both sides of the equation. For this reaction: $2 Cl^- (aq)$

   3. Examine the remaining species to see if a reduction of <u>every</u> coefficient is possible:

   $Zn (s) + 2 H^+ (aq) \rightarrow H_2 (g) + Zn^{2+} (aq)$

   b. $Mg(OH)_2 (s) + 2 HCl (aq) \rightarrow MgCl_2 (aq) + 2 H_2O (\ell)$

   1. $Mg(OH)_2(s) + 2 H^+ (aq) + 2 Cl^- (aq) \rightarrow Mg^{2+} (aq) + 2 Cl^- (aq) + 2 H_2O(\ell)$
   2. Spectator ions: $2 Cl^- (aq)$
   3. $Mg(OH)_2 (s) + 2 H^+ (aq) \rightarrow Mg^{2+} (aq) + 2 H_2O (\ell)$

   c. $2 HNO_3 (aq) + CaCO_3 (s) \rightarrow Ca(NO_3)_2 (aq) + H_2O (\ell) + CO_2 (g)$

   1. $2 H^+ (aq) + 2 NO_3^- (aq) + CaCO_3 (s) \rightarrow$
      $Ca^{2+} (aq) + 2 NO_3^- (aq) + H_2O (\ell) + CO_2 (g)$
   2. Spectator ions: $2 NO_3^- (aq)$
   3. $2 H^+ (aq) + CaCO_3 (s) \rightarrow Ca^{2+} (aq) + H_2O (\ell) + CO_2 (g)$

39. Balance the following equations, and then write the net ionic equation:

   a. $Ba(OH)_2 (s) + 2 HNO_3 (aq) \rightarrow Ba(NO_3)_2 (aq) + 2 H_2O (\ell)$
      $Ba(OH)_2 (s) + 2 H^+ (aq) \rightarrow 2 H_2O(\ell) + Ba^{2+}(aq)$

   b. $BaCl_2 (aq) + Na_2CO_3 (aq) \rightarrow BaCO_3 (s) + 2 NaCl (aq)$
      $Ba^{2+} (aq) + CO_3^{2-} (aq) \rightarrow BaCO_3 (s)$

   c. $2 Na_3PO_4(aq) + 3 Ni(NO_3)_2 (aq) \rightarrow Ni_3(PO_4)_2 (s) + 6 NaNO_3 (aq)$
      $2 PO_4^{3-} (aq) + 3 Ni^{2+} (aq) \rightarrow Ni_3(PO_4)_2 (s)$

41

Types of Reactions in Aqueous Solution

41. Acid-base reactions (AB) usually produce a salt and water. Precipitation reactions (PR) always form a salt which is insoluble (usually in water). Gas-forming reactions (GF) produce a gas. A chart of solubility rules may come in handy if you haven't already learned the rules.

   a. K_2CO_3 (aq) + $Cu(NO_3)_2$ (aq) $\rightarrow$ $CuCO_3$ (s) + 2 KNO_3 (aq)   PR

   b. $Pb(NO_3)_2$ (aq) + 2 HCl (aq) $\rightarrow$ $PbCl_2$ (s) + 2 HNO_3 (aq)   PR

   c. $MgCO_3$ (s) + 2 HCl (aq) $\rightarrow$ $MgCl_2$ (aq) + H_2O (ℓ) + CO_2(g)   GF

43. Acid-Base (AB), Precipitation (PR) or Gas-Forming (GF)

   a. $MnCl_2$(aq) + Na_2S(aq) $\rightarrow$ MnS(s) + 2 NaCl(aq)   PR

   b. K_2CO_3(aq) + $ZnCl_2$(aq) $\rightarrow$ $ZnCO_3$(s) + 2 KCl(aq)   PR

   c. K_2CO_3(aq) + 2 $HClO_4$(aq) $\rightarrow$ 2 $KClO_4$(aq) + CO_2(g) + H_2O(ℓ)   GF

   Net ionic equations:
   a. Mn^{2+}(aq) + S^{2-}(aq) $\rightarrow$ MnS(s)
   b. CO_3^{2-}(aq) + Zn^{2+}(aq) $\rightarrow$ $ZnCO_3$(s)
   c. CO_3^{2-}(aq) + 2 H^+(aq) $\rightarrow$ CO_2(g) + H_2O(ℓ)

Precipitation Reactions

45. $CdCl_2$(aq) + 2 NaOH(aq) $\rightarrow$ $Cd(OH)_2$ (s) + 2 NaCl (aq)
   Net ionic equation: Cd^{2+}(aq) + 2 OH^-(aq) $\rightarrow$ $Cd(OH)_2$ (s)

47. Balanced equations for precipitation reactions:

   a. $NiCl_2$(aq) + $(NH_4)_2S$(aq) $\rightarrow$ NiS(s) + 2 NH_4Cl(aq)

   b. 3 $Mn(NO_3)_2$(aq) + 2 Na_3PO_4(aq) $\rightarrow$ $Mn_3(PO_4)_2$(s) + 6 $NaNO_3$(aq)

49. $Pb(NO_3)_2$(aq) + 2 KOH(aq) $\rightarrow$ $Pb(OH)_2$(s) + 2 KNO_3(aq)
   lead(II)   potassium   lead (II)   potassium
   nitrate   hydroxide   hydroxide   nitrate

Acid-Base Reactions

51. a. $2\ CH_3CO_2H(aq)\ +\ Mg(OH)_2(s)\ \rightarrow\ Mg(CH_3CO_2)_2(aq)\ +\ 2\ H_2O(\ell)$

        acetic   magnesium   magnesium   water
        acid   hydroxide   acetate

    b. $HClO_4(aq)\ +\ NH_3(aq)\ \rightarrow\ NH_4ClO_4(aq)$

      perchloric   ammonia   ammonium
        acid   perchlorate

53. Write and balance the equation:

    $Ba(OH)_2\ (s)\ +\ 2\ HNO_3\ (aq)\ \rightarrow\ Ba(NO_3)_2\ (aq)\ +\ 2\ H_2O\ (\ell)$

Gas-Forming Reactions

55. $MnCO_3\ (s)\ +\ 2\ HCl\ (aq)\ \rightarrow\ MnCl_2\ (aq)\ +\ H_2O\ (\ell)\ +\ CO_2(g)$

  manganese(II)   hydrochloric   manganese(II)   water   carbon
    carbonate   acid   chloride   dioxide

Oxidation Numbers

57. For questions on oxidation number, read the symbol (x) as "the oxidation number of x."

        a. BrO_3^- $(Br) + 3(O)$ $= -1$

Since oxygen almost always has an oxidation number of -2, we can substitute this value and solve for the oxidation number of Br.

                        $(Br) + 3(-2)$ $= -1$
                        (Br) $= +5$

        b. $C_2O_4^{2-}$ $2\ (C) + 4\ (O)$ $= -2$
                        $2\ (C) + 4\ (-2)$ $= -2$
                        $2\ (C) + -8$ $= -2$
                        $2\ (C)$ $= +6$
                        (C) $= +3$

        c. F_2 The oxidation number for any free element is zero.

d. CaH_2 (Ca) + 2 (H)   = 0
                    (Ca) + 2 (-1)   = 0
                    (Ca)   = +2

e. H_4SiO_4 4(H) + (Si) + 4(O)   = 0
                    4(+1) + (Si) + 4(-2) = 0
                          (Si)   = +4

f. SO_4^{2-} (S) + 4(O) = -2
                    (S) + 4(-2) = -2
                    (S)   = +6

59. Oxidation number of atoms in the equation:

$$2 NaI(s) + 2 H_2SO_4(aq) + MnO_2(s) \rightarrow Na_2SO_4(aq) + MnSO_4(aq) + I_2(g) + 2 H_2O(\ell)$$

Na = +1	H = +1	Mn = +4	Na = +1	Mn = +2	I = 0	H = +1
I = -1	S = +6	O = -2	S = +6	S = +6		O = -2
	O = -2		O = -2	O = -2		

Oxidation-Reduction Reactions

61. Identify the oxidation-reduction (OR) reactions:

    a. $CdCl_2(aq) + Na_2S(aq) \rightarrow CdS(s) + 2 NaCl(aq)$ PR

    b. $2 Ca(s) + O_2(g) \rightarrow 2 CaO(s)$ OR
    The oxidation state of Ca changes from $0 \rightarrow +2$, indicating a loss of electrons (oxidation).
    The oxidation state of O changes from $0 \rightarrow -2$, indicating a gain of electrons (reduction).
    c. $Ca(OH)_2(s) + 2 HCl(aq) \rightarrow CaCl_2(aq) + 2 H_2O(\ell)$ AB

63. Determine which reactant is oxidized and which is reduced:
    a. $2 Mg (s) + O_2 (g) \rightarrow 2 MgO (s)$

	ox. number			
specie	before	after	has experienced	functions as the
Mg	0	+2	oxidation	(Mg) reducing agent
O	0	-2	reduction	(O_2) oxidizing agent

44

b. C_2H_4 (g) + 3 O_2 (g) → 2 CO_2 (g) + 2 H_2O (g)

ox. number

specie	before	after	has experienced	functions as the	
C	-2	+4	oxidation	(C_2H_4)	reducing agent
H	+1	+1	no change		
O	0	-2	reduction	(O_2)	oxidizing agent

c. Si (s) + 2 Cl_2 (g) → $SiCl_4$ (ℓ)

ox. number

specie	before	after	has experienced	functions as the	
Si	0	+4	oxidation	(Si)	reducing agent
Cl	0	-1	reduction	(Cl_2)	oxidizing agent

General Questions

65. $MgCO_3$ (s) + 2 H^+(aq) + 2 Cl^-(aq) → CO_2(g) + Mg^{2+}(aq) + 2 Cl^-(aq) + H_2O (ℓ)

1. Write species as they exist in aqueous solution, and name the spectator ions.

$MgCO_3$ (s) + 2 H^+(aq) + 2 Cl^-(aq) → CO_2(g) + Mg^{2+}(aq) + 2 Cl^-(aq) + H_2O (ℓ)

spectator ion (chloride ion)

2. Remove any species which appear in exactly the same form on both sides of the equation. For this reaction: 2 Cl^- (aq).

$MgCO_3$ (s) + 2 H^+(aq) → CO_2(g) + Mg^{2+}(aq) + H_2O (ℓ)

3. This is a gas-forming reaction.

67.   a. A balanced equation:

$(NH_4)_2S$(aq) + $Hg(NO_3)_2$(aq) → HgS(s) + 2 NH_4NO_3(aq)

b. compounds named:

ammonium	mercury(II)	mercury(II)	ammonium
sulfide	nitrate	sulfide	nitrate

c. This is a **precipitation** reaction.

69. $Cu_3(CO_3)_2(OH)_2(s) + HCl(aq) \rightarrow$?

  Note that azurite is a basic carbonate, that is it contains both OH^- and CO_3^{2-} moieties.

  Since both these groups react with the strong acid HCl, we anticipate the following:

$$2\,OH^- + 2\,HCl \rightarrow 2\,H_2O + 2\,Cl^-$$
$$2\,CO_3^{2-} + 4\,HCl \rightarrow 2\,H_2O + 2\,CO_2 + 4\,Cl^-$$

  So, the overall reaction would involve:

$$Cu_3(CO_3)_2(OH)_2(s) + 6\,HCl(aq) \rightarrow 4\,H_2O(\ell) + 6\,Cl^-(aq) + 2\,CO_2(g) + 3\,Cu^{2+}(aq)$$

71.   a. $MnCl_2$ (aq) + Na_2S (aq) $\rightarrow$ 2 NaCl (aq) + MnS (s)   PR
          net: Mn^{2+}(aq) + S^{2-}(aq) $\rightarrow$ MnS(s)

      b. K_2CO_3(aq) + $ZnCl_2$(aq) $\rightarrow$ $ZnCO_3$(s) + 2 KCl(aq)   PR
          net: CO_3^{2-}(aq) + Zn^{2+}(aq) $\rightarrow$ $ZnCO_3$(s)

      c. K_2CO_3(aq) + 2 $HClO_4$(aq) $\rightarrow$ 2 $KClO_4$ (aq) + CO_2(g) + H_2O (ℓ)   GF
          net: CO_3^{2-}(aq) + 2 H^+(aq) $\rightarrow$ CO_2(g) + H_2O (ℓ)

          Acid-base (AB), Precipitation reactions (PR), or Gas-forming (GF)

73. $C_6H_8O_6$(aq) + Br_2(aq) $\rightarrow$ 2 HBr(aq) + $C_6H_6O_6$(aq)

Element oxidized:	C (oxidation state +4/6 in $C_6H_8O_6$; +1 in $C_6H_6O_6$)
Element reduced:	Br (oxidation state 0 in Br_2 ; -1 in HBr)
Oxidizing agent:	Br_2 — causes oxidation of C
Reducing agent:	$C_6H_8O_6$ — causes reduction of Br

Conceptual Questions

75. A precipitation reaction to form barium sulfate:
      $BaCl_2$ (aq) + Na_2SO_4 (aq) $\rightarrow$ $BaSO_4$ (s) + 2 NaCl (aq)

  A gas-forming reaction to form barium sulfate:
      $BaCO_3$ (s) + H_2SO_4 (aq) $\rightarrow$ $BaSO_4$ (s) + CO_2(g) + $H_2O(\ell)$

Summary Question

78. a. Balanced equation:

    $$CaF_2 (s) + H_2SO_4 (aq) \rightarrow CaSO_4 (s) + 2 HF (g)$$

    compounds named:

calcium	sulfuric	calcium	hydrogen
fluoride	acid	sulfate	fluoride

   b. No element changes oxidation state, so this is **NOT** an oxidation-reduction reaction. Since a base is defined as a substance that increases the concentration of OH^- in solution, this reaction contains no base. A solid ($CaSO_4$) is formed in this reaction, so it is a precipitation reaction. Additionally HF is formed, so it is also a gas-forming reaction.

   c. Supply names

    $$2 HF(g) + CCl_4(\ell) \rightarrow CCl_2F_2(g) + 2 HCl(g)$$

carbon		hydrogen
tetrachloride		chloride

   d. The empirical formula for the compound that is 8.74% C, 77.43% Cl, and 13.83% F:

   Assuming that we have 100 g of this compound, calculate the moles of each atom involved:

   $$8.74 \text{ g C} \cdot \frac{1 \text{ mol C}}{12.011 \text{ g C}} = 0.728 \text{ mol C}$$

   $$77.43 \text{ g Cl} \cdot \frac{1 \text{ mol Cl}}{35.45 \text{ g Cl}} = 2.184 \text{ mol Cl}$$

   $$13.83 \text{ g F} \cdot \frac{1 \text{ mol F}}{18.998 \text{ g F}} = 0.728 \text{ mol F}$$

   Establish the ratio of moles of C : Cl : F atoms by dividing each of the three by the smallest (0.728) giving: $C_1Cl_3F_1$. The empirical formula would be CCl_3F.

Chapter 5
Stoichiometry

General Stoichiometry

1. Since we're told that 26.6 g of oxygen is the exact amount of oxygen needed to react with 10.0 g C, the mass of CO_2 obtainable is (10.0 + 26.6) or 36.6 g CO_2.

3. The formula weight for GaAs is 144.6 g/mol

The % of Ga is $\dfrac{69.72}{144.6}$ or 48.2 % and the % of As is (100.0 - 48.2) or 51.8 %.

So 1.45 g GaAs $\cdot$ $\dfrac{69.72 \text{ g Ga}}{144.6 \text{ g GaAs}}$ = 0.699 g Ga

and the amount of As needed is (1.45 g - 0.699 g) or 0.75 g As

5. Balance the equation:

$$Co(s) + 2\, HCl(aq) \longrightarrow CoCl_2(aq) + H_2(g)$$

$\dfrac{2.56 \text{ g Co}}{1}$ $\cdot$ $\dfrac{1 \text{ mol Co}}{58.93 \text{ g Co}}$ $\cdot$ $\dfrac{129.8 \text{ g CoCl}_2}{1 \text{ mol CoCl}_2}$ = 5.64 g $CoCl_2$

Noting that the ratio of H_2 : Co is 1:1, the mass of H_2 can be calculated:

$\dfrac{2.56 \text{ g Co}}{1}$ $\cdot$ $\dfrac{1 \text{ mol Co}}{58.93 \text{ g Co}}$ $\cdot$ $\dfrac{1 \text{ mol H}_2}{1 \text{ mol Co}}$ $\cdot$ $\dfrac{2.02 \text{ g H}_2}{1 \text{ mol H}_2}$ = 0.0876 g H_2

7. Moles and Mass of O_2:

$\dfrac{2.2 \text{ mol NO}}{1}$ $\cdot$ $\dfrac{1 \text{ mol O}_2}{2 \text{ mol NO}}$ = 1.1 mol O_2

and 1.1 mol of O_2 have a mass of : $\dfrac{32 \text{ g O}_2}{1 \text{ mol O}_2}$ $\cdot$ $\dfrac{1.1 \text{ mol O}_2}{1}$ = 35 g O_2

Mass of NO_2 produced:

2.2 mol NO $\cdot$ $\dfrac{2 \text{ mol NO}_2}{2 \text{ mol NO}}$ $\cdot$ $\dfrac{46.0 \text{ g NO}_2}{1 \text{ mol NO}_2}$ = 101.2 g NO_2 or 1.0 x 10^2 g NO_2 (2 sf)

9. $Fe_2O_3(s) + 3\, CO(g) \longrightarrow 2\, Fe(s) + 3\, CO_2(g)$

a. Mass of Fe obtained from 454 g Fe_2O_3:

454 g Fe_2O_3 $\cdot$ $\dfrac{1 \text{ mol Fe}_2O_3}{159.7 \text{ g Fe}_2O_3}$ $\cdot$ $\dfrac{2 \text{ mol Fe}}{1 \text{ mol Fe}_2O_3}$ $\cdot$ $\dfrac{55.847 \text{ g Fe}}{1 \text{ mol Fe}}$ = 318 g Fe

b. Mass of CO to consume 454 g Fe_2O_3:

$$454 \text{ g Fe}_2\text{O}_3 \cdot \frac{1 \text{ mol Fe}_2\text{O}_3}{159.7 \text{ g Fe}_2\text{O}_3} \cdot \frac{3 \text{ mol CO}}{1 \text{ mol Fe}_2\text{O}_3} \cdot \frac{28.01 \text{ g CO}}{1 \text{ mol CO}} = 239 \text{ g CO}$$

11. $2 \text{ SO}_2\text{(g)} + 2 \text{ CaCO}_3\text{(s)} + \text{O}_2\text{(g)} \longrightarrow 2 \text{ CaSO}_4\text{(s)} + 2 \text{ CO}_2\text{(g)}$

    a. Compounds involved:

sulfur dioxide	calcium carbonate	calcium sulfate	carbon dioxide

    b. Mass of $CaCO_3$ to remove 150. g SO_2:

$$150. \text{ g SO}_2 \cdot \frac{1 \text{ mol SO}_2}{64.07 \text{ g SO}_2} \cdot \frac{2 \text{ mol CaCO}_3}{2 \text{ mol SO}_2} \cdot \frac{100.1 \text{ g CaCO}_3}{1 \text{ mol CaCO}_3} = 234 \text{ g CaCO}_3$$

    c. Mass of $CaSO_4$ formed:

$$2.34 \text{ mol SO}_2 \cdot \frac{2 \text{ mol CaSO}_4}{2 \text{ mol SO}_2} \cdot \frac{136.1 \text{ g CaSO}_4}{1 \text{ mol CaSO}_4} = 319 \text{ g CaSO}_4$$

13. Decomposition of NH_4NO_3 :

    a. Balanced equation: $NH_4NO_3\text{(s)} \longrightarrow N_2O\text{(g)} + 2 \text{ H}_2\text{O}(\ell)$

    b. Mass of N_2O and H_2O from 10.0 g of NH_4NO_3:

$$10.0 \text{ g NH}_4\text{NO}_3 \cdot \frac{1 \text{ mol NH}_4\text{NO}_3}{80.04 \text{ g NH}_4\text{NO}_3} = 0.125 \text{ mol NH}_4\text{NO}_3$$

$$0.125 \text{ mol NH}_4\text{NO}_3 \cdot \frac{1 \text{ mol N}_2\text{O}}{1 \text{mol NH}_4\text{NO}_3} \cdot \frac{44.01 \text{ g N}_2\text{O}}{1 \text{ mol N}_2\text{O}} = 5.50 \text{ g N}_2\text{O}$$

$$\text{and } 0.125 \text{ mol NH}_4\text{NO}_3 \cdot \frac{2 \text{ mol H}_2\text{O}}{1 \text{mol NH}_4\text{NO}_3} \cdot \frac{18.02 \text{ g H}_2\text{O}}{1 \text{ mol H}_2\text{O}} = 4.50 \text{ g H}_2\text{O}$$

Note: Given that N_2O and H_2O are the only products, once the mass of one (say N_2O) is determined, The Law of Conservation of Mass would provide a quick calculation for the mass of the other product: 10.0 g - 5.50 g N_2O = 4.5 g H_2O

Limiting Reagent

15. $2 Al(s) + 3 Cl_2(g) \longrightarrow 2 AlCl_3(s)$

    a. Limiting reagent:

    $$2.70 \text{ g Al} \cdot \frac{1 \text{ mol Al}}{26.98 \text{ g Al}} = 0.100 \text{ mol Al}$$

    $$4.05 \text{ g Cl}_2 \cdot \frac{1 \text{ mol Cl}_2}{70.91 \text{ g Cl}_2} = 0.0571 \text{ mol Cl}_2$$

    Since each mol of Al requires 1.5 mol Cl_2, we clearly have a deficiency of Cl_2, so **Cl_2 is the Limiting Reagent**.

    b. Mass of $AlCl_3$ possible:

    $$0.0571 \text{ mol Cl}_2 \cdot \frac{2 \text{ mol AlCl}_3}{3 \text{ mol Cl}_2} \cdot \frac{133.3 \text{ g AlCl}_3}{1 \text{ mol AlCl}_3} = 5.08 \text{ g AlCl}_3$$

    c. Mass of Al remaining:

    $$0.0571 \text{ mol Cl}_2 \cdot \frac{2 \text{ mol Al}}{3 \text{ mol Cl}_2} \cdot \frac{26.98 \text{ g Al}}{1 \text{ mol Al}} = 1.03 \text{ g Al consumed}$$

    Remaining = (2.70 - 1.03) = 1.67 g Al

17. The limiting reagent can be determined by calculating the moles-available and moles-required ratios for the equation:

    $$CO (g) + 2 H_2 (g) \rightarrow CH_3OH (\ell)$$

    Moles of each reactant present:

    $$12.0 \text{ g H}_2 \cdot \frac{1 \text{ mol H}_2}{2.016 \text{ g H}_2} = 5.95 \text{ mol H}_2$$

    $$74.5 \text{ g CO} \cdot \frac{1 \text{ mol CO}}{28.01 \text{ g CO}} = 2.66 \text{ mol CO}$$

    moles-required ratio: $\dfrac{2 \text{ mol H}_2}{1 \text{ mol CO}}$

    moles-available ratio: $\dfrac{5.95 \text{ mol H}_2}{2.66 \text{ mol CO}} = \dfrac{2.24 \text{ mol H}_2}{1 \text{ mol CO}}$

    Since we require 2 moles of hydrogen per mole of CO, and we have available 2.24 moles of hydrogen per mole of CO, **CO is the limiting reagent** and **H_2 is present in excess**.

To determine the mass of excess reagent remaining after the reaction is complete, calculate the amount needed:

$$2.66 \text{ mol CO} \cdot \frac{2 \text{ mol H}_2}{1 \text{ mol CO}} \cdot \frac{2.016 \text{ g H}_2}{1 \text{ mol H}_2} = 10.7 \text{ g H}_2 \text{ needed}$$

Excess H_2 = 12.0 - 10.7 = 1.3 g

We calculate the theoretical yield of CH_3OH from the amount of limiting reagent:

$$2.66 \text{ mol CO} \cdot \frac{1 \text{ mol CH}_3\text{OH}}{1 \text{ mol CO}} \cdot \frac{32.04 \text{ g CH}_3\text{OH}}{1 \text{ mol CH}_3\text{OH}} = 85.2 \text{ g } CH_3OH$$

19.   The limiting reagent can be determined by calculating the mole-available and mole-needed ratios for the equation:

$$CaO \text{ (s)} + 2 \text{ NH}_4\text{Cl (s)} \rightarrow 2 \text{ NH}_3 \text{ (g)} + \text{H}_2\text{O (g)} + \text{CaCl}_2 \text{ (s)}$$

Calculate the moles of CaO and of NH_4Cl :

$$112 \text{ g CaO} \cdot \frac{1 \text{ mol CaO}}{56.08 \text{ g CaO}} = 2.00 \text{ mol CaO}$$

$$224 \text{ g NH}_4\text{Cl} \cdot \frac{1 \text{ mol NH}_4\text{Cl}}{53.49 \text{ g NH}_4\text{Cl}} = 4.19 \text{ mol NH}_4\text{Cl}$$

moles-required ratio: $\dfrac{2 \text{ mol NH}_4\text{Cl}}{1 \text{ mol CaO}}$

moles-available ratio: $\dfrac{4.19 \text{ mol NH}_4\text{Cl}}{2.00 \text{ mol CaO}} = \dfrac{2.10 \text{ mol NH}_4\text{Cl}}{1.00 \text{ mol CaO}}$

CaO is the limiting reagent, and will determine the maximum amount of products obtainable:

$$112 \text{ g CaO} \cdot \frac{1 \text{ mol CaO}}{56.08 \text{ g CaO}} \cdot \frac{2 \text{ mol NH}_3}{1 \text{ mol CaO}} \cdot \frac{17.03 \text{g NH}_3}{1 \text{ mol NH}_3} = 68.0 \text{ g NH}_3$$

The balanced equation shows that for each mole of CaO, 2 moles of NH_4Cl are required. So 2.00 mol of CaO would require 4.00 mol of NH_4Cl, leaving (4.19 - 4.00) 0.19 mol of NH_4Cl in excess. This number of moles would have a mass of:

$$0.19 \text{ mol NH}_4\text{Cl} \cdot \frac{53.49 \text{ g NH}_4\text{Cl}}{1 \text{ mol NH}_4\text{Cl}} = 10.3 \text{ g NH}_4\text{Cl}$$

Percent Yield

21. Percent yield of NH_3:

$$\frac{\text{actual}}{\text{theoretical}} = \frac{100.\text{ g } NH_3}{136 \text{ g } NH_3} \times 100 = 73.5 \text{ % yield}$$

23. Theoretical yield of $ZnCl_2$:

Calculate the % of Zn in $ZnCl_2$: $\frac{65.39 \text{ g Zn}}{136.30 \text{ g } ZnCl_2}$

So: 35.5 g Zn $\cdot \dfrac{136.30 \text{ g } ZnCl_2}{65.39 \text{ g Zn}} = 74.0 \text{ g } ZnCl_2$

% yield of $ZnCl_2$:

$$\frac{\text{actual}}{\text{theoretical}} = \frac{65.2 \text{ g } ZnCl_2}{74.0 \text{ g } ZnCl_2} \times 100 = 88.1 \text{ % yield}$$

Chemical Analysis

25. Weight percent of $CuSO_4 \cdot 5\,H_2O$ in the mixture:

Mass of H_2O = 1.245 g - 0.832 g = 0.413 g H_2O

Since this water was a part of the hydrated salt, let's calculate the mass of that salt present: In 1 mol of $CuSO_4 \cdot 5\,H_2O$ there are 90.08 g H_2O and 159.61 g $CuSO_4$ or 249.69 g $CuSO_4 \cdot 5\,H_2O$. So:

$$0.413 \text{ g } H_2O \cdot \frac{249.69 \text{ g } CuSO_4 \cdot 5\,H_2O}{90.08 \text{ g } H_2O} = 1.14 \text{ g hydrated salt}$$

$$\text{% hydrated salt} = \frac{1.14 \text{ g hydrated salt}}{1.245 \text{ g mixture}} \times 100 = 91.6\%$$

27. The moles of benzene present are:

$$0.951 \text{ g } C_6H_6 \cdot \frac{1 \text{ mol } C_6H_6}{78.11 \text{ g } C_6H_6} = 1.22 \times 10^{-2} \text{ mol } C_6H_6$$

The amount of $Al(C_6H_5)_3$ present is:

$$1.22 \times 10^{-2} \text{ mol } C_6H_6 \cdot \frac{1 \text{ mol } Al(C_6H_5)_3}{3 \text{ mol } C_6H_6} \cdot \frac{258.30 \text{ g } Al(C_6H_5)_3}{1 \text{ mol} Al(C_6H_5)_3} =$$

$$1.05 \text{ g } Al(C_6H_5)_3$$

The percent of $Al(C_6H_5)_3$ in the sample is: $\dfrac{1.05 \text{ g } Al(C_6H_5)_3}{1.25 \text{ g sample}} \cdot 100 = 83.9\%$

Determination of Empirical Formulas

29. The basic equation is:

$$C_xH_y + O_2 \longrightarrow x\,CO_2 + \frac{y}{2}\,H_2O$$

Without balancing the equation, one can see that all the C in CO_2 comes from the styrene as does all the H in H_2O. Let's use the percentage of C in CO_2 to determine the mass of C in styrene, and the percentage of H in H_2O to provide the mass of H in styrene.

$$1.481 \text{ g } CO_2 \cdot \frac{12.01 \text{ g C}}{44.01 \text{ g } CO_2} = 0.404 \text{ g C}$$

Similarly :

$$0.303 \text{ g } H_2O \cdot \frac{2.02 \text{ g H}}{18.02 \text{ g } H_2O} = 0.0340 \text{ g H}$$

Alternatively, the mass of H could be determined by subtracting the mass of C from the 0.438 g styrene

Mass H = 0.438 g styrene - 0.404 g C

Establish the ratio of C atoms to H atoms

$$0.0340 \text{ g H} \cdot \frac{1 \text{ mol H}}{1.008 \text{ g H}} = 0.0337 \text{ mol H}$$

$$0.404 \text{ g C} \cdot \frac{1 \text{ mol C}}{12.011 \text{ g C}} = 0.0336 \text{ mol C}$$

This number of H and C atoms indicates an empirical formula for styrene of 1:1 or C_1H_1.

31. Let's follow the model we used in Question 29. Note the difference here - propanoic acid contains C, H, **and** O. So we can determine the masses of C and H and, by difference, determine the mass of O.

$$0.421 \text{ g } CO_2 \cdot \frac{12.011 \text{ g C}}{44.01 \text{ g } CO_2} = 0.115 \text{ g C}$$

$$0.172 \text{ g } H_2O \cdot \frac{2.02 \text{ g H}}{18.02 \text{ g } H_2O} = 0.0193 \text{ g H}$$

The mass of O present is then:

$$
\begin{aligned}
&0.236 \text{ g acid} \\
&\underline{-0.115 \text{ g C}} \\
&\underline{-0.0193 \text{ g H}} \\
&0.102 \text{ g O}
\end{aligned}
$$

Now determine the moles of C, H, O atoms present:

$$0.115 \text{ g C} \cdot \frac{1 \text{ mol C}}{12.011 \text{ g C}} = 0.00957 \text{ mol C}$$

$$0.0193 \text{ g H} \cdot \frac{1 \text{ mol H}}{1.0079 \text{ g H}} = 0.01949 \text{ mol H}$$

$$0.102 \text{ g O} \cdot \frac{1 \text{ mol O}}{15.9994 \text{ g O}} = 0.006375 \text{ mol O}$$

Dividing each of these by the smallest (0.006375) indicates an empirical formula for propanoic acid of 1.5 : 3 : 1 or $C_3H_6O_2$.

33. Calculate the mass of Si in the compounds; using the % of Si in SiO_2:

$$11.64 \text{ g SiO}_2 \cdot \frac{28.09 \text{ g Si}}{60.084 \text{ g SiO}_2} = 5.442 \text{ g Si or } 0.1937 \text{ mol Si}$$

and the mass of H :

$$6.980 \text{ g H}_2\text{O} \cdot \frac{2.016 \text{ g H}}{18.02 \text{ g H}_2\text{O}} = 0.7809 \text{ g H or } 0.7747 \text{ mol H}$$

The ratio of Si:H is $\dfrac{0.7747 \text{ mol H}}{0.1937 \text{ mol Si}} = 4$

The empirical formula is then SiH_4.

Solution Concentration

35. Molarity of Na_2CO_3 solution:

$$6.73 \text{ g Na}_2\text{CO}_3 \cdot \frac{1 \text{ mol Na}_2\text{CO}_3}{106.0 \text{ g Na}_2\text{CO}_3} = 0.0635 \text{ mol Na}_2\text{CO}_3$$

$$\text{Molarity} \equiv \frac{\# \text{ mol}}{L} = \frac{0.0635 \text{ mol Na}_2\text{CO}_3}{0.250 \text{ L}} = 0.254 \text{ M Na}_2\text{CO}_3$$

Concentration of Na^+ and CO_3^{2-} ions:

$$\frac{0.254 \text{ mol Na}_2\text{CO}_3}{L} \cdot \frac{2 \text{ mol Na}^+}{1 \text{ mol Na}_2\text{CO}_3} = 0.508 \text{ M Na}^+$$

$$\frac{0.254 \text{ mol Na}_2\text{CO}_3}{L} \cdot \frac{1 \text{ mol CO}_3^{2-}}{1 \text{ mol Na}_2\text{CO}_3} = 0.254 \text{ M CO}_3^{2-}$$

37. Mass of $KMnO_4$:

$$\frac{0.0125 \text{ mol } KMnO_4}{L} \cdot \frac{0.250 \text{ L}}{1} \cdot \frac{158.0 \text{ g } KMnO_4}{1 \text{ mol } KMnO_4} = 0.494 \text{ g } KMnO_4$$

39. Volume of 0.123 M NaOH to contain 25.0 g NaOH:

Calculate moles of NaOH in 25.0 g:

$$\frac{25.0 \text{ g NaOH}}{1} \cdot \frac{1 \text{ mol NaOH}}{40.00 \text{ g NaOH}} = 0.625 \text{ mol NaOH}$$

The volume of 0.123 M NaOH that contains 0.625 mol NaOH:

$$0.625 \text{ mol NaOH} \cdot \frac{1 \text{ L}}{0.123 \text{ mol NaOH}} \cdot \frac{1 \times 10^3 \text{ mL}}{1L} = 5.08 \times 10^3 \text{ mL}$$

41. Molarity of Cu(II) sulfate in the diluted solution:

We can calculate the molarity if we know the number of moles of $CuSO_4$ in the 10.0 mL solution

1. Moles of $CuSO_4$ in 4.00 mL of 0.0250 M $CuSO_4$:

$$M \times V = \frac{0.0250 \text{ mol } CuSO_4}{L} \cdot \frac{4.00 \times 10^{-3} \text{ L}}{1} = 1.00 \times 10^{-4} \text{ mol } CuSO_4$$

2. When that number of moles is distributed in 10.0 mL:

$$\frac{1.00 \times 10^{-4} \text{ mol } CuSO_4}{10.0 \times 10^{-3} \text{ L}} = 0.0100 \text{ M } CuSO_4$$

Perhaps a shorter way to solve this problem is to note the number of moles (found by multiplying the original molarity times the volume) is distributed in a given volume, resulting in the diluted molarity. Mathematically: $M_1 \times V_1 = M_2 \times V_2$

43. Using the formula discussed in 41, we can calculate the concentration of H_2SO_4 in the following:

a. $\dfrac{1.25 \text{ mol } H_2SO_4}{L} \cdot 36.0 \times 10^{-3} \text{ L} = M_2 \cdot 1.00 \text{ L}$

$\dfrac{0.0450 \text{ mol } H_2SO_4}{1.00 \text{ L}} = 0.0450 \text{ M} = M_2$

b. $\dfrac{6.00 \text{ mol } H_2SO_4}{L} \cdot 20.8 \times 10^{-3} \text{ L} = M_2 \cdot 1.00 \text{ L}$

$\dfrac{0.125 \text{ mol } H_2SO_4}{1.00 \text{ L}} = 0.125 \text{ M} = M_2$

c. $\dfrac{3.00 \text{ mol H}_2\text{SO}_4}{L} \cdot 50.0 \times 10^{-3}\text{ L} = M_2 \cdot 1.00\text{ L}$

$\dfrac{0.150 \text{ mol H}_2\text{SO}_4}{1.00 \text{ L}} = 0.150\text{ M} = M_2$

d. $0.500 \text{ mol H}_2\text{SO}_4 \cdot 0.500\text{ L} = M_2 \cdot 1.00\text{ L}$

$\dfrac{0.250 \text{ mol H}_2\text{SO}_4}{1.00 \text{ L}} = 0.250\text{ M} = M_2$

Method (b) would provide 1.00 L of 0.125 M H_2SO_4

45. Concentration of ions found in:

a. 0.12 M $BaCl_2$ 0.12 M Ba^{2+} and 0.24 M Cl^-
b. 0.0125 M $CuSO_4$ 0.0125 M Cu^{2+} and 0.0125 M SO_4^{2-}
c. 0.146 M $AlCl_3$ 0.146 M Al^{3+} and 0.438 M Cl^-
d. 0.500 M $K_2Cr_2O_7$ 1.000 M K^+ and 0.500 M $Cr_2O_7^{2-}$

Stoichiometry of Reactions in Solution

47. Calculate:

1. moles of HNO_3 present
2. moles of Na_2CO_3 that react with that amount of HNO_3
3. mass of Na_2CO_3 corresponding to that number of moles

The balanced equation is:

$$2 \text{ HNO}_3 \text{ (aq)} + \text{Na}_2\text{CO}_3 \text{ (aq)} \rightarrow 2\text{ NaNO}_3 \text{ (aq)} + \text{CO}_2 \text{ (g)} + \text{H}_2\text{O} \text{ (}\ell\text{)}$$

$25.0 \text{ mL} \cdot \dfrac{1\text{ L}}{1000\text{ mL}} \cdot \dfrac{0.155 \text{ mol HNO}_3}{1\text{ L}} \cdot \dfrac{1 \text{ mol Na}_2\text{CO}_3}{2 \text{ mol HNO}_3}$

$\cdot \dfrac{106.0 \text{ g Na}_2\text{CO}_3}{1 \text{ mol Na}_2\text{CO}_3} = 0.205 \text{ g Na}_2\text{CO}_3$

49. Mass of NaOH formed from 10.0 L of 0.15 M NaCl:

$\dfrac{0.15 \text{ mol NaCl}}{1\text{ L}} \cdot \dfrac{10.0\text{ L}}{1} \cdot \dfrac{2 \text{ mol NaOH}}{2 \text{ mol NaCl}} \cdot \dfrac{40.0 \text{ g NaOH}}{1 \text{ mol NaOH}} = 60. \text{ g NaOH}$

Mass of Cl_2 obtainable:

$$\frac{0.15 \text{ mol NaCl}}{1 \text{ L}} \cdot \frac{10.0 \text{ L}}{1} \cdot \frac{1 \text{ mol } Cl_2}{2 \text{ mol NaCl}} \cdot \frac{70.9 \text{ g } Cl_2}{1 \text{ mol } Cl_2} = 53 \text{ g } Cl_2$$

51. Calculate

    1. mol of AgBr

    2. mol of $Na_2S_2O_3$ needed to react (balanced equation)

    3. volume of 0.0138 M $Na_2S_2O_3$ containing that number of moles.

$$0.250 \text{ g AgBr} \cdot \frac{1 \text{ mol AgBr}}{187.8 \text{ g AgBr}} \cdot \frac{2 \text{ mol } Na_2S_2O_3}{1 \text{ mol AgBr}} \cdot \frac{1 \text{ L}}{0.0138 \text{ mol } Na_2S_2O_3}$$

$$\cdot \frac{1000 \text{ mL}}{1 \text{ L}} = 193 \text{ mL } Na_2S_2O_3$$

53. The balanced equation:

$$Pb(NO_3)_2 + 2 \text{ NaCl} \rightarrow PbCl_2 + 2 \text{ NaNO}_3$$

Volume of 0.750 M $Pb(NO_3)_2$ needed:

$$\frac{2.25 \text{ mol NaCl}}{1 \text{ L}} \cdot \frac{1.00 \text{ L}}{1} \cdot \frac{1 \text{ mol } Pb(NO_3)_2}{2 \text{ mol NaCl}} \cdot \frac{1 \text{ L}}{0.750 \text{ mol } Pb(NO_3)_2} \cdot \frac{1000 \text{ mL}}{1 \text{ L}}$$

$$= 1500 \text{ mL or } 1.50 \times 10^3 \text{ mL}$$

55. The balanced equation:

$$AgNO_3 + \text{NaCl} \rightarrow AgCl + \text{NaNO}_3$$

Calculate moles of each reactant:

$$\frac{0.025 \text{ mol } AgNO_3}{1 \text{ L}} \cdot \frac{0.0500 \text{ L}}{1} = 1.25 \times 10^{-3} \text{ mol } AgNO_3$$

$$\frac{0.025 \text{ mol NaCl}}{1 \text{ L}} \cdot \frac{0.1000 \text{ L}}{1} = 2.50 \times 10^{-3} \text{ mol NaCl}$$

Reagent in excess: NaCl

Maximum mass of AgCl is limited by $AgNO_3$ present:

$$1.25 \times 10^{-3} \text{ mol } AgNO_3 \cdot \frac{1 \text{ mol AgCl}}{1 \text{ mol } AgNO_3} \cdot \frac{143.3 \text{ g AgCl}}{1 \text{ mol AgCl}} = 0.18 \text{ g AgCl (2 sf)}$$

Concentration of NaCl remaining:

Moles of NaCl remaining $= (2.50 \times 10^{-3} - 1.25 \times 10^{-3}) = 1.25 \times 10^{-3}$ mol NaCl

$$\frac{\text{moles}}{\text{volume}} = \frac{1.25 \times 10^{-3} \text{ mol NaCl}}{0.150 \text{ L}} = 8.3 \times 10^{-3} \text{ M NaCl} \quad (2 \text{ sf})$$

Titrations

57. To calculate the volume of HCl needed, we calculate the moles of NaOH in 1.33 g, then use the stoichiometry of the balanced equation:

$$HCl \text{ (aq)} + NaOH \text{ (aq)} \rightarrow NaCl \text{ (aq)} + H_2O \text{ (}\ell\text{)}$$

$$1.33 \text{ g NaOH} \cdot \frac{1 \text{ mol NaOH}}{40.00 \text{ g NaOH}} \cdot \frac{1 \text{ mol HCl}}{1 \text{ mol NaOH}} \cdot \frac{1 \text{ L}}{0.812 \text{ mol HCl}} \cdot \frac{1000 \text{ mL}}{1 \text{ L}}$$

$$= 40.9 \text{ mL HCl}$$

59. Calculate:

1. moles of Na_2CO_3 corresponding to 2.152 g Na_2CO_3

2. moles of HCl that react with that number of moles (using the balanced equation)

3. the volume of HCl containing that number of moles of HCl

The balanced equation is:

$$Na_2CO_3 \text{ (aq)} + 2 \text{ HCl (aq)} \rightarrow 2 \text{ NaCl (aq)} + H_2O \text{ (}\ell\text{)} + CO_2 \text{ (g)}$$

$$2.152 \text{ g Na}_2CO_3 \cdot \frac{1 \text{ mol Na}_2CO_3}{106.0 \text{ g Na}_2CO_3} \cdot \frac{2 \text{ mol HCl}}{1 \text{ mol Na}_2CO_3} \cdot \frac{1 \text{ L}}{0.955 \text{ mol HCl}} \cdot \frac{1000 \text{ mL}}{1 \text{ L}}$$

$$= 42.5 \text{ mL HCl}$$

61. Mass of citric acid per 100. mL of soft drink:

Calculate:

1. mol NaOH used in the neutralization

2. mol citric acid that react with that amount of NaOH (balanced equation)

3. mass of citric acid corresponding to that number of moles.

$$\frac{0.0102 \text{ mol NaOH}}{1 \text{ L}} \cdot \frac{0.03351 \text{ L}}{1} \cdot \frac{1 \text{ mol citric acid}}{3 \text{ mol NaOH}} \cdot \frac{192.1 \text{ g citric acid}}{1 \text{ mol citric acid}}$$

$$= 0.0219 \text{ g citric acid}$$

63. To determine the acid's identity, calculate the moles of each acid corresponding to 0.956 g.

$$0.956 \text{ g} \cdot \frac{1 \text{ mol citric acid}}{192.1 \text{ g citric acid}} = 4.98 \times 10^{-3} \text{ mol citric acid}$$

$$0.956 \text{ g} \cdot \frac{1 \text{ mol tartaric acid}}{150.1 \text{ g tartaric acid}} = 6.37 \times 10^{-3} \text{ mol tartaric acid}$$

Calculate the number of moles of NaOH present:

$$\frac{0.513 \text{ mol NaOH}}{1 \text{ L}} \cdot \frac{0.0291 \text{ L}}{1} = 1.49 \times 10^{-2} \text{ mol NaOH}$$

From the balanced equations note that each mole of citric acid requires 3 moles NaOH while each mole of tartaric acid requires 2 moles NaOH. Using the moles of citric acid and tartaric acid calculated above, calculate the amount of NaOH needed for each of the acids.

$$4.98 \times 10^{-3} \text{ mol citric acid} \cdot \frac{3 \text{ mol NaOH}}{1 \text{ mol citric acid}} = 1.49 \times 10^{-2} \text{ mol NaOH}$$

$$6.37 \times 10^{-3} \text{ mol tartaric acid} \cdot \frac{2 \text{ mol NaOH}}{1 \text{ mol tartaric acid}} = 1.27 \times 10^{-2} \text{ mol NaOH}$$

From this latter calculation we see that the solid acid is **citric acid**.

65. Mass of vitamin C in a 1.00 g tablet :

$$\frac{0.102 \text{ mol Br}_2}{1 \text{ L}} \cdot \frac{0.02785 \text{ L}}{1} \cdot \frac{1 \text{ mol C}_6\text{H}_8\text{O}_6}{1 \text{ mol Br}_2} \cdot \frac{176.1 \text{ g vitamin C}}{1 \text{ mol C}_6\text{H}_8\text{O}_6}$$
$$= 0.500 \text{ g vitamin C}$$

General Questions:

67.   a. Balanced equation: $2 \text{ Fe(s)} + 3 \text{ Cl}_2\text{(g)} \rightarrow 2 \text{ FeCl}_3\text{(s)}$

   b. 1. Mass of Cl_2 to react with 10.0 g iron:

$$10.0 \text{ g Fe} \cdot \frac{1 \text{ mol Fe}}{55.85 \text{ g Fe}} \cdot \frac{3 \text{ mol Cl}_2}{2 \text{ mol Fe}} \cdot \frac{70.91 \text{ g Cl}_2}{1 \text{ mol Cl}_2} = 19.0 \text{ g Cl}_2$$

   2. Amount of $FeCl_3$ produced :

$$10.0 \text{ g Fe} \cdot \frac{1 \text{ mol Fe}}{55.85 \text{ g Fe}} \cdot \frac{2 \text{ mol FeCl}_3}{2 \text{ mol Fe}} = 0.179 \text{ mol FeCl}_3$$

$$0.179 \text{ mol FeCl}_3 \cdot \frac{162.2 \text{ g FeCl}_3}{1 \text{ mol FeCl}_3} = 29.0 \text{ g FeCl}_3$$

    c. Percent yield of $FeCl_3$:

$$\frac{\text{Actual}}{\text{Theoretical}} \times 100 \quad \text{or} \quad \frac{18.5 \text{ g FeCl}_3}{29.0 \text{ g FeCl}_3} \times 100 = 63.8 \%$$

69. 1.056 g MCO_3 produced MO + 0.376 g CO_2

    The MO had a mass of (1.056 - 0.376) 0.680 g

    According to the equation given, CO_2 and MO are produced in equimolar amounts.

$$0.376 \text{ g CO}_2 \cdot \frac{1 \text{ mol CO}_2}{44.0 \text{ g CO}_2} = 8.54 \times 10^{-3} \text{ mol CO}_2$$

So the metal oxide (0.680 g) must correspond to 8.54×10^{-3} mol of metal oxide. The molar mass of metal oxide is then:

$$\frac{0.680 \text{ g}}{8.54 \times 10^{-3} \text{ mol}} = 79.6 \text{ g/ mol}$$

If the oxide contains one mol of O atoms per mol of M atoms, we can deduce the molar mass of M

        MO = 79.6 g/mol

        79.6 = M g/mol + 16.0 g O/mol

        63.6 g/mol = M

The metal with the atomic weight close to 63.6 is Cu.

71. When 15.5 g of $(NH_4)_2PtCl_4$ combine with 225 mL of 0.75 M NH_3:

    a. The reactant in excess:

$$\frac{0.75 \text{ mol NH}_3}{L} \cdot 0.225 \text{ L} = 0.169 \text{ mol NH}_3 \text{ (to 3 sf)}$$

$$15.5 \text{ g (NH}_4)_2\text{PtCl}_4 \cdot \frac{1 \text{ mol (NH}_4)_2\text{PtCl}_4}{373.0 \text{ g (NH}_4)_2\text{PtCl}_4} = 0.0416 \text{ mol (to 3 sf)}$$

Since 1 mol of the Pt compound requires 2 mol of NH_3, NH_3 is in excess, and $(NH_4)_2PtCl_4$ is the limiting reagent:

$$0.0416 \text{ mol(NH}_4)_2\text{PtCl}_4 \cdot \frac{2 \text{ mol NH}_3}{1 \text{ mol(NH}_4)_2\text{PtCl}_4} = 0.0831 \text{ mol NH}_3 \text{ needed}$$

    [You get 0.0832 if you've rounded already.]

    Excess NH_3 = (0.169 - 0.0831) = 0.086 mol NH_3

                available - needed

b. Mass of cisplatin possible:

$$0.0416 \text{ mol } (NH_4)_2PtCl_4 \cdot \frac{1 \text{ mol } Pt(NH_3)_2Cl_2}{1 \text{ mol } (NH_4)_2PtCl_4} \cdot \frac{300.0 \text{ g } Pt(NH_3)_2Cl_2}{1 \text{ mol } Pt(NH_3)_2Cl_2}$$

$$= 12.5 \text{ g } Pt(NH_3)_2Cl_2$$

c. Mass of NH_3 remaining:

$$0.086 \text{ mol } NH_3 \cdot \frac{17.03 \text{ g } NH_3}{1 \text{ mol } NH_3} = 1.46 \text{ g } NH_3$$

73. Balance the equation: $CS_2 + 3 Cl_2 \rightarrow S_2Cl_2 + CCl_4$

Theoretical yield of CCl_4 :

1. Determine the limiting reagent.

$$125 \text{ g } CS_2 \cdot \frac{1 \text{ mol } CS_2}{76.14 \text{ g } CS_2} = 1.64 \text{ mol } CS_2$$

$$435 \text{ g } Cl_2 \cdot \frac{1 \text{ mol } Cl_2}{70.91 \text{ g } Cl_2} = 6.13 \text{ mol } Cl_2$$

Note that each mole of CS_2 requires 3 mol Cl_2. So 1.64 mol CS_2 requires 4.92 mol Cl_2. Since we have more Cl_2 than required, CS_2 is the limiting reagent.

$$1.64 \text{ mol } CS_2 \cdot \frac{1 \text{ mol } CCl_4}{1 \text{ mol } CS_2} \cdot \frac{153.8 \text{ g } CCl_4}{1 \text{ mol } CCl_4} = 252 \text{ g } CCl_4$$

2. Mass of Cl_2 in excess:

The excess Cl_2 is 6.13 - 4.92 or 1.21 mol Cl_2

which would have a mass of:

$$1.21 \text{ mol } Cl_2 \cdot \frac{70.91 \text{ g } Cl_2}{1 \text{ mol } Cl_2} = 85.8 \text{ g } Cl_2$$

75. Empirical formula of B_xH_y :

Since **all** the B in B_2O_3 came from the B_xH_y, calculate the mass of that boron:

$$0.422 \text{ g } B_2O_3 \cdot \frac{21.62 \text{ g } B}{69.62 \text{ g } B_2O_3} = 0.131 \text{ g } B$$

The original sample contains **only** B and H, so the H present is:

$$0.148 \text{ g } B_xH_y - 0.131 \text{ g } B = 0.017 \text{ g } H$$

Calculate the moles of B, H present in the sample:

$$0.131 \text{ g } B \cdot \frac{1 \text{ mol } B}{10.81 \text{ g } B} = 0.0121 \text{ mol } B$$

$$0.017 \text{ g } H \cdot \frac{1 \text{ mol } H}{1.00 \text{ g } H} = 0.017 \text{ mol } H$$

The ratio of B:H is :   0.0121 mol B : 0.017 mol H

                                              or   1 : 1.4

Since we prefer integral subscripts for compounds, we need to determine a multiplier that provides <u>integers</u> for both B and H. Trial and error indicates that multiplying by 5 yields B_5H_7.

77. Diluting 10.0 mL of 2.56 M HCl to 250. mL results in a concentration of HCl of:

moles HCl : $\dfrac{2.56 \text{ mol HCl}}{1 \text{ L}} \cdot 0.0100 \text{ L} = 2.56 \times 10^{-2}$ mol HCl

Molarity HCl $= \dfrac{2.56 \times 10^{-2} \text{ mol HCl}}{0.250 \text{ L}}$ or   0.102 M HCl (to 3 sf)

Since HCl is a strong acid, we anticipate total dissociation into H^+ and Cl^- ions.
So 0.102 M HCl $\Rightarrow$ 0.102 M H^+(aq) and 0.102 M Cl^- (aq).

79. The preparation of diborane, B_2H_6, may be written:

$2 \text{ NaBH}_4(aq) + H_2SO_4(aq) \rightarrow 2 H_2(g) + Na_2SO_4(aq) + B_2H_6(g)$

The yield of B_2H_6 can be calculated by determining the limiting regent:

$0.0875 \text{ M } H_2SO_4 \cdot 0.250 \text{ L} = 0.0219 \text{ mol } H_2SO_4$

$1.55 \text{ g NaBH}_4 \cdot \dfrac{1 \text{ mol NaBH}_4}{37.83 \text{ g NaBH}_4} = 0.0410 \text{ mol NaBH}_4$

Since each mol of H_2SO_4 requires 2 mol $NaBH_4$,

$0.0219 \text{ mol } H_2SO_4 \cdot \dfrac{2 \text{ mol NaBH}_4}{1 \text{ mol } H_2SO_4} = 0.0438 \text{ mol NaBH}_4$ (needed)

$NaBH_4$ will limit the amount of B_2H_6 that we can make.

$0.0410 \text{ mol NaBH}_4 \cdot \dfrac{1 \text{ mol } B_2H_6}{2 \text{ mol NaBH}_4} \cdot \dfrac{27.67 \text{ g } B_2H_6}{1 \text{ mol } B_2H_6} = 0.567 \text{ g } B_2H_6$

81.   The balanced equation is:

$PbO_2 \text{ (s)} + 4 H^+ \text{ (aq)} + 2 I^- \text{ (aq)} \rightarrow Pb^{2+} \text{ (aq)} + I_2 \text{ (aq)} + 2 H_2O \text{ (}\ell\text{)}$

$\dfrac{0.0500 \text{ mol Na}_2S_2O_3}{L} \cdot \dfrac{0.03523 \text{ L}}{1} \cdot \dfrac{1 \text{ mol } I_2}{2 \text{ mol } S_2O_3{}^{2-}} \cdot \dfrac{1 \text{ mol } Pb^{2+}}{1 \text{ mol } I_2} \cdot \dfrac{1 \text{ mol Pb}}{1 \text{ mol } Pb^{2+}}$

$\cdot \dfrac{207.2 \text{ g Pb}}{1 \text{ mol Pb}} = 0.182 \text{ g Pb}$

Weight percent of lead in the ore $= \dfrac{0.182 \text{ g Pb}}{0.576 \text{ g ore}} \times 100 = 31.7\% \text{ Pb}$

62

83. The mixture can be described if we know the amounts of acid and base added.

$$\frac{1.50 \text{ mol } HNO_3}{1 \text{ L}} \cdot 0.0250 \text{ L} = 0.0375 \text{ mol } HNO_3$$

$$\frac{2.50 \text{ mol } NaOH}{1 \text{ L}} \cdot 0.0500 \text{ L} = 0.125 \text{ mol } NaOH$$

The reaction can be represented: $HNO_3 + NaOH \rightarrow NaNO_3 + H_2O$

Since the reaction entails 1 mol HNO_3 / mol $NaOH$, HNO_3 is the limiting reagent.

Following reaction, there will be **0 M HNO_3** .

The **NaOH** should be (0.125 - 0.0375 mol) in (25.0 + 50.0 mL)

$$\frac{0.0875 \text{ mol } NaOH}{0.0750 \text{ L}} = 1.17 \text{ M } NaOH$$

The $NaNO_3$ formed (0.0375 mol) in (25.0 + 50.0 mL) :

$$\frac{0.0375 \text{ mol } NaNO_3}{0.0750 \text{ L}} = 0.500 \text{ M } NaNO_3$$

So (b) is the best answer.

85. The amount of $C_2O_4^{2-}$ from the iron-containing compound:

$$\frac{0.108 \text{ mol } MnO_4^-}{1 \text{ L}} \cdot \frac{0.03450 \text{ L}}{1} \cdot \frac{5 \text{ mol } H_2C_2O_4}{2 \text{ mol } MnO_4^-} \cdot \frac{5 \text{ mol } C_2O_4^{2-}}{5 \text{ mol } H_2C_2O_4}$$

$$= 9.32 \times 10^{-3} \text{ mol } C_2O_4^{2-}$$

Calculate the moles of each compound based upon the two possible formulations:

$K_3[Fe(C_2O_4)_3]$ has a formula weight of 437.2 while $K[Fe(C_2O_4)_2(H_2O)_2]$ has a formula weight of 307.0.

$$1.356 \text{ g cpd} \cdot \frac{1 \text{ mol } K_3[Fe(C_2O_4)_3]}{437.2 \text{ g cpd}} = 3.10 \times 10^{-3} \text{ mol } K_3[Fe(C_2O_4)_3]$$

For the second compound:

$$1.356 \text{ g cpd} \cdot \frac{1 \text{ mol } K[Fe(C_2O_4)_2(H_2O)_2]}{307.0 \text{ g cpd}} = 4.42 \times 10^{-3} \text{ mol } K[Fe(C_2O_4)_2(H_2O)_2]$$

Each mol of this second compound has 2 mol of oxalate ion or 8.83×10^{-3} mol $C_2O_4^{-2}$, while for the first compound—3 mol of oxalate ions or 9.30×10^{-3} mol $C_2O_4^{2-}$.

The compound must be $K_3[Fe(C_2O_4)_3]$.

87. We need to determine the stoichiometric relationship between Cl_2 and $KClO_4$.

Equation 1 :   1 mol Cl_2 : 1 mol $KClO$

Equation 2 :   3 mol $KClO$: 1 mol $KClO_3$

Equation 3 :   4 mol $KClO_3$: 3 mol $KClO_4$

Starting with equation 3, we can determine the relation between Cl_2 and $KClO_4$

$$\frac{3 \text{ mol } KClO_4}{4 \text{ mol } KClO_3} \cdot \frac{4 \text{ mol } KClO_3}{12 \text{ mol } KClO} \cdot \frac{12 \text{ mol } KClO}{12 \text{ mol } Cl_2} \qquad \text{or} \qquad \frac{1 \text{ mol } KClO_4}{4 \text{ mol } Cl_2}$$

$$1.0 \times 10^3 \text{ g } KClO_4 \cdot \frac{1 \text{ mol } KClO_4}{138.5 \text{ g } KClO_4} \cdot \frac{4 \text{ mol } Cl_2}{1 \text{ mol } KClO_4} \cdot \frac{70.9 \text{ g } Cl_2}{1 \text{ mol } Cl_2} = 2.0 \times 10^3 \text{ g } Cl_2$$

$$\text{or } 2.0 \text{ kg } Cl_2$$

89. $\qquad TiO_2 + H_2 \rightarrow H_2O + Ti_xO_y$

$\qquad$ 1.598 g $\qquad\qquad\qquad$ 1.438 g

The moles of TiO_2 initially present:

$$1.598 \text{ g } TiO_2 \cdot \frac{1 \text{ mol } TiO_2}{79.879 \text{ g } TiO_2} = 0.02000 \text{ mol } TiO_2 \text{ (and Ti)}$$

The mass of Ti present in the TiO_2 :

$$1.598 \text{ g } TiO_2 \cdot \frac{47.88 \text{ g } Ti}{79.879 \text{ g } TiO_2} = 0.9579 \text{ g } Ti$$

Since all the Ti in the unknown compound, Ti_xO_y, originates in the TiO_2, 0.9579 g of the 1.438 g of the new oxide is Ti, leaving (1.438 - 0.9579) g of O. The number of moles of O is:

$$0.480 \text{ g } O \cdot \frac{1 \text{ mol } O}{16.00 \text{ g } O} = 0.03000 \text{ mol } O$$

The new oxide is then $Ti_{0.02000}O_{0.03000}$ or Ti_2O_3

91. If 1.2 g A react with 3.2 g O to form AO_x it follows that 2.4 g A would react with 6.4 g O to form AO_x as well. 6.4 g O represents:

$$6.4 \text{ g } O \cdot \frac{1 \text{ mol } O}{16.0 \text{ g } O} = 0.4 \text{ mol } O$$

a. For the oxide AO_y, 2.4 g A react with 3.2 g O (0.2 mol O)

$\qquad$ The ratio $\dfrac{x}{y}$ is then $\qquad \dfrac{0.4 \text{ mol } O}{0.2 \text{ mol } O}$ or $\dfrac{2}{1}$

b. If x = 2 identify A

For the oxide AO_x (AO_2), 0.4 mol O would combine with 0.2 mol A. Then 2.4 g A represents 0.2 mol or 12 g/mol. Element A must be **Carbon.**

64

Conceptual Questions

93. According to the graph 2.4 g of Fe combine to produce 12.9 g of compound Fe_xBr_y.

$$2.4 \text{ g Fe} \cdot \frac{1 \text{ mol Fe}}{55.85 \text{ g Fe}} = 0.043 \text{ mol Fe}$$

$$10.5 \text{ g Br} \cdot \frac{1 \text{ mol Br}}{79.9 \text{ g Br}} = 0.131 \text{ mol Br}$$

The ratio of $\frac{Br}{Fe}$ is $\frac{0.131}{0.043}$ or 3. The empirical formula is $FeBr_3$.

The balanced equation : $2 \text{ Fe} + 3 \text{ Br}_2 \rightarrow 2 \text{ FeBr}_3$

The product is $FeBr_3$: iron(III) bromide

96. To decide the relative concentrations, calculate the dilutions. Let's assume that the HCl has a concentration of 0.100 M. Then calculate the diluted concentrations in each case.

Student 1 :

20.0 mL of 0.100 M HCl is diluted to 40.0 mL total

The diluted molarity is then :

$$\frac{0.100 \text{ mol HCl}}{1 \text{ L}} \cdot 0.0200 \text{ L} = 0.400 \text{ L} \cdot M$$

$$0.050 \frac{\text{mol HCl}}{L} = M$$

Student 2 :

20.0 mL of 0.100 M HCl is diluted to 80.0 mL total

The diluted molarity is :

$$\frac{0.100 \text{ mol HCl}}{1L} \cdot 0.0200 \text{ L} = 0.0800 \text{ L} \cdot M$$

$$0.025 \frac{\text{mol HCl}}{L} = M$$

So the second student's molarity is half the concentration of the first student's. Noting that the volumes of the two solutions differ by a factor of 2, and their concentrations differ by a factor of 2, when the two students calculate **the concentration of the original solution, they will find identical concentrations** (e)—since they both used equal volumes (and equal number of moles) of the orginal HCl solution.

98. a. Product of the reactions of Mg, Ca, and Sr with Br_2 :

   $MgBr_2$ magnesium bromide

   $CaBr_2$ calcium bromide

   $SrBr_2$ strontium bromide

   b. The balanced equations:

   $Mg + Br_2 \rightarrow MgBr_2$

   $Ca + Br_2 \rightarrow CaBr_2$

   $Sr + Br_2 \rightarrow SrBr_2$

   c. These reactions involve the combination of good reducing agents, (Mg, Ca, and Sr), with a good oxidizing agent (Br_2). This is an **oxidation-reduction reaction**.

   d. For magnesium   1.50 g Mg form 11.5 g compound

   For calcium   2.60 g Ca form 12.5 g compound

   For strontium   5.50 g Sr form 15.5 g compound

   Mg: 11.5 g - 1.50 g   =   10.0 g Br

   Ca: 12.5 g - 2.60 g   =   9.9 g Br

   Sr: 15.5 g - 5.50 g   =   10.0 g Br

   From the calculation of the amount of the metals in the compounds:

   The empirical formula of the magnesium bromide is:

   $$1.50 \text{ g Mg} \cdot \frac{1 \text{ mol Mg}}{24.3 \text{ g Mg}} = 0.062 \text{ mol Mg}$$

   $$10.0 \text{ g Br} \cdot \frac{1 \text{ mol Br}}{79.9 \text{ g Br}} = 0.125 \text{ mol Br}$$

   $$\frac{\text{mol Br}}{\text{mol Mg}} = \frac{0.125 \text{ mol Br}}{0.062 \text{ mol Mg}} \text{ or approximately } \frac{2}{1}$$

   Note that 2.60 g Ca and 5.50 g Sr correspond to approximately 0.066 mole of the metals. The plots become level at different masses of metals and products owing to the different masses of the alkaline earth metals per mole of metal. Note, however, that the masses of metals and bromine correspond to equivalent numbers of moles of bromine per mole of metal.

Chapter 6
Energy and Chemical Reactions

Energy Units

11. Express 1670 kJ in Calories:

$$1670 \text{ kJ} \cdot \frac{1000 \text{ J}}{1 \text{ kJ}} \cdot \frac{1 \text{ cal}}{4.184 \text{ J}} \cdot \frac{1 \text{ Calorie}}{1000 \text{ cal}} = 399 \text{ Calories}$$

13. Energy striking the parking lot:

Area of the lot = 300. m x 50.0 m or 1.50×10^4 m^2

$$2.6 \times 10^7 \frac{\text{J}}{\text{m}^2 \cdot \text{day}} \cdot 1.50 \times 10^4 \text{ m}^2 = 3.9 \times 10^{11} \frac{\text{J}}{\text{day}}$$

Specific Heat

15. Heat capacity of silver :

$$\text{Specific heat capacity (C)} = \frac{\text{quantity of heat supplied}}{(\text{mass of object}) \cdot (\text{temperature change})}$$

We could convert both 27.0 °C and 10.0 °C to their Kelvin equivalents, by adding 273.2.

The difference in temperature (T_{final} - $T_{initial}$) will be 17.0 K.

$$C = \frac{74.8 \text{ J}}{(18.69 \text{ g})(17.0 \text{ K})} = 0.235 \frac{\text{J}}{\text{g} \cdot \text{K}}$$

17. Heat energy to warm a liquid from 22 °C to 85 °C:

$$\text{Heat} = \text{mass x heat capacity x } \Delta T$$

$$\text{For water} = (50.0 \text{ g})(\frac{4.184 \text{ J}}{\text{g} \cdot \text{K}})(63 \text{ K})$$

$$= 1.32 \times 10^4 \text{ J or } 13.2 \text{ kJ}$$

$$\text{For ethylene glycol} = (100. \text{ g})(\frac{2.39 \text{ J}}{\text{g} \cdot \text{K}})(63 \text{ K})$$

$$= 1.51 \times 10^4 \text{ J or } 15.1 \text{ kJ}$$

Note that the temperature difference is 63 K whether you calculate the temperature difference in Celsius (85 °C - 20 °C) or Kelvin (358 K - 295 K).

19. Using the specific heat for aluminum (0.902 $\frac{J}{g \cdot K}$) and the change in T (255 °C - 25 °C), calculate the energy added:

$$q_{Al} = (mass)(heat\ capacity)(\Delta T)$$
$$= (500.\ g)(0.902\ \frac{J}{g \cdot K})(230.\ K)$$
$$= 104000\ J\ \ or\ \ 104\ kJ$$

21. While you could calculate the heat energy needed using the equation, q = (mass)(heat capacity)(ΔT), you can easily answer this question by noting the relative magnitudes of the heat capacity for ethylene glycol (2.42 $\frac{J}{g \cdot K}$) and water (4.184 $\frac{J}{g \cdot K}$). **Water would require more heat energy**.

23. Final T of copper-water mixture:

We must **assume** that **no energy** will be transferred to or from the beaker containing the water. Then the **magnitude** of energy lost by the hot copper and the energy gained by the cold water will be equal (but opposite in sign).

$$q_{copper} = -q_{water}$$

Using the heat capacities of H_2O and copper, and expressing the temperatures in Kelvin we can write:

mass of water

$$(192\ g)(0.385\ \frac{J}{g \cdot K})(T_{final} - 373.2\ K) = -(750.\ mL)(1.00\frac{g}{mL})(4.184\ \frac{J}{g \cdot K})(T_{final} - 277.2\ K)$$

Simplifying each side gives:

$$73.92\ \frac{J}{K} \cdot T_{final} - 27{,}600\ J = -3138\ \frac{J}{K} \cdot T_{final} + 870{,}000\ J$$
$$3212\ \frac{J}{K} \cdot T_{final} = 897600\ J$$
$$T_{final} = 279.4\ K\ \ or\ \ 6.2\ °C$$

Don't forget: **Round numbers only at the end**. To show the steps, I have rounded some of these intermediate numbers.

25. Initial temperature of gold sample:

This problem is solved almost exactly like question 23. The difference is that we know T_{final} for the gold.

$$q_{gold} = -q_{water}$$

$$(182 \text{ g})(0.129 \frac{J}{g \cdot K})(300.7 \text{ K} - T_{initial}) = -(22.1 \text{ g})(4.184 \frac{J}{g \cdot K})(2.5 \text{ K})$$

$$7060. \text{ J} - 23.5 \frac{J}{K} \cdot T_{initial} = -231 \text{ J}$$

rearranging:

$$7291 \text{ J} = 23.5 \text{ J/K} \cdot T_{initial}$$

$$311 \text{ K} = T_{initial}$$

or 38 °C

27. Example 6.3 in your text is a good template for this problem.

$$q_{metal} = -q_{water}$$

$$(150.0 \text{ g})(C_{metal})(296.5 \text{ K} - 353.2 \text{ K}) = -(150.0 \text{ g})(4.184 \frac{J}{g \cdot K})(296.5 \text{ K} - 293.2 \text{ K})$$

$$-8505 \text{ g} \cdot K(C_{metal}) = -2071.1 \text{ J}$$

$$C_{metal} = 0.24 \frac{J}{g \cdot K}$$

Changes of State

29. Quantity of energy to melt 16 ice cubes at 0 °C.

The mass of ice involved:

$$16 \text{ ice cubes} \cdot \frac{62.0 \text{ g ice}}{1 \text{ ice cube}} = 992 \text{ g ice}$$

To melt 992 g ice: $992 \text{ g ice} \cdot \frac{333 \text{ J}}{1.000 \text{ g ice}} = 330. \times 10^3 \text{ J or } 330. \text{ kJ}$

31. To calculate the quantity of heat for the process described, think of the problem in three steps:

1. melt ice at 0°C to liquid water at 0°C

2. warm liquid water from 0°C to 100°C

3. convert liquid water at 100 °C to gaseous water at 100 °C

1. The energy to melt 60.1 g of ice at 0 °C is:

$$60.1 \text{ g ice} \cdot \frac{333 \text{ J}}{\text{g ice}} = 2.00 \times 10^4 \text{ J}$$

2. The energy required to warm the liquid water from 0°C to 100 °C ($\Delta T = 100$ K) is:

$$(4.18 \text{ J/g} \cdot K) \cdot 60.1 \text{ g} \cdot 100 \text{ K} = 2.51 \times 10^4 \text{ J}$$

3. To convert liquid water at 100 °C to gaseous water at 100 °C:

$$2260 \text{ J/g} \cdot 60.1 \text{ g} = 13.6 \times 10^4 \text{ J}$$

The total energy required is: $[2.00 \times 10^4 + 2.51 \times 10^4 \text{ J} + 13.6 \times 10^4 \text{J}] = 1.81 \times 10^5 \text{ J}$

33. To accomplish the process, one must:

1. heat the tin from 25.0 °C to 231.9 °C ($\Delta T = 206.9$ K)

2. melt the tin at 231.9 °C

Using the specific heat for tin, the energy for the first step is:

$$(0.227 \frac{\text{J}}{\text{g} \cdot \text{K}})(454 \text{ g})(206.9 \text{ K}) = 21,300 \text{ J}$$

To melt the tin at 231.9 °C, we need:

$$59.2 \frac{\text{J}}{\text{g}} \cdot 454 \text{ g} = 26,900 \text{ J}$$

The total heat energy needed (in J) is $(21,300 + 26,900) = 48,200$ J

Enthalpy

Note that in this chapter, I have left negative signs with the value for heat released

(heat released = - ; heat absorbed = +)

35. For a process in which the $\Delta H°$ is negative, that process is **exothermic**.

To calculate heat released when 1.25 g NO react, note that the energy shown (-114.1 kJ) is released when **2** moles of NO react, so we'll need to account for that:

$$1.25 \text{ g NO} \cdot \frac{1 \text{ mol NO}}{30.01 \text{ g NO}} \cdot \frac{-114.1 \text{ kJ}}{2 \text{ mol NO}} = -2.38 \text{ kJ}$$

37. The combustion of isooctane (IO) is **exothermic**. The molar mass of IO is: 114.2 g/mol.

The heat evolved is:

$$1.00 \text{ L of IO} \cdot \frac{0.6878 \text{ g IO}}{1 \text{mL}} \cdot \frac{1 \times 10^3 \text{ mL}}{1 \text{ L}} \cdot \frac{1 \text{ mol IO}}{114.2 \text{ g IO}} \cdot \frac{-10922 \text{ kJ}}{2 \text{ mol IO}} = -32,900 \text{ kJ}$$

39. The molar mass of CH_3OH is 32.04 g/mL

$$0.115 \text{ g } CH_3OH \cdot \frac{1 \text{ mol } CH_3OH}{32.04 \text{ g } CH_3OH} = 3.59 \times 10^{-3} \text{ mol } CH_3OH$$

The enthalpy change will be

$$\frac{-1110 \text{ J}}{3.59 \times 10^{-3} \text{ mol } CH_3OH} = -309,000 \frac{\text{J}}{\text{mol}} \text{ or } -309 \frac{\text{kJ}}{\text{mol}} \text{ (molar heat of combustion)}$$

The $\Delta H°\text{rxn}$ for the equation shown would be -618 kJ.

Hess's Law

41. The molar enthalpy of formation for benzene may be calculated by describing **several processes** that **when added** are:

$$6\ C(s) + 3\ H_2(g) \longrightarrow C_6H_6\ (\ell)$$

Begin with the equation given for the combustion of benzene.

Reversing it gives:

$$12\ CO_2(g) + 6\ H_2O(\ell) \longrightarrow 2\ C_6H_6(\ell) + 15\ O_2(g) \qquad \Delta H = +6534.8\ kJ$$

Use the equation for the formation of CO_2, noting that we need 12 $CO_2(g)$

$$12\ C(s) + 12\ O_2(g) \longrightarrow 12\ CO_2(g) \qquad \Delta H = (-393.509)(12)\ kJ$$

and the equation for the formation of H_2O, noting that we need 6 $H_2O(\ell)$

$$6\ H_2(g) + 3\ O_2(g) \longrightarrow 6\ H_2O(\ell) \qquad \Delta H = (-285.830)(6)\ kJ$$

Adding these 3 equations gives

$$12C(s) + 6\ H_2(g) \longrightarrow 2\ C_6H_6\ (\ell) \qquad \Delta H = +97.9\ kJ$$

So, halving all the coefficients provides the desired equation with a $\Delta H = +48.9\ kJ$

43. The desired equation is: $Pb(s) + 1/2\ O_2(g) \longrightarrow PbO(s)$

$$\begin{array}{lll} 1.\ Pb(s) + CO(g) \longrightarrow PbO(s) + C(s) & \Delta H° = -106.8\ kJ \\ \underline{1/2\ x\ 2.\ \ C(s) + 1/2\ O_2(g) \longrightarrow CO(g)} & \underline{\Delta H° = (-221.0\ kJ)(0.5)} \\ \qquad Pb(s) + 1/2\ O2(g) \longrightarrow PbO(s) & \Delta H° = -217.3\ kJ \end{array}$$

The process is **exothermic**.

The heat evolved when 250 g Pb react:

$$250\ g\ Pb \cdot \frac{1\ mol\ Pb}{207.2\ g\ Pb} \cdot \frac{-217.3kJ}{1\ mol\ Pb} = -260\ kJ\ (2\ sf)$$

Standard Enthalpies of Formation

45. The equation requested requires that we form **one** mol of product (chromium(III) oxide) from its elements—each in their standard state.

Begin by writing a balanced equation:

$$4\ Cr(s) + 3O_2(g) \longrightarrow 2Cr_2O_3(s)$$

Now express the reaction so that you form one mole of Cr_2O_3—divide all coefficients by 2

$$2\ Cr(s) + 3/2O_2(g) \longrightarrow Cr_2O_3(s)$$

47.   a. The $\Delta H°$ is **negative**; the formation of glucose from its elements is **exothermic**.

b. The balanced equation:
$$6 \text{ C(graphite)} + 3 O_2(g) + 6 H_2(g) \longrightarrow C_6H_{12}O_6(s)$$

49. Use Hess's Law to obtain the overall $\Delta H°_{rxn}$: Note that we'll need to reverse the first equation shown:

$PbCl_2(s) \longrightarrow Pb(s) + Cl_2(g)$	$-\Delta H°_f = +359.4$ kJ
$Pb(s) + 2 Cl_2(g) \longrightarrow PbCl_4(\ell)$	$\Delta H°_f = -329.3$ kJ
$PbCl_2(s) + 2 Cl_2(g) \longrightarrow PbCl_4(\ell)$	$\Delta H°_f = + 30.1$ kJ

51. The enthalpic change for SO_3 formation:
$$SO_2(g) + 1/2 O_2(g) \longrightarrow SO_3(g)$$

$$\begin{aligned}
\Delta H°_{rxn} &= \Delta H°_f\, SO_3 - [\Delta H°_f\, SO_2 + 1/2\, \Delta H°_f\, O_2] \\
&= (-395.7 \text{ kJ/mol})(1 \text{ mol}) - [(-296.8 \text{ kJ/mol})(1 \text{ mol}) + (0 \text{ kJ/mol})(1/2 \text{ mol})] \\
&= -395.7 \text{ kJ} + 296.8 \text{ kJ} \\
&= -98.9 \text{ kJ}
\end{aligned}$$

Since the enthalpic change is negative, the reaction is **exothermic**.

53. a. The enthalpy change for the reaction:

	$4 NH_3(g)$	$+ 5 O_2(g)$	$\longrightarrow$	$4 NO(g)$	$+ 6 H_2O(g)$
$\Delta H°_f$(kJ/mol)	-46.1	0		+90.3	-241.8

$$\begin{aligned}
\Delta H°_{rxn} &= [\,(4 \text{ mol})(+90.3\, \tfrac{kJ}{mol}) + (6 \text{ mol})(-241.8\, \tfrac{kJ}{mol})\,] - \\
&\qquad\qquad [\,(4 \text{ mol})(-46.1\, \tfrac{kJ}{mol}) + (5 \text{ mol})(0)] \\
&= (-1089.6 \text{ kJ}) - (-184.4 \text{ kJ}) \\
&= -905.2 \text{ kJ} \qquad \text{The reaction is exothermic.}
\end{aligned}$$

b. Heat evolved when 10.0 g NH_3 react:

The balanced equation shows that 4 mol NH_3 result in the release of 905.2 kJ.

$$10.0 \text{ g } NH_3 \cdot \frac{1 \text{ mol } NH_3}{17.03 \text{ g } NH_3} \cdot \frac{-905.2 \text{ kJ}}{4 \text{ mol } NH_3} = -133 \text{ kJ}$$

55. a. The enthalpy change for the reaction:

$$WO_3(s) + 3\ H_2(g) \longrightarrow W(s) + 3\ H_2O(\ell)$$

$\Delta H°_f$(kJ/mol)   -842.9   0   0   -285.8

$$\Delta H°_{rxn} = [\ (3\ mol)(-285.8\ \frac{kJ}{mol})] - [(1\ mol)(-842.9\ \frac{kJ}{mol})]$$

$$= (-857.4\ kJ) - (-842.9\ kJ)$$

$$= -14.5\ kJ$$

b. The heat **evolved** when 1.00 g WO_3 reacts:

$$1.00\ g\ WO_3 \cdot \frac{1\ mol\ WO_3}{231.8\ g\ WO_3} \cdot \frac{-14.5\ kJ}{1 mol\ WO_3} = -0.0626\ kJ$$

57. The molar enthalpy of formation of naphthalene can be calculated since we're given the enthalpic change for the reaction:

$$C_{10}H_8(s) + 12\ O_2(g) \longrightarrow 10\ CO_2(g) + 4\ H_2O(\ell)$$

$\Delta H°_f$(kJ/mol)   ?   0   -393.5   -285.8

$\Delta H°_{rxn}$ $= \sum \Delta H°_f$ products $- \sum \Delta H°_f$ reactants

-5156.1 kJ $= [(10\ mol)(-393.5\ \frac{kJ}{mol}) + (4\ mol)(-285.8\ \frac{kJ}{mol})] - [\Delta H°_f\ C_{10}H_8]$

-5156.1 kJ $= (-5078.2\ kJ) - \Delta H°_f\ C_{10}H_8$

- 77.9 kJ $= - \Delta H°_f\ C_{10}H_8$

77.9 kJ $= \Delta H°_f\ C_{10}H_8$

Calorimetry

59.   Heat evolved = Heat absorbed by bomb and water

$-q_{rxn}$ = q_{bomb} + q_{water}

q_{bomb} = $C \cdot \Delta t$ = (650 J/ K)(3.33 K) = 2.16×10^3 J

(where C = heat capacity of bomb)

q_{water} = $S \cdot m \cdot \Delta t$ = (4.184 J/g $\cdot$ K)(320. g)(3.33 K) = 4.46×10^3 J

(where S = specific heat of water)

$- q_{rxn}$ = 2.16×10^3 J + 4.46×10^3 J = 6.62×10^3 J

q_{rxn} = $- 6.62 \times 10^3$ J   or -6.62 kJ

61.   Heat evolved by reaction = - Heat absorbed by surroundings

                                = - (Heat absorbed by bomb and water)

                                = - (qbomb + qwater)

Δt   = (27.38°C - 25.00°C) = 2.38 °C or 2.38 K

qbomb = (893 J/K) • 2.38 K   = 2130 J

qwater = (775 g)(4.18 J/g •K) • 2.38 K   = 7710 J

Heat absorbed by bomb + water = 9840 J or 9.84 kJ

The heat evolved by reaction of 0.300 g C is then: -9.84 kJ

Expressing this on a molar basis:

$$\frac{-\ 9.84\ kJ}{0.300\ g\ C}\ \bullet\ \frac{12.01\ g\ C}{1\ mol\ C}\ =\ \frac{-394\ kJ}{1\ mol\ C}$$

63.   100. mL of 0.200 M CsOH and 50.0 mL of 0.400 M HCl each supply 0.0200 moles of
       base and acid respectively. If we assume the specific heat capacities of the solutions are
       4.20 J/g • K, the **heat evolved** for 0.200 moles of CsOH is:

       q = (4.20 J/g • K)(150. g)(24.28 °C - 22.50°C) [and since 1.78°C = 1.78 K]

       q = (4.20 J/g • K)(150. g)(1.78 K)

       q = 1120 J

The molar enthalpy of neutralization is : $\frac{-\ 1120\ J}{0.200\ mol\ CsOH}$ = - 56100 J/mol

                                                                or -56.1 kJ/mol

65. Heat absorbed by the ice : $\frac{333\ J}{1.00\ g\ ice}$ • 7.33 g ice = 2440 J (to 3 sf)

     Since this energy is released by the metal, we can calculate the heat capacity of the metal:

          heat   = heat capacity x mass x ΔT

       -2440 J   = C x 50.0 g x (273.2 K - 373 K)

     0.489 $\frac{J}{g\ \bullet\ K}$ = C

     Note that the heat released has a negative sign.

67. Greater quantity of heat released :

To calculate heat transferred:

$$q = (mass)(heat\ capacity)(\Delta T)$$

$$q_{H_2O} = (50.0\ g)(\frac{4.184\ J}{g \cdot K})(283\ K - 323\ K)$$

$$= -8400\ J \quad (to\ 2\ sf)$$

$$q_{C_2H_5OH} = (100.0\ g)(\frac{2.46\ J}{g \cdot K})(283\ K - 323\ K)$$

$$= -9800\ J \quad (to\ 2\ sf)$$

Ethanol releases (q is negative) more heat.

69. The enthalpy change for the reaction:

$$Mg(s) + 2\ H_2O(\ell) \longrightarrow Mg(OH)2(s) + H_2(g)$$

$\Delta H°_f(kJ/mol)$ 0   -285.8   -924.5   0

$$\Delta H°_{rxn} = (1\ mol)(-924.5\ \frac{kJ}{mol}) - (2\ mol)(-285.8\ \frac{kJ}{mol})$$

$$= -352.9\ kJ \quad or\ -3.529\ x\ 10^5\ J$$

Each mole of magnesium releases 352.9 kJ of heat energy.

Calculate the heat required to warm 25 mL of water from 25 to 85 °C.

$$heat = heat\ capacity\ x\ mass\ x\ \Delta T$$

$$= (4.184\ \frac{kJ}{mol})(25\ mL)(\frac{1.00\ g}{1\ mL})(60\ K)$$

$$= 6276\ or\ 6300\ J \quad or\ 6.3\ kJ \quad (to\ 2\ sf)$$

Magnesium required:

$$6.3\ kJ \cdot \frac{1\ mol\ Mg}{352.9\ kJ} \cdot \frac{24.3\ g\ Mg}{1\ mol\ Mg} = 0.43\ g\ Mg$$

71. The molar heat capacities for Al, Fe, Cu, Au:

Convert the heat capacities ($\frac{J}{g \cdot K}$) to the molar capacity ($\frac{J}{mol \cdot K}$)

$$Al: \quad 0.902\ \frac{J}{g \cdot K} \cdot \frac{26.98\ g\ Al}{1\ mol\ Al} = 24.3\ \frac{J}{mol \cdot K}$$

$$Fe: \quad 0.451\ \frac{J}{g \cdot K} \cdot \frac{55.85\ g\ Fe}{1\ mol\ Fe} = 25.2\frac{J}{mol \cdot K}$$

$$Cu: \quad 0.385\ \frac{J}{g \cdot K} \cdot \frac{63.55\ g\ Cu}{1\ mol\ Cu} = 24.5\ \frac{J}{mol \cdot K}$$

Au: $0.128 \dfrac{J}{g \cdot K} \cdot \dfrac{197.0 \text{ g Au}}{1 \text{ mol Au}} = 25.2 \dfrac{J}{mol \cdot K}$

The molar heat capacities are remarkably similar.

Estimate the heat capacity of Ag:

One might well imagine the heat capacity of Ag as an **average** of the heat capacities of Cu and Au ($\sim 24.9 \dfrac{J}{mol \cdot K}$) or $24.9 \dfrac{J}{mol \cdot K} \cdot \dfrac{1 \text{ mol Ag}}{107.9 \text{ g Ag}} = 0.231 \dfrac{J}{g \cdot K}$

73. Calculate the ΔH°_f for $B_2H_6(g)$:

$$B_2H_6(g) + 3\,O_2(g) \longrightarrow B_2O_3(s) + 3\,H_2O(g)$$

ΔH°_f(kJ/mol)   ?   0   -1271.9   -241.8

Given that the enthalpic change for the reaction is $-1941 \dfrac{kJ}{mol\ B_2H_6}$

ΔH°_{rxn} $= [(1 \text{ mol})(-1271.9 \dfrac{kJ}{mol}) + (3 \text{ mol})(-241.8 \dfrac{kJ}{mol})] - \Delta H^\circ_f\ B_2H_6(g)$

-1941 kJ $= [-1997.3 \text{ kJ}] - \Delta H^\circ_f\ B_2H_6(g)$

56 kJ $= -\Delta H^\circ_f\ B_2H_6(g)$

-56 kJ $= \Delta H^\circ_f\ B_2H_6(g)$

75. Calculate the ΔH°_f for $N_2H_4(\ell)$

$$N_2H_4(\ell) + O_2(g) \longrightarrow N_2(g) + 2\,H_2O(g)$$

ΔH°_f(kJ/mol)   ?   0   0   -241.8

Given that the ΔH°_{rxn} is -534 kJ.

ΔH°_{rxn} $= [(2 \text{ mol})(-241.8 \dfrac{kJ}{mol})] - \Delta H^\circ_f\ N_2H_4(\ell)$

-534 kJ $= -483.6 \text{ kJ} - \Delta H^\circ_f\ N_2H_4(\ell)$

-50. kJ $= -\Delta H^\circ_f\ N_2H_4(\ell)$

50. kJ $= \Delta H^\circ_f\ N_2H_4(\ell)$

77. a. Enthalpy change for:

1. $2\,C_2H_2(g) + 5\,O_2(g) \longrightarrow 4\,CO_2(g) + 2\,H_2O(g)$

ΔH°_f(kJ/mol)   +226.7   0   -393.5   -241.8

ΔH°_{rxn} $= [(4 \text{ mol})(-393.5 \dfrac{kJ}{mol}) + (2 \text{ mol})(-241.8 \dfrac{kJ}{mol})] - [(2 \text{ mol})(+226.7 \dfrac{kJ}{mol})]$

$= (-2057.6 \text{ kJ}) - (+453.4 \text{ kJ})$

$= -2511.0 \text{ kJ}$ **exothermic**

2. $2\ CO(g)\ +\ O_2(g)\ \longrightarrow\ 2\ CO_2(g)$

$\Delta H°_f$(kJ/mol) -110.5   0   -393.5

$\Delta H°_{rxn}$ $=\ [(2\ mol)(-393.5\ \frac{kJ}{mol})\ ]\ -[(2\ mol)(-110.5\ \frac{kJ}{mol})]$

              $=\ (-787.0\ kJ)\ -\ (-221.0\ kJ)$

              $=\ -566.0\ kJ$ **exothermic**

3. $2\ C_2H_2(g)\ +\ 10\ NO(g)\ \longrightarrow\ 4\ CO_2(g)\ +\ 2\ H_2O(g)\ +\ 5\ N_2(g)$

$\Delta H°_f$(kJ/mol) +226.7   +90.3   -393.5   -241.8   0

$\Delta H°_{rxn}$ $=\ [(4\ mol)(-393.5\ \frac{kJ}{mol})\ +\ (2\ mol)(-241.8\ \frac{kJ}{mol})]\ -$

              $[(2\ mol)(+226.7\ \frac{kJ}{mol})\ +\ (10\ mol)(+90.3\ \frac{kJ}{mol})]$

              $=\ (-2057.6\ kJ)\ -\ (+1356.4\ kJ)$

              $=\ -3414.0\ kJ$ **exothermic**

4. $2\ NO(g)\ +\ 2\ CO(g)\ \longrightarrow\ N_2(g)\ +\ 2\ CO_2(g)$

$\Delta H°_f$(kJ/mol) +90.3   -110.5   0   -393.5

$\Delta H°_{rxn}$ $=\ [(2\ mol)(-393.5\ \frac{kJ}{mol})\ ]\ -\ [(2\ mol)(+90.3\ \frac{kJ}{mol})\ +\ (2\ mol)(-110.5\ \frac{kJ}{mol})]$

              $=\ (-787.0\ kJ)\ -\ (-40.4\ kJ)$

              $=\ -746.6\ kJ$ **exothermic**

b. Quantity of heat evolved when 15.0 g of C_2H_2 reacts in the third equation.

$15.0\ g\ C_2H_2\ \cdot\ \dfrac{1\ mol\ C_2H_2}{26.04\ g\ C_2H_2}\ \cdot\ \dfrac{-3414.0\ kJ}{2\ mol\ C_2H_2}\ =\ -983\ kJ$

An error in the problem statement in the text book prohibits answering the question posed.
Instead I have calculated the heat evolved in the first equation.

Quantity of heat evolved when 15.0 g of C_2H_2 reacts in the first equation.

$15.0\ g\ C_2H_2\ \cdot\ \dfrac{1\ mol\ C_2H_2}{26.04\ g C_2H_2}\ \cdot\ \dfrac{-2511.0\ kJ}{2\ mol\ C_2H_2}\ =\ -723\ kJ$

79. Calculate $\Delta H°_{rxn}$ for each of the steps:

Step 1 $SO_2(g)\ +\ 2\ H_2O(g)\ +\ Br_2(g)\ \longrightarrow\ H_2SO_4(\ell)\ +\ 2\ HBr(g)$

$\Delta H°_f$(kJ/mol) -296.8   -241.8   +30.907   -813.989   -36.4

$\Delta H°_{rxn}$ $=\ [(1\ mol)(-813.989\ \frac{kJ}{mol})\ +\ (2\ mol)(-36.4\ \frac{kJ}{mol})]\ -$

              $[(1\ mol)(-296.8\ \frac{kJ}{mol})\ +\ (2\ mol)(-241.8\ \frac{kJ}{mol})\ +\ (1\ mol)(+30.907\ \frac{kJ}{mol})]$

$$= (-886.8 \text{ kJ}) - (-749.5 \text{ kJ})$$
$$= -137.3 \text{ kJ}$$

Step 2 $H_2SO_4(\ell) \longrightarrow H_2O(g) + SO_2(g) + 1/2\,O_2(g)$

$\Delta H°_f(\text{kJ/mol})$ -813.989   -241.8   -296.8   0

$\Delta H°_{rxn}$ $= [(1 \text{ mol})(-241.8 \frac{\text{kJ}}{\text{mol}}) + (1 \text{ mol})(-296.8 \frac{\text{kJ}}{\text{mol}})] - [(1 \text{ mol})(-813.989 \frac{\text{kJ}}{\text{mol}})]$

$= (-538.6 \text{ kJ}) + (813.989 \text{ kJ})$

$= +275.4 \text{ kJ}$

Step 3 $2\,HBr(g) \longrightarrow H_2(g) + Br_2(g)$

$\Delta H°_f(\text{kJ/mol})$ -36.4   0   +30.907

$\Delta H°_{rxn}$ $= [(1 \text{ mol})(+30.907 \frac{\text{kJ}}{\text{mol}})] - [(2 \text{ mol})(-36.4 \frac{\text{kJ}}{\text{mol}})]$

$= (30.907 \text{ kJ}) - (72.8 \text{ kJ})$

$= -41.9 \text{ kJ}$

The equation for the overall process is found by adding the 3 steps:

$$H_2O(g) \longrightarrow 1/2\,O_2(g) + H_2(g)$$

The overall enthalpic change is found by summing each of the step-wise changes.

$\Delta H°_{overall}$ $= (-137.3 \text{ kJ}) + (+275.4 \text{ kJ}) + (-41.9 \text{ kJ})$

$= 96.2 \text{ kJ}$

The overall process is **endothermic**.

81.   For the combustion of C_8H_{18}:

$$C_8H_{18}(\ell) + 25/2\,O_2(g) \rightarrow 8\,CO_2(g) + 9\,H_2O(\ell)$$

$\Delta H°_{rxn} = [(8 \text{ mol})(-393.5 \frac{\text{kJ}}{\text{mol}}) + (9 \text{ mol})(-285.8 \frac{\text{kJ}}{\text{mol}})] - [(1 \text{ mol})(-259.2 \frac{\text{kJ}}{\text{mol}}) + 0]$

$\Delta H°_{rxn} = -5461.0 \text{ kJ}$

Expressed on a gram basis

$-5461.0 \frac{\text{kJ}}{\text{mol}} \cdot \frac{1 \text{ mol } C_8H_{18}}{114.2 \text{ g } C_8H_{18}} = -47.81 \text{ kJ/g}$

For the combustion of CH_3OH:

$$2\,CH_3OH(\ell) + 3\,O_2(g) \rightarrow 4\,H_2O(\ell) + 2\,CO_2(g)$$

$\Delta H°_{rxn} = [(-393.5 \text{ kJ/mol})(2 \text{ mol}) + (-285.8 \text{ kJ/mol})(4 \text{ mol})] -$

$[(-238.66 \text{ kJ/mol})(2 \text{ mol}) + 0]$

$$= [(- 787.0) + (- 967.2)] + 477.4 \text{ kJ}$$
$$= - 1452.9 \text{ kJ}$$

Express this on a per mol and per gram basis:

$$\frac{- 1452.9 \text{ kJ}}{2 \text{ mol CH}_3\text{OH}} \cdot \frac{1 \text{ mol CH}_3\text{OH}}{32.04 \text{ g CH}_3\text{OH}} = - 22.67 \text{ kJ/g}$$

On a per gram basis, **octane liberates the greater amount** of heat energy.

83. $N_2H_4 \, (\ell) + O_2 \, (g) \rightarrow N_2 \, (g) + 2 \, H_2O \, (g)$

$\Delta H_f^\circ \, N_2H_4 \, (\ell) = \; + 50.6 \text{ kJ/mol}$

$\Delta H_{rxn} \, N_2H_4 \, (\ell) = [2 \, \Delta H_f^\circ \, H_2O(g) + \Delta H_f^\circ \, N_2(g)] - [\Delta H_f^\circ \, N_2H_4(\ell) + \Delta H_f^\circ \, O_2(g)]$

$$= - 534.2 \text{ kJ}$$

On a per gram basis this corresponds to:

$$\frac{- 534.2 \text{ kJ}}{1 \text{ mol N}_2\text{H}_4} \cdot \frac{1 \text{ mol N}_2\text{H}_4}{32.05 \text{ g N}_2\text{H}_4} = - 16.67 \text{ kJ/g}$$

Similarly for the dimethylhydrazine:

$N_2H_2(CH_3)_2 \, (\ell) + 4 \, O_2 \, (g) \rightarrow 2 \, CO_2 \, (g) + 4 \, H_2O \, (g) + N_2 \, (g)$

$\Delta H_f^\circ \, N_2H_2(CH_3)_2 \, (l) = \; + 48.9 \text{ kJ/mol}$

$\Delta H_{rxn} \, N_2H_2(CH_3)_2 \, (\ell) = [2 \, \Delta H_f^\circ \, CO_2 \, (g) + 4 \, \Delta H_f^\circ \, H_2O \, (g) + \Delta H_f^\circ \, N_2 \, (g)]$

$$- [\Delta H_f^\circ \, N_2H_2(CH_3)_2 \, (\ell) + 4\Delta H_f^\circ \, O_2 \, (g)] = - 1803 \text{ kJ}$$

On a per gram basis this corresponds to:

$$\frac{- 1803 \text{ kJ}}{1 \text{ mol N}_2\text{H}_2(\text{CH}_3)_2} \cdot \frac{1 \text{ mol N}_2\text{H}_2(\text{CH}_3)_2}{60.10 \text{ g N}_2\text{H}_2(\text{CH}_3)_2} = - 30.00 \text{ kJ/g}$$

Dimethylhydrazine gives more heat per gram.

85. Express BTU/ft^2 as J/m^2:

A conversion factor relating feet to meters is : 3.281 ft = 1 m.

Area is (Length)^2, so squaring both sides of this equality provides the conversion factor :
10.76 ft^2 = 1 m^2

$$\frac{2080 \text{ BTU}}{1 \text{ ft}^2} \cdot \frac{1055 \text{ J}}{1 \text{ BTU}} \cdot \frac{10.76 \text{ ft}^2}{1 \text{ m}^2} = 2.36 \times 10^7 \frac{\text{J}}{\text{m}^2} \quad (\text{ to 3 sf})$$

87. a. The Enthalpy diagram for the isomers of butene:

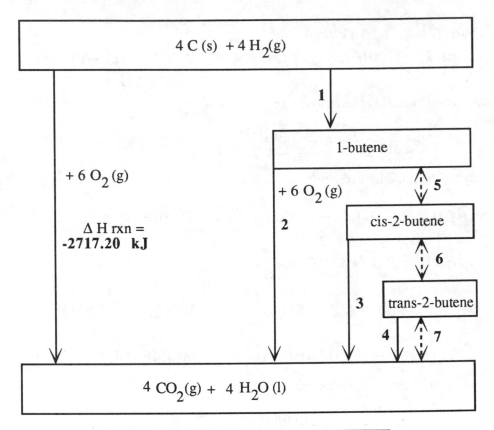

Step	ΔH (in kJ/mol)
1	- 20.5
2	-2696.7
3	-2687.5
4	-2684.2

b. The enthalpy change from *cis*-2-butene to *trans*-2-butene:

This energy difference corresponds to step 6 in the diagram. Note that this difference corresponds to the difference in the Enthalpies of Combustion of steps 3-4 or: $\Delta H = (-2684.2) - (-2687.5) = -3.3$ kJ

c. The enthalpies of formation for both *cis*-2-butene and *trans*-2-butene:

Step 1 corresponds to the ΔH_f for 1-butene. Step 1 + step 5 corresponds to the ΔH_f for *cis*-2-butene. Step 5 corresponds to the difference in the Enthalpies of Combustion of steps 2-3 or:

$\Delta H = (-2687.5) - (-2696.7) = -9.2$ kJ

Step 1 + Step 5 = $(-20.5) + (-9.2) = -29.7$ kJ (ΔH_f for *cis*-2-butene)

Using the same logic as above, step 1 + step 5 + step 6 corresponds to the ΔH_f for *trans*-2-butene. The magnitude for step 6 corresponds to the difference in the Enthalpies of Combustion of steps 3-4, as calculated in part b above, or:

$\Delta H = (-2684.2) - (-2687.5) = -3.3$ kJ

Step 1 + Step 5 + Step 6 = $(-20.5) + (-9.2) + (-3.3) = -33.0$ kJ

(ΔH_f for *trans*2-butene)

Conceptual Questions

90. This is a losing battle. To extract heat from the inside of the refrigerator, work has to be done. That work (by the condenser and motor) releases heat to the environment (your room). So while the temporary relief of cool air from the inside of the refrigerator is pleasant, the motor has to do work—and heats your room.

Summary Question

92.   a. Mass of SO_2 contained in 440 million gallons of wine that is 100. ppm in SO_2:

$$440 \times 10^6 \text{ gal} \cdot \frac{4 \text{ qt}}{1 \text{ gal}} \cdot \frac{0.9463 \text{ L}}{1 \text{ qt}} \cdot \frac{1 \times 10^3 \text{ cm}^3}{1 \text{ L}} \cdot \frac{1.00 \text{ g wine}}{1 \text{ cm}^3}$$

$$\cdot \frac{100. \text{ g } SO_2}{1 \times 10^6 \text{ g wine}} = 1.7 \times 10^8 \text{ g } SO_2$$

The number of moles to which this corresponds is:

$$1.7 \times 10^8 \text{ g SO}_2 \cdot \frac{1 \text{ mol SO}_2}{64.06 \text{ g SO}_2} = 2.6 \times 10^{10} \text{ mol SO}_2$$

b. For the consumption of 20. million tons of SO_2 by MgO:

$$20. \times 10^6 \text{ T SO}_2 \cdot \frac{1 \text{ mol SO}_2}{64.06 \text{ g SO}_2} \cdot \frac{1 \text{ mol MgO}}{1 \text{ mol SO}_2} \cdot \frac{40.30 \text{ g MgO}}{1 \text{ mol MgO}}$$

$$= 1.3 \times 10^7 \text{ T MgO } (1.1 \times 10^{13} \text{ g MgO})$$

The quantity of $MgSO_4$ produced:

$$1.3 \times 10^7 \text{ T MgO} \cdot \frac{120.4 \text{ g MgSO}_4}{40.30 \text{ g MgO}} = 3.8 \times 10^7 \text{ T MgSO}_4 \, (3.4 \times 10^{13} \text{ g MgSO}_4)$$

c. The heat energy change per mole of $MgSO_4$ by the reaction:

$$\text{MgO (s)} + \text{SO}_2 \text{ (g)} + \tfrac{1}{2} \text{ O}_2 \text{ (g)} \rightarrow \text{MgSO}_4 \text{ (s)}$$

ΔH_f° - 601.70   - 296.8   0   - 2817.5

$\Delta H_{rxn}^\circ = \Sigma \, \Delta H_f^\circ$ products - $\Sigma \, \Delta H_f^\circ$ reactants

$\qquad = (- 2817.5 \text{ kJ/mol})(1 \text{ mol})$

$\qquad\qquad - [(-601.70 \text{ kJ/mol})(1 \text{ mol}) + (- 296.8 \text{ kJ/mol})(1 \text{ mol}) + 0]$

$\qquad = -1919.0 \text{ kJ}$ **Heat evolved**

d. Heat evolved by production of 750. T H_2SO_4 per day.

The overall ΔH_{rxn} is the sum of the three reactions given:

$\Delta H_{overall} = (- 296.8 \text{ kJ}) + (- 98.9 \text{ kJ}) + (-130.0 \text{ kJ}) = - 525.7 \text{ kJ/mol H}_2\text{SO}_4$

The number of moles of H_2SO_4 corresponding to 750. T is:

$$750 \text{ T H}_2\text{SO}_4 \cdot \frac{9.08 \times 10^5 \text{ g H}_2\text{SO}_4}{1 \text{ T H}_2\text{SO}_4} \cdot \frac{1 \text{ mol H}_2\text{SO}_4}{98.08 \text{ g H}_2\text{SO}_4} = 6.94 \times 10^6 \text{ mol}$$

The overall energy produced per day is then:

$$6.94 \times 10^6 \text{ mol H}_2\text{SO}_4 \cdot \frac{- 525.7 \text{ kJ}}{\text{mol H}_2\text{SO}_4} = 3.65 \times 10^9 \text{ kJ}$$

Chapter 7:
Atomic Structure

Electromagnetic Radiation

19.   a **Red, orange, and yellow** light have less energy than green light.

   b. **Blue** light photons are of greater energy than yellow light.

   c. **Blue** light has greater frequency than green light.

21. Since frequency • wavelength = speed of light (c)

$$76 \text{ s}^{-1} \bullet \text{ l } = 3.0 \times 10^8 \text{ m/s}$$

$$\text{l } = 3.9 \times 10^6 \text{ m}$$

$$\text{wavelength in miles } = 3.9 \times 10^6 \text{ m} \bullet \frac{1 \text{ km}}{1.0 \times 10^3 \text{ m}} \bullet \frac{1 \text{ mi}}{1.61 \text{ km}} = 2.5 \times 10^3 \text{ mi}$$

23. $\text{frequency } = \dfrac{\text{speed of light}}{\text{wavelength}} = \dfrac{2.9979 \times 10^8 \text{ m/s}}{5.0 \times 10^2 \text{ nm}} \bullet \dfrac{1.00 \times 10^9 \text{ nm}}{1.00 \text{ m}}$

$$= 6.0 \times 10^{14} \text{ s}^{-1}$$

The energy of a photon of this light may be determined :
$$E = h\upsilon$$

Planck's constant, h, has a value of 6.626×10^{-34} J • s • photons^{-1}

$$E = (6.626 \times 10^{-34} \text{ J} \bullet \text{s} \bullet \text{photons}^{-1})(6.0 \times 10^{14} \text{ s}^{-1})$$
$$= 3.98 \times 10^{-19} \text{ J} \bullet \text{photon}^{-1} \text{ or } 4.0 \times 10^{-19} \text{ J} \bullet \text{photon}^{-1} \text{ to 2 sf}$$

Energy of 1.00 mol of photons $= 3.98 \times 10^{-19}$ J • photon$^{-1} \bullet \dfrac{6.0221 \times 10^{23} \text{ photons}}{1.00 \text{ mol photons}}$

$$= 2.4 \times 10^5 \text{ J/mol photon}$$

25. The frequency of the line at 396.15 nm:

$\text{frequency } = \dfrac{\text{speed of light}}{\text{wavelength}} = \dfrac{2.9979 \times 10^8 \text{ m/s}}{3.9615 \times 10^2 \text{ nm}} \bullet \dfrac{1.00 \times 10^9 \text{ nm}}{1.00 \text{ m}}$

$$= 7.5676 \times 10^{14} \text{ s}^{-1}$$

The energy of a photon of this light may be determined : $E = h\upsilon$

Planck's constant, h, has a value of 6.626×10^{-34} J • s • photons^{-1}

$$E = (6.626 \times 10^{-34} \text{ J} \cdot \text{s} \cdot \text{photons}^{-1})(7.5676 \times 10^{14} \text{ s}^{-1})$$
$$= 5.0143 \times 10^{-19} \text{ J} \cdot \text{photon}^{-1}$$

Energy of 1.00 mol of photons = 5.0143×10^{-19} J • photon^{-1} • $\dfrac{6.0221 \times 10^{23} \text{ photons}}{1.00 \text{ mol photons}}$

$$= 3.0197 \times 10^5 \text{ J/mol photon}$$

27.   a. The **most energetic light** would be represented by the light of **shortest wavelength** (253.652 nm).

b. The frequency of this light is :
$$\frac{2.9979 \times 10^8 \text{ m/s}}{253.652 \text{ nm}} \cdot \frac{1.00 \times 10^9 \text{ nm}}{1.00 \text{ m}} = 1.18190 \times 10^{15} \text{ s}^{-1}$$

The energy of 1 photon with this wavelength is:
$$E = h\upsilon = (6.62608 \times 10^{-34} \frac{\text{J} \cdot \text{s}}{\text{photon}})(1.18190 \times 10^{15} \text{ s}^{-1})$$
$$= 7.83139 \times 10^{-19} \frac{\text{J}}{\text{photon}}$$

c. The line emission spectrum of mercury shows the visible region between $\approx$ 400 and 750 nm. The lines at 404 and 435 nm are present while the lines at 253 nm, 365 nm and 1013 nm lie outside the visible region. The 404 nm line is violet, while the 435 nm line is blue.

29. Since energy is proportional to frequency ($E = h\upsilon$), we can arrange the radiation in order of increasing energy per photon by listing the types of radiation in increasing frequency (or decreasing wavelength).

$$\rightarrow \qquad \text{Energy increasing} \qquad \rightarrow$$

FM music   microwave   yellow light   x-rays

$$\rightarrow \quad \text{Frequency } (\upsilon) \text{ increasing} \quad \rightarrow$$
$$\leftarrow \quad \text{Wavelength } (\lambda) \text{ increasing} \quad \leftarrow$$

Photoelectric Effect

31. Energy $= 2.0 \times 10^2$ kJ/mol $\cdot \dfrac{1 \text{ mol}}{6.0221 \times 10^{23} \text{ photons}} \cdot \dfrac{1.00 \times 10^3 \text{ J}}{1.00 \text{ kJ}}$

$= 3.3 \times 10^{-19}$ J $\cdot$ photons $^{-1}$

$E = h\upsilon = \dfrac{hc}{\lambda}$ or $\lambda = \dfrac{hc}{E} = \dfrac{(6.626 \times 10^{-34} \text{ J} \cdot \text{s} \cdot \text{photons}^{-1})(2.9979 \times 10^8 \text{ m} \cdot \text{s}^{-1})}{3.3 \times 10^{-19} \text{ J} \cdot \text{photons}^{-1}}$

$= 6.0 \times 10^{-7}$ m or 6.0×10^2 nm

Radiation of this wavelength--in the **visible** region of the electromagnetic spectrum-- would appear **orange**.

Atomic Spectra and The Bohr Atom

33. a. | Transitions from | to |
| --- | --- |
| n = 4 | n = 3, 2, or 1 |
| n = 3 | n = 2 or 1 |
| n = 2 | n = 1 |

Six transitions are possible from these four quantum levels, providing 6 emission lines.

b. Photons of the highest energy will be emitted in a transition from the level with **n = 4** to the level **n = 1**. This is easily seen with the aid of the equation

$$\Delta E = Rhc\left(\dfrac{1}{n^2_f} - \dfrac{1}{n^2_i}\right).$$

Since R, h, and c are constant for any transition, inspection shows that the largest change in energy results if $n_f = 1$ and $n_i = 4$.

c. The emission line having the **longest wavelength** also has the **lowest frequency** and the **lowest energy**. A transition from $n_i = 4$ to $n_f = 3$ would provide the longest wavelength line.

35. One can calculate the wavelength of any transition by applying the relationship:

$$\frac{1}{\lambda} = R\left(\frac{1}{n^2_f} - \frac{1}{n^2_i}\right).$$

For any line with a wavelength greater than 102.6 nm, the left side of the equation ($\frac{1}{\lambda}$) would be smaller than that for the 102.6 nm line. This will be true if the term ($\frac{1}{n^2_f} - \frac{1}{n^2_i}$) is smaller than 8/9.

For $n_f = 1$ and $n_i = 3$: $\frac{1}{1^2} - \frac{1}{3^2} = 8/9$ or 0.89

Substituting the suggested transitions into the term ($\frac{1}{n^2_f} - \frac{1}{n^2_i}$) yields:

	transition between	($\frac{1}{n^2_f} - \frac{1}{n^2_i}$)
a.	2 and 4	0.19
b.	1 and 4	0.94
c.	1 and 5	0.96
d.	3 and 5	0.07

The transition between **2 and 4** and between **3 and 5** would produce a line of longer wavelength than 102.6 nm (lower energy). The remaining transitions would produce lines of shorter wavelength (and higher energy) .

37. The wavelength of emitted light for the transition n = 3 to n = 1.

$$\frac{1}{\lambda} = R\left(\frac{1}{1^2} - \frac{1}{3^2}\right) = (1.0974 \times 10^7 \text{ m}^{-1})(8/9) = 9.7547 \times 10^6 \text{ m}^{-1}$$

$\lambda = 1.0252 \times 10^{-7}$ m or approximately 103 nm (far ultraviolet)

39.   The energy needed to move an electron from n = 1 to n = 5 will be the same as the amount emitted as the electron relaxed from n = 5 to n = 1, i.e. 2.093×10^{-18} J.

De Broglie and Matter Waves

41. Mass of an electron: 9.11×10^{-31} kg

    Planck's constant: 6.626×10^{-34} J • s • photon^{-1}

    Velocity of the electron: 2.5×10^8 cm • s^{-1} or 2.5×10^6 m • s^{-1}

$$\lambda = \frac{h}{m \cdot v} = \frac{6.626 \times 10^{-34} \text{ J} \cdot \text{s}}{(9.11 \times 10^{-31} \text{ kg} \cdot 2.5 \times 10^6 \text{ m} \cdot \text{s}^{-1})}$$

$$= 2.9 \times 10^{-10} \text{ m} = 2.9 \text{ Angstroms} = 0.29 \text{ nm}$$

43. The wavelength can be determined exactly as in Question 41:

$$\lambda = \frac{h}{m \cdot v} = \frac{6.626 \times 10^{-34} \text{ J} \cdot \text{s}}{(1.0 \times 10^{-1} \text{ kg} \cdot 30. \text{ m} \cdot \text{s}^{-1})}$$

$$= 2.2 \times 10^{-34} \text{ m} \quad \text{or} \quad 2.2 \times 10^{-25} \text{ nm}$$

    Velocity to have a wavelength of 5.6×10^{-3} nm

$$5.6 \times 10^{-3} \text{ nm} \cdot \frac{1 \text{ m}}{1 \times 10^9 \text{ nm}} = 5.6 \times 10^{-12} \text{ m}$$

    Rewriting the above equation

$$V = \frac{h}{m \cdot \lambda} = \frac{6.626 \times 10^{-34} \text{ J} \cdot \text{s}}{(1.0 \times 10^{-1} \text{ kg} \cdot 5.6 \times 10^{-12} \text{ m})} = 1.2 \times 10^{-21} \frac{\text{m}}{\text{s}}$$

Quantum Mechanics

45. Complete the following table: (Answers are emboldened.)

QUANTUM NUMBER	ATOMIC PROPERTY DETERMINED
n	orbital size
m$_\ell$	relative orbital orientation
ℓ	orbital shape

47.   a. $n = 4$ possible ℓ values $= 0,1,2,3$ $(\ell = 0,1,... (n - 1))$
      b. $\ell = 2$ possible m_ℓ values $= -2,-1,0,+1,+2$ $(-\ell ..., 0,....+\ell)$
      c. orbital $= 4s$ $n = 4$; $\ell = 0$; $m_\ell = 0$
      d. orbital $= 4f$ $n = 4$; $\ell = 3$; $m_\ell = -3,-2,-1,0,+1,+2,+3$

49. An electron in a 4p orbital must have $n = 4$ and $\ell = 1$. The possible m_ℓ values give rise to the
      following sets of n, ℓ, and m_ℓ

n	ℓ	m_ℓ	
4	1	-1	Note that the **three values** of m describe
4	1	0	**three orbital orientations**.
4	1	+1	

51. Subshells in the electron shell with $n = 4$:
      There are 4 : s, p, d, and f sublevels corresponding to $\ell = 0, 1, 2$, and 3 respectively.

53. Explain why each of the following is not a possible set of quantum numbers for an electron in
      an atom.
      a. $n = 2$, $\ell = 2$, $m_\ell = 0$ For $n = 2$, maximum value of ℓ is one (1).
      b. $n = 3$, $\ell = 0$, $m_\ell = -2$ For $\ell = 0$, possible values of m_ℓ are $\pm \ell$ and 0.
      c. $n = 6$, $\ell = 0$, $m_\ell = 1$ For $\ell = 0$, possible values of m_ℓ are $\pm \ell$ and 0.

55.   <u>quantum number designation</u>   <u>maximum number of orbitals</u>
      a. $n = 4$; $\ell = 3$ 7 ("**f**" orbitals)

      b. $n = 5$; $\ell = 0$ 1 ("s" orbital)
              $\ell = 1$ 3 ("**p**" orbitals)
              $\ell = 2$ 5 ("**d**" orbitals)
              $\ell = 3$ 7 ("**f**" orbitals)
              $\ell = 4$ <u>9 ("**g**" orbitals)</u>
                                        25 orbitals

      c. $n = 2$; $\ell = 2$ none; for $n = 2$
                                        the max number of $\ell = 1$

      d. $n = 3$; $\ell = 1$; $m_\ell = -1$ 1 (a 3 "**p**" orbital)

57. The number of planar nodes possessed by each of the following:

	orbital	number of planar nodes
a.	2s	0
b.	5d	2
c.	5f	3

59. Which of the following orbitals cannot exist and why:

2s exists	$n = 2$ permits ℓ values as large as 1
2d cannot exist	$\ell = 2$ is not permitted for $n < 3$
3p exists	$n = 3$ permits ℓ values as large as 2
3f cannot exist	$\ell = 3$ is not permitted for $n < 4$
4f exists	$\ell = 4$ permits ℓ values as large as 3
5s exists	$n = 5$ permits ℓ values as large as 4

61. The complete set of quantum numbers for :

		$\underline{n}$	$\underline{\ell}$	$\underline{m_\ell}$	
a.	2p	2	1	-1, 0, +1	(3 orbitals)
b.	3d	3	2	-2, -1, 0, +1, +2	(5 orbitals)
c.	4f	4	3	-3, -2, -1, 0, +1, +2, +3	(7 orbitals)

63. All orbitals except **the s orbital** may have magnetic quantum numbers, $m_\ell = -1$.
The s orbital has as its only value for the magnetic quantum number, $m_\ell = 0$.

General Questions

65. The frequency 6.00×10^5 s^{-1} has a wavelength of:

$$\lambda = \frac{c}{\upsilon} = \frac{2.998 \times 10^8 \text{ m} \cdot \text{s}^{-1}}{6.00 \times 10^5 \text{ s}^{-1}} = 5.00 \times 10^2 \text{ m}$$

The energy of 1 photon of frequency 6.00×10^5 s^{-1} :

$$E = h\upsilon = (6.626 \times 10^{-34} \text{ J} \cdot \text{s} \cdot \text{photon}^{-1})(6.00 \times 10^5 \text{ s}^{-1}) = 3.98 \times 10^{-28} \text{J} \cdot \text{photon}^{-1}$$

Compared to a photon of red light of $\lambda = 685$ nm :

$$E = \frac{hc}{\lambda} = (6.626 \times 10^{-34} \text{ J} \cdot \text{s} \cdot \text{photon}^{-1}) \frac{2.9979 \times 10^8 \text{ m} \cdot \text{s}^{-1}}{785 \times 10^{-9} \text{ m}}$$

$$= 2.90 \times 10^{-19} \text{ J} \cdot \text{photon}^{-1}$$

67. Number of photons ($\lambda = 12$ cm) to raise the temperature of eye by 3.0 °C:

  1. Determine the energy needed :

$$\begin{aligned} \text{Energy} &= \text{mass} \times \text{heat capacity} \times \Delta t \\ &= (10.0 \text{ g})(4.0 \frac{J}{g \cdot K})(3.0 \text{ K}) \\ &= 120 \text{ J} \end{aligned}$$

  2. Determine the energy of a photon of light with $\lambda = 12$ cm:

$$\text{Energy} = \frac{hc}{\lambda} = (6.626 \times 10^{-34} \text{ J} \cdot \text{s} \cdot \text{photon}^{-1})\frac{2.9979 \times 10^8 \text{ m} \cdot \text{s}^{-1}}{12 \times 10^{-2} \text{ m}}$$

  [Note the expression of 12 cm in units of meters.]

$$= 1.66 \times 10^{-24} \text{ J} \cdot \text{photon}^{-1}$$

  3. The number of photons needed to furnish 120 J:

$$1.20 \times 10^2 \text{ J} \cdot \frac{1 \text{ photon}}{1.66 \times 10^{-24} \text{ J}} = 7.2 \times 10^{25} \text{ photons} \quad (2 \text{ sf})$$

69. Time for Voyager I's signal to travel 2.7×10^9 miles:

If light travels at 2.9979×10^8 m $\cdot$ s^{-1}, we can calculate the time:

$$2.7 \times 10^9 \text{ mi} \cdot \frac{1.61 \text{ km}}{1 \text{ mi}} \cdot \frac{1 \times 10^3 \text{ m}}{1 \text{ km}} \cdot \frac{1 \text{ s}}{2.9979 \times 10^8 \text{ m}} = 1.4 \times 10^4 \text{ s}$$

Given there are 3600 s in 1 hr,

$$1.4 \times 10^4 \text{ s} \cdot \frac{1 \text{ hr}}{3600 \text{ s}} = 4.0 \text{ hr}$$

71. Example 7.3 in your text illustrates the calculation of the ionization energy for H's electron

$$E = \frac{-Z^2 Rhc}{n^2} = -2.179 \times 10^{-18} \text{ J/atom} \implies -1312 \text{ kJ/mol}$$

For He$^+$ the calculation yields

$$E = \frac{-(2)^2(1.097 \times 10^7 \text{ m}^{-1})(6.626 \times 10^{-34} \text{ J} \cdot \text{s})(2.998 \times 10^8 \text{ m} \cdot \text{s}^{-1})}{(1)^2}$$

$$= -8.717 \times 10^{-18} \text{ J/ion and expressing this energy for a mol of ions}$$

$$= \frac{-8.717 \times 10^{-18} \text{ J}}{\text{ion}} \cdot \frac{6.0221 \times 10^{23} \text{ atoms}}{\text{mol}} \cdot \frac{1 \text{ kJ}}{1000 \text{ J}} = -5249 \frac{\text{kJ}}{\text{mol}}$$

The energy to remove the electron is then 5249 kJ/mol of ions. Note that this energy is four times that for H.

73. The shortest wavelength photon from an excited H atom is the photon of highest energy ($E = h\upsilon$). That energy would be associated with the interval of $n = 1$ to $n = \infty$. As you can see from Figure 7.13, that transition ($n = \infty$ to $n = 1$) would release light with a wavelength of 93.8 nm.

75. The maximum probability density for the 2p orbital occurs at a distance of $2a_o$ from the nucleus (or $2 \cdot 0.0529$ nm) or 0.1058 nm.

77. Orbitals in an atom that can be specified by:

	orbital designation	maximum number of orbitals
a.	3p	3 orbitals ($m_\ell = +1, 0, -1$; the p_x, p_y, and p_z orbitals)
b.	4p	3 orbitals ($m_\ell = +1, 0, -1$; the p_x, p_y, and p_z orbitals)
c.	$4p_x$	1 orbital
d.	$n = 5$; $\ell = 0$	1 ("**s**" orbital)
	$\ell = 1$	3 ("**p**" orbitals)
	$\ell = 2$	5 ("**d**" orbitals)
	$\ell = 3$	7 ("**f**" orbitals)
	$\ell = 4$	9 ("**g**" orbitals)
		25 orbitals
e.	6d	5 orbitals ($m_\ell = +2, +1, 0, -1, -2,$)
f.	5d	5 orbitals ($m_\ell = +2, +1, 0, -1, -2,$)
g.	5f	7 orbitals ($m_\ell = +3, +2, +1, 0, -1, -2, -3$)
h.	7s	1 orbital ($m_\ell = 0,$)

79.   a. The quantum number n describes the **size (and energy)** of an atomic orbital and the quantum number ℓ describes its **shape or type**.

b. When n = 3, the possible values of ℓ are 0,1, and 2 [0.....(n-1)].

c. The **f** orbital corresponds to $\ell = 3$.

d. For a 4d orbital, the value of n is **4**, the value of ℓ is **2**, and a possible value of m_ℓ is
$$0, \pm\ 1,\ \text{or}\ \pm 2.$$

e. The orbitals pictured have the following characteristics:

letter =	d	s	p
ℓ value =	2	0	1
nodal planes =	2	0	1

f. An atomic orbital with 3 nodal planes is **f**.

g. According to modern quantum theory, the **2d** and **4g** orbitals cannot exist.
Maximum ℓ value for n = 1 is 0 (s) and for n = 4 (f).

h. The set **n = 2, ℓ = 1, and m_ℓ = 2** is not a valid set of quantum numbers (maximum value of $m_\ell = +\ell$)

i. The maximum number of orbitals associated with the following sets of quantum numbers is: orbitals

    i. n = 2 and ℓ = 1   3 (the p orbitals)

    ii. n = 3   1 s, 3 p's, and 5 d's = 9 total

    iii n = 3 and ℓ = 3   none (max value of ℓ = (n - 1)

    iv. n = 2, ℓ = 1, and m_ℓ = 0   1 (of the three 2p orbitals)

81. The $2p_z$ orbital is oriented along the z axis, with minimal density along the x (and y) axis.
Hence the density 0.05 nm from the nucleus on the x axis would be **lower** compared to the z axis.

Conceptual Questions

86. For the first three electron shells:

N = 1	N = 2	N = 3
L = 1	L = 2	L = 3
M = ±1,0	M = ±1,0	M = ±1,0

There would be 3 orbitals in each of the 3 shells for a **total of 9 orbitals**.

90.   a. The spherical symmetry of the 1s orbital means that the probability of finding the electron at Y = C is equal to that at x = d. Given the spherical nature of the s orbital, the electron probability will **increase** at a distance closer to the nucleus.

b. Given the symmetry of the p_x orbital, the probability on the y and z axes should be the same (1×10^{-3}).

Summary Question

91.   a. Technetium is in period 5, group 7B

b. A set of quantum numbers for a 5s electron :   n = 5, $\ell = 0$, and $m_\ell = 0$

c. Wavelength and frequency of a photon with energy of 0.141 MeV

$$E = h\upsilon$$

$$0.141 \text{ MeV} \cdot \frac{1 \times 10^6 \text{ eV}}{1 \text{MeV}} \cdot \frac{9.6485 \times 10^4 \text{ J/mol}}{1 \text{ eV}} \cdot \frac{1 \text{ mol e}^-}{6.0221 \times 10^{23} \text{ e}^-}$$

$$= 6.626 \times 10^{-34} \text{ J} \cdot s\upsilon$$

$$2.26 \times 10^{-14} \text{ J} = 6.626 \times 10^{-34} \text{ J} \cdot s\upsilon$$
$$3.41 \times 10^{19} \text{ s}^{-1} = \upsilon$$
$$\upsilon\lambda = c \quad \text{so} \quad \lambda = \frac{2.9979 \times 10^8 \text{ m} \cdot s^{-1}}{3.41 \times 10^{19} \text{ s}^{-1}}$$

$$\lambda = 8.79 \times 10^{-12} \text{ m or } 8.79 \times 10^{-3} \text{ nm}$$

d. (i) $HTcO_4$ (aq) + NaOH(aq) $\longrightarrow$ $NaTcO_4$ (aq) + $H_2O(\ell)$
Mass of $NaTcO_4$ possible:

$$4.5 \text{ mg Tc} \cdot \frac{185 \text{ g NaTcO}_4}{98 \text{ g Tc}} = 8.5 \text{ mg NaTcO}_4$$

$$8.5 \times 10^{-3} \text{ g NaTcO}_4 \cdot \frac{1 \text{ mol NaTcO}_4}{185 \text{ g NaTcO}_4} \cdot \frac{1 \text{ mol NaOH}}{1 \text{ mol NaTcO}_4} \cdot \frac{40.0 \text{ g NaOH}}{1 \text{ mol NaOH}}$$

$$= 1.8 \times 10^{-3} \text{ g NaOH}$$

Chapter 8:
Atomic Electron Configuration and Chemical Periodicity

Writing Electron Configurations

13. The electron configurations for Mg and Cl using both the orbital box and spectroscopic notations:

   Orbital box notation

   Mg [↑↓] [↑↓] [↑↓][↑↓][↑↓] [↑↓]

   Cl [↑↓] [↑↓] [↑↓][↑↓][↑↓] [↑↓] [↑↓][↑↓][↑]

   Spectroscopic notation

   $1s^2 2s^2 2p^6 3s^2$

   $1s^2 2s^2 2p^6 3s^2 3p^5$

15. Vanadium's electron configuration: $1s^2 2s^2 2p^6 3s^2 3p^6 3d^3 4s^2$

17. Germanium's electron configuration:
   Spectroscopic notation: $1s^2 2s^2 2p^6 3s^2 3p^6 3d^{10} 4s^2 4p^2$
   Noble gas notation: $[Ar] 3d^{10} 4s^2 4p^2$

19. Electron configurations, using the Noble gas notation:

   a. Sr: $[Kr] 5s^2$ b. Zr: $[Kr] 4d^2 5s^2$
   c. Rh: $[Kr] 4d^8 5s^1$ d. Sn: $[Kr] 4d^{10} 5s^2 5p^2$

   The spectroscopic notation for each of these is obtained by replacing [Kr] with
   $1s^2 2s^2 2p^6 3s^2 3p^6 3d^{10} 4s^2 4p^2 4p^6$

21. Electron configuration for:

   a. Eu: $1s^2 2s^2 2p^6 3s^2 3p^6 3d^{10} 4s^2 4p^6 4d^{10} 4f^7 5s^2 5p^6 6s^2$
      $[Xe] 4f^7 6s^2$

   b. Yb: $1s^2 2s^2 2p^6 3s^2 3p^6 3d^{10} 4s^2 4p^6 4d^{10} 4f^{14} 5s^2 5p^6 6s^2$
      $[Xe] 4f^{14} 6s^2$

23. Electron configuration for:

    a. Pu: $1s^2\ 2s^2\ 2p^6\ 3s^2\ 3p^6\ 3d^{10}\ 4s^2\ 4p^6\ 4d^{10}\ 4f^{14}\ 5s^2\ 5p^6\ 5d^{10}\ 5f^6\ 6s^2\ 6p^6\ 7s^2$
       [Rn] $5f^6\ 7s^2$

    b. Es: $1s^2\ 2s^2\ 2p^6\ 3s^2\ 3p^6\ 3d^{10}\ 4s^2\ 4p^6\ 4d^{10}\ 4f^{14}\ 5s^2\ 5p^6\ 5d^{10}\ 5f^{11}\ 6s^2\ 6p^6\ 7s^2$
       [Rn] $5f^{11}\ 7s^2$

25. The orbital box representations for the following ions:

                       Orbital box notation

    a. Na^+

    b. Al^{3+}

    c. Cl^-

27.   Electron configurations of:

    a. Ti [Ar]   [Ar] $3d^2\ 4s^2$

    b. Ti^{2+} [Ar]   [Ar] $3d^2$

    c. Ti^{4+} [Ar]   [Ar]

    Note that the Ti^{2+} ion contains two unpaired electrons, and is therefore paramagnetic.

29.   Element 25 is manganese.

    a. Mn   [Ar]   [Ar] $3d^5 4s^2$

    b. Mn^{2+} [Ar]   [Ar] $3d^5$

    c. Having five (5) unpaired electrons, Mn $^{2+}$ is **paramagnetic.**

31. For ruthenium :

    a. The predicted and actual electron configuration are:

    Predicted : Ru   [Kr] $4d^6 5s^2$ or $1s^2 2s^2 2p^6 3s^2 3p^6 3d^{10} 4s^2 4p^6 4d^6 5s^2$

    Actual :   Ru   [Kr] $4d^7 5s^1$ or $1s^2 2s^2 2p^6 3s^2 3p^6 3d^{10} 4s^2 4p^6 4d^7 5s^1$

    b. The formation of the Ru^{3+} ion can be arrived at (from the predicted configuration) by loss of the two 5s electrons and one 4d electron. From the actual configuration, we see the loss of two 4d electrons and the 5s electron.

    Ru^{3+} [Kr]   4d [↑|↑|↑|↑|↑]   5s [ ]   [Kr] $4d^5$

33.   The electron configuration for Sm and Ho and their ions using the orbital box method:

    a.   Sm: [Xe]   4f [↑|↑|↑|↑|↑|↑|]   6s [↑↓]   [Xe] $4f^6 6s^2$

    Sm:$^{3+}$ [Xe]   [↑|↑|↑|↑|↑| |]   [ ]   [Xe] $4f^5$

    b.   Ho [Xe]   [↑↓|↑↓|↑↓|↑↓|↑|↑|↑]   [↑↓]   [Xe] $4f^{11} 6s^2$

    Ho^{3+} [Xe]   [↑↓|↑↓|↑↓|↑|↑|↑|↑]   [ ]   [Xe] $4f^{10}$

35.   The electron configuration for Ti^{2+} is : [Ar] $3d^2$ hence this ion has 2 unpaired electrons. The electron configuration for Co^{3+} is : [Ar] $3d^6$ giving this ion 4 unpaired electrons. Both ions therefore are paramagnetic (contain unpaired electrons).

37.   The 2+ ions for Ti through Zn are:

Ion	Unpaired electrons	Ion	Unpaired electrons
Ti $^{2+}$	2	Co $^{2+}$	3
V $^{2+}$	3	Ni $^{2+}$	2
Cr $^{2+}$	4	Cu $^{2+}$	1

Ion	Unpaired electrons		Ion	Unpaired electrons
Mn^{2+}	5		Zn^{2+}	0 **diamagnetic**
Fe^{2+}	4			

Electron Configurations and Quantum Numbers

39. The electron configuration for Mg using the orbital box method:

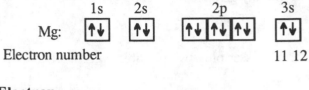

Electron number   11 12

Electron number:	n	ℓ	m_ℓ	m_s
11	3	0	0	+ 1/2
12	3	0	0	- 1/2

41.   The electron configuration for Ti using the orbital box method:

        3d   4s

Ti [Ar] | ↑ | ↑ | | | | | ↑↓ | [Ar] $3d^2 4s^2$

Electron Number   3   4   12

Electron number:	n	ℓ	m_ℓ	m_s
1	4	0	0	+ 1/2
2	4	0	0	- 1/2
3	3	2	-2	+ 1/2
4	3	2	-1	+ 1/2

43. Maximum number of electrons associated with the following sets of quantum numbers:

	Characterized as	Maximum number of electrons
a. n = 2 and ℓ = 1	2p electrons	6
b. n = 3	3s, 3p, 3d electrons	18
c. n = 3 and ℓ = 3	[Maximum ℓ = n-1]	none

d. $n = 4$, $\ell = 1$, $m_\ell = -1$,   4p orbital   1 (4 q. n. completely

    and $m_s = -1/2$ describe 1 electron)

e. $n = 5$, $\ell = 0$, $m_\ell = +1$ [m_ℓ range = $-\ell$...,0,... $+\ell$] none

45. Explain why the following sets of quantum numbers are not valid:

a. $n = 2$, $\ell = 2$, $m_\ell = 0$, $m_s = +1/2$:

    The maximum value of ℓ for any given n value is (n - 1).

b. $n = 2$, $\ell = 1$, $m_\ell = -1$, $m_s = 0$:

    The spin quantum number possesses only two possible values: +1/2 or -1/2.

c. $n = 3$, $\ell = 1$, $m_\ell = +1$, $m_s = +1/2$

    Oxygen has only 8 electrons, therefore it has <u>no</u> electrons in the third shell (n = 3) in the ground state.

Periodic Properties

47. a. Estimate E-Cl bond distances: [The atomic radius for Cl = 100 pm.]

Element	Atomic Radius (pm)	E—Cl Bond Distances (pm)
N	70	170
P	110	210
As	120	220
Sb	140	240
Bi	150	250

49. Elements arranged in order of increasing size: C < B < Al < Na < K

    Radii from Figure 8.10 (in pm)   77 < 85 < 143 < 186 < 227

51.   The specie in each pair with the larger radius:

    a. Cl^- is larger than Cl -- The ion has more electrons/proton than the atom.

    b. Al is larger than N -- Al is in period 3, while N is in period 2.

    c. In is larger than Sn -- Atomic radii decrease, in general, across a period.

53.   The group of elements with correctly ordered increasing ionization energy (IE):

c. Li < Si < C < Ne.

Neon would have the greatest IE. Silicon, being slightly larger in atomic radius than carbon, has a lesser IE. Lithium, the largest atom of this group, would have the smallest IE.

55. For the elements :   Li, K, C, and N

Increasing ionization energy: K < Li < C < N

57. For the elements Li, K, C, and N:

a. The largest atomic radius: K

b. The largest electron affinity: C (most negative value for EA)

c. Increasing ionization energy: K < Li < C < N

59.   a. Increasing ionization energy: S < O < F

Ionization energy is inversely proportional to atomic size.

b. Largest ionization energy of O, S, or Se: O

Oxygen is the smallest of these Group 6A elements.

c. Largest electron affinity of Se, Cl, or Br: Cl

Chlorine is the smallest of these three elements. EA tends to increase (become more negative) on a diagonal from the lower left of the periodic table to the upper right.

d. Largest radius of O^{2-}, F^-, F: O^{2-}

The oxide ion has the largest electron : proton ratio.

Lattice Energies

61.

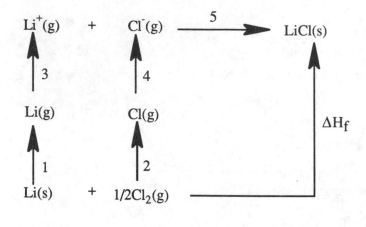

From Hess's Law we know that the ΔH's for steps 1 - 5 must equal the ΔH_f for LiCl(s).

step	ΔH (kJ/mol)
1	+159.4
2	+121.68
3	+520.
4	-349
5	-852
	-400.

The experimental value for ΔH_f for LiCl is -409 kJ/mol.

63. For a given alkali metal cation, lattice energies tend to decrease (become less negative) with heavier halide ions.

For an estimate of the lattice energy for RbBr, note that the **difference** in lattice energies between the chloride and bromide are 37 kJ/mol (for Li), 34 kJ/mol (for Na), and 28 kJ/mol (for K). An estimate of between 21 and 24 kJ difference for Rb would give an approximate lattice energy range of -674 to -671 kJ/mol for RbBr.

General Questions

65. The electronic configuration for element 109 is: [Rn] $5f^{14} 6d^7 7s^2$
    Or using the spectroscopic notation: $1s^2 2s^2 2p^6 3s^2 3p^6 3d^{10} 4s^2 4p^6 4d^{10} 4f^{14} 5s^2$
    $5p^6 5d^{10} 5f^{14} 6s^2 6p^6 6d^7 7s^2$

Iridium would be another element in the same group.

67. The set of quantum numbers which is incorrect is (b). The maximum value of ℓ is (n - 1). With an ℓ =1, this value could not exceed 0.

69. The number of complete electron shells in element 71 is four (4).
    Element 71 is Lu, which has all subshells of the fourth shell filled.

71. Possible quantum numbers for an electron in a 4p orbital:

n	ℓ	m_ℓ	m_s
4	1	1	+1/2
4	1	1	-1/2
4	1	0	+1/2
4	1	0	-1/2
4	1	-1	+1/2
4	1	-1	-1/2

73. Elements containing four unpaired electrons in the 3^+ ion: Tc and Rh.

75. For element A = [Ar] $4s^2$ and B = [Ar] $3d^{10}4s^24p^5$
    a. Element A is a metal (two electrons in the s sublevel).
    b. Element B is a nonmetal (seven electrons in the outer shell).
    c. Since B is "nonmetallic", it should have the larger ionization energy.
    d. B would be smaller as a result of the larger effective nuclear charge for B's outer
       electrons when compared to A's outer electrons.

77. These 5 species are isoelectronic (18 electrons).

Radius (pm)	170	167	152	114	98
	S^{2-} >	Cl^- >	K^+ >	Ca^{2+} >	Ar

79. An examination of the electron configurations for Group 4A and 5A shows that the gain of
    one electron by a Group 4A element would result in a half-filled p subshell. The 5A
    elements have a half-filled p subshell naturally, and the addition of an electron would result
    in increased electron-electron repulsions--resulting in lower electron affinities for Group
    5A.

81. Since $\underline{Ca}$ is smaller than $\underline{K}$, we would expect the first IE of $\underline{Ca}$ to be greater than that of $\underline{K}$. Once $\underline{K}$ has lost its "first" electron, it possess an [Ar] core. Removal of an additional electron (the second) requires much energy. $\underline{Ca}$, on the other hand, can lose a "second" electron to obtain the stable noble gas configuration with a much smaller amount of energy (smaller IE).

83. For the elements Na, B, Al, and C:

   a. The largest atomic radius: Na (186 pm)

   b. The largest electron affinity: C (most negative value for EA)

   c. Increasing ionization energy: Na < Al < B < C

85.   a. With 22 electrons, the element is **titanium**.

   b. Titanium is in period 4 (electrons in 4s) and group IV B (4 electrons in the 3d and 4s shells).

   c. With a 3d shell partially occupied, the element is a **transition element**.

   d. With **two unpaired electrons**, titanium is **paramagnetic**.

   e. Quantum numbers for electrons 1, 2, and 4.

Electron	n	ℓ	m_ℓ	m_s
1	3	2	-2	+1/2
2	3	2	-1	+1/2
4	4	0	0	-1/2

   f. When two electrons are removed to form the Ti^{2+} ion, the "electrons of highest n" (4s) are removed—leaving the paramagnetic Ti^{2+} ion.

Conceptual Questions

87. The decrease in atomic size across a period is attributable to the increasing nuclear charge with an increasing number of protons. Given that the electrons are in the same outer energy level, the nuclear attraction for the electrons results in a diminishing atomic radius.

89. The slight decrease in atomic radius of the transition metals is a result of increased repulsions of $(n-1)d$ electrons for $(n)s$ electrons. Thus repulsion reduces the effects of the increasing nuclear charge across a period.

91. The element with the greatest difference between the first and second IE's is sodium. Of the four elements shown, Na is in the first group. The loss of the first electron results in an ion with a filled outer shell—a very stable configuration. Removal of a second electron would require a much larger amount of energy.

93. Ions likely to be formed:
    Cs^+ and Se^{2-}; These ions both have the noble gas configuration.

95. A plot of the atomic radii of the elements Ca—Cu shows a decrease. With the mass of these elements increasing from Ca through Cu, the density is expected to increase.

97. The electron configuration for Co^{2+} and Co^{3+} are:

Co^{2+} 25 e⁻   [Ar]   [↑↓|↑↓|↑ |↑ |↑]

Co^{3+} 24 e⁻   [Ar]   [↑↓|↑ |↑ |↑ |↑]

One could measure the compounds' magnetic susceptibilities. The Co^{2+} ion would have 3 unpaired electrons while Co^{3+} would have 4 unpaired electrons.

100.   a. Orbital box notation for sulfur:

    1s   2s   2p   3s   3p
    [↑↓]   [↑↓]   [↑↓|↑↓|↑↓]   [↑↓]   [↑↓|↑ |↑]

b. Quantum numbers for the "last electron": $n = 3$, $\ell = 1$, $m\ell = +1$ (or 0 or -1), $m_s = -1/2$

c. Element with the smallest ionization energy : S
   Element with the smallest radius: O

d. S: Negative ions are always larger than the element from which they are derived.

e. Grams of Cl_2 to make 675 g of $OSCl_2$:

$$675 \text{ g } OSCl_2 \cdot \frac{1 \text{ mol } OSCl_2}{119.0 \text{ g } OSCl_2} \cdot \frac{2 \text{ mol } Cl_2}{1 \text{ mol } OSCl_2} \cdot \frac{70.91 \text{ g } Cl_2}{1 \text{ mol } Cl_2} = 804 \text{ g } Cl_2$$

f. The theoretical yield of $OSCl_2$ if 10.0 g of SO_2 and 20.0 g of Cl_2 are used:

$$\text{Moles of } SO_2 = 10.0 \text{ g } SO_2 \cdot \frac{1 \text{ mol } SO_2}{64.06 \text{ g } SO_2} = 0.156 \text{ moles } SO_2$$

$$\text{Moles of } Cl_2 = 20.0 \text{ g } Cl_2 \cdot \frac{1 \text{ mol } Cl_2}{70.91 \text{ g } Cl_2} = 0.282 \text{ moles } Cl_2$$

Moles-available ratio: $\dfrac{0.156 \text{ moles } SO_2}{0.282 \text{ moles } Cl_2} = \dfrac{0.553 \text{ moles } SO_2}{1 \text{ mol } Cl_2}$

Moles-required ratio: $\dfrac{1 \text{ mol } SO_2}{2 \text{ mol } Cl_2} = \dfrac{0.5 \text{ mol } SO_2}{1 \text{ mol } Cl_2}$

Cl_2 is the limiting reagent.

$$0.282 \text{ moles } Cl_2 \cdot \frac{1 \text{ mol } OSCl_2}{2 \text{ moles } Cl_2} \cdot \frac{119.0 \text{ g } OSCl_2}{1 \text{ mol } OSCl_2} = 16.8 \text{ g } OSCl_2$$

g. For the reaction:

$$SO_2 (g) + 2 Cl_2 (g) \rightarrow OSCl_2 (g) + Cl_2O (g) \qquad \Delta H°_{rxn} = +164.6 \text{ kJ}$$

$$\Delta H°_{rxn} = [1 \cdot \Delta H°_f \, OSCl_2 (g) + 1 \cdot \Delta H°_f \, Cl_2O (g)]$$
$$- [1 \cdot \Delta H°_f \, SO_2(g) + 2 \cdot \Delta H°_f \, Cl_2 (g)]$$

$+ 164.6 \text{ kJ} = [\Delta H°_f \, OSCl_2 (g) + 80.3 \text{ kJ}] - [-296.8 \text{ kJ} + 2 \cdot (0)]$

$+ 164.6 \text{ kJ} = \Delta H°_f \, OSCl_2 (g) + 377.1 \text{ kJ}$

$- 212.5 \text{ kJ} = \Delta H°_f \, OSCl_2 (g)$

Chapter 9:
Bonding and Molecular Structure: Fundamental Concepts

Valence Electrons

26.

Element		Number of Valence Electrons	Group Number
a.	N	5	5A
b.	B	3	3A
c.	Na	1	1A
d.	Mg	2	2A
e.	F	7	7A
f.	S	6	6A

The Octet Rule

28.

Group Number	Number of Bond Pairs
1A	1
2A	2
3A	3
4A	4
5A	3 (or 4 in species such as NH_4^+)
6A	2 (or 3 as in H_3O^+)
7A	1
8A	0

Lewis Electron Dot Structures

30. a. NF_3 : $[1(5) + 3(7)]$ = 26 valence electrons

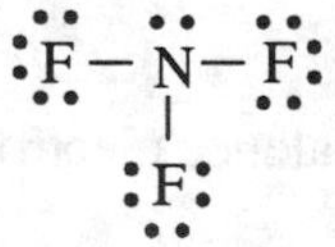

$$:\ddot{F} - \ddot{N} - \ddot{F}:$$
$$|$$
$$:\ddot{F}:$$

b. ClO_3^- : $[1(7) + 3(6) + 1]$ = 26 valence electrons

$\uparrow$

ion charge

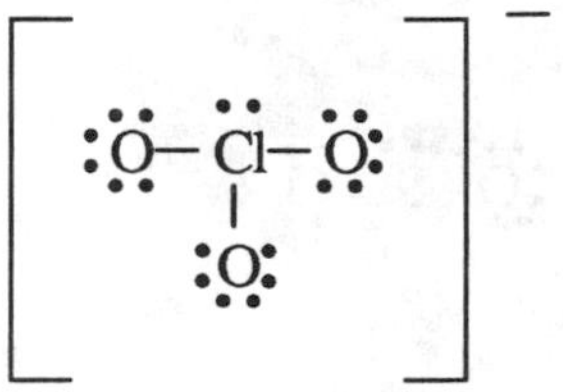

$$\left[:\ddot{O} - \ddot{Cl} - \ddot{O}: \right]^-$$
$$|$$
$$:\ddot{O}:$$

c. HOBr: [1(1) + 1(6) + 1(7)] = 14 valence electrons

$$H-\ddot{O}-\ddot{B}r\!:$$

d. SO_3^{2-} : [1(6) + 3(6) + 2] = 26 valence electrons

ion charge

$$\left[\;:\!\ddot{O}\!-\;\ddot{S}\!-\!\ddot{O}\!:\atop\quad\;\;:\!\ddot{O}\!:\right]^{2-}$$

32.   a. $CHClF_2$: [1(4) +1(1) + 1(7) + 2(7)] = 26 valence electrons

$$:\!\ddot{F}-\overset{\overset{\displaystyle H}{|}}{\underset{\underset{\displaystyle :\!\ddot{Cl}\!:}{|}}{C}}-\ddot{F}\!:$$

b. HCOOH: [1(1) + 1(4) + 2(6) + 1(1)] = 18 valence electrons

$$H-\overset{\overset{\displaystyle :\!\ddot{O}\!:}{\|}}{C}-\ddot{O}-H$$

c. H_3CCN: [3(1) + 2(4) + 1(5)] = 16 valence electrons

$$H-\overset{\overset{\displaystyle H}{|}}{\underset{\underset{\displaystyle H}{|}}{C}}-C\equiv N\!:$$

d. H_3COH: [3(1) + 1(4) + 1(6) + 1(1)] = 14 valence electrons

$$H-\overset{\overset{\displaystyle H}{|}}{\underset{\underset{\displaystyle H}{|}}{C}}-\ddot{O}-H$$

34.   Resonance structures for:

a. SO_2:

$$\ddot{O}=\ddot{S}-\ddot{O}\!:\quad\longleftrightarrow\quad:\!\ddot{O}-\ddot{S}=\ddot{O}$$

b. SO_3:

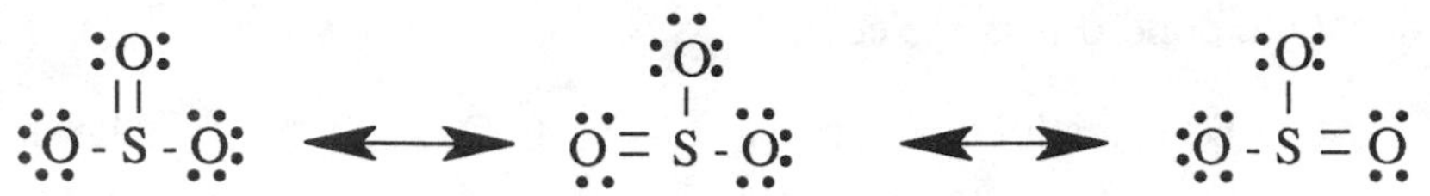

$$:\ddot{O}: \quad \quad :\ddot{O}: \quad \quad :\ddot{O}:$$

$$:\ddot{O}-S-\ddot{O}: \longleftrightarrow \ddot{O}=S-\ddot{O}: \longleftrightarrow :\ddot{O}-S=\ddot{O}$$

c. SCN^- :

$$\left[:\ddot{N}-C\equiv S: \right]^- \longleftrightarrow \left[\ddot{N}=C=\ddot{S}: \right]^- \longleftrightarrow \left[:N\equiv C-\ddot{S}: \right]^-$$

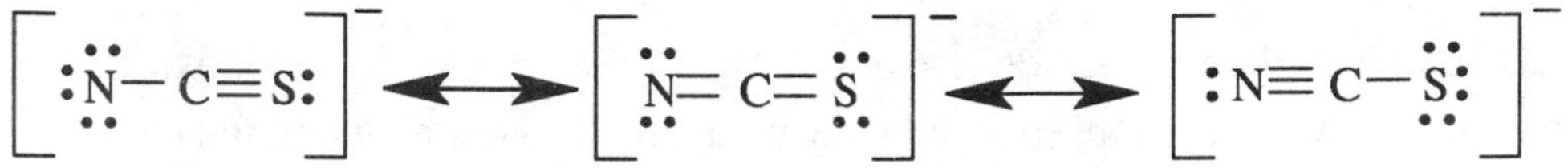

36.   a. BrF_3 : $[1(7) + 3(7)] = 28$ valence electrons

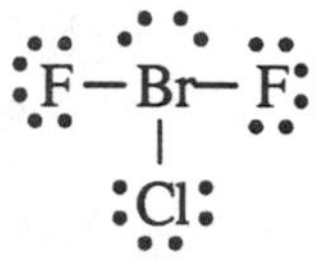

$$:\ddot{F}-Br-\ddot{F}:$$
$$|$$
$$:\ddot{Cl}:$$

b. I_3^- : $[3(7) + 1] = 22$ valence electrons

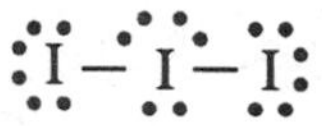

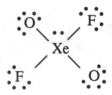

c. XeO_2F_2 : $[1(8) + 2(6) + 2(7)] = 34$ valence electrons

$$:\ddot{O}\cdot \quad \cdot\ddot{F}:$$
$$\diagdown :: \diagup$$
$$Xe$$
$$\diagup \diagdown$$
$$:\ddot{F}\cdot \quad \cdot\ddot{O}:$$

Bond Properties

38.

Specie	Number of bonds	Bond Order : Bonded Atoms
a. H_2CO	3	1 : CH 2: C = O
b. SO_3^{2-}	3	1 : SO
c. NO_2^+	2	2 : NO

40. In each case the shorter bond length should be between the atoms with smaller radii--if we assume that the bond orders are equal.

a. B-Cl   B is smaller than Ga   b. C-O   C is smaller than Sn
c. P-O   O is smaller than S   d. C=O   O is smaller than C

42. The CO bond in carbon monoxide is shorter. The CO bond in carbon monoxide is a triple bond, thus it requires more energy to break than the CO double bond in H_2CO.

44. The bond order for NO_2^+ is 2, while the bond order for NO_3^- is 4/3. The Lewis dot structure for the NO_2^+ ion indicates that both NO bonds are double, while in the nitrate ion, any resonance structure (there are three) shows one π bond and three σ bonds. Hence the NO bonds in nitrate will be longer than those in the NO_2^+ ion.

Bond Energies and Reaction Enthalpies

46. Estimate the enthalpy of the hydrogenation reaction for butene:

$$H_3C\text{-}CH_2\text{-}C{=}C\text{-}H \quad + \quad H_2 \longrightarrow \quad H_3C\text{-}CH_2\text{-}\overset{\overset{H}{|}\ \overset{H}{|}}{\underset{\underset{H}{|}\ \underset{H}{|}}{C\text{-}C}}\text{-}H$$

Energy input: 1 mol C=C = 1 mol • 611 kJ/mol =   611 kJ
                 1 mol H-H = 1 mol • 436 kJ/mol =   436 kJ
                              Total input   = 1047 kJ

Energy release: 1 mol C-C = 1 mol • 347 kJ/mol = 347 kJ
                 2 mol C-H = 2 mol • 414 kJ/mol = 828 kJ
                              Total released   = 1175 kJ
Energy change: 1047 kJ - 1175 kJ = - 128 kJ

48. Heat of reaction for : CO (g) + Cl_2 (g) $\rightarrow$ Cl_2CO (g)

Energy input: 1 mol C≡O = 1 mol • 1075 kJ/mol = 1075 kJ
                 1 mol Cl-Cl = 1 mol • 243 kJ/mol = 243 kJ
                              Total input   = 1318 kJ

Energy release: 2 mol C-Cl = 2 mol • 330 kJ/mol =   660 kJ

1 mol C=O = 1 mol • 745 kJ/mol =   745 kJ

Total released   = 1405 kJ

Energy change: 1318 kJ - 1405 kJ = - 87 kJ

50. OF_2 (g) + H_2O (g) → O_2 (g) + 2 HF (g)   ΔH = - 318 kJ

Energy input :   2 mol O-F = 2 x (where x = O-F bond energy)

2 mol O-H = 2 mol • 464 kJ/mol = 928 kJ

Total input   =   (928 + 2x) kJ

Energy release: 1 mol O=O = 1 mol • 498 kJ/mol =   498 kJ

2 mol H-F = 2 mol • 569 kJ/mol = 1138 kJ

Total release   = 1636 kJ

- 318 kJ   = 928 kJ + 2x - 1636 kJ

390 kJ   = 2x

195 kJ/mol   = O-F bond energy

Electronegativity and Bond Polarity

52. Indicate the more polar bond (Arrow points toward the more negative atom in the dipole).

a. C-O >   C-N
    →   →

b. P-O >   P-S
    →   →

c. P-H <   P-N
    not polar   →

d. B-H <   B-I
    →   →

54.   For the bonds in acrolein the polarities are as follows:

	H-C	C-C	C=O
$\dfrac{\Delta\chi}{\Sigma\chi}$	0.09	0	0.16

a. The C-C bonds are nonpolar, the C-H bonds are slightly polar, and the C=O bond
        is polar.

b. The most polar bond in the molecule is the C=O bond, with the oxygen atom being the
        negative end of the dipole.

Oxidation Numbers and Atom Formal Charges

56. Using the Lewis structures, the oxidation numbers of the atoms in the following:

a. water   b. hydrogen peroxide   c. sulfur dioxide

$H - \overset{..}{\underset{..}{O}} - H$ $H - \overset{..}{\underset{..}{O}} - \overset{..}{\underset{..}{O}} - H$ $\overset{..}{\underset{..}{O}} = \overset{..}{S} - \overset{..}{\underset{..}{O}}:$

$H = +1; O = -2$ $H = +1; O = -1$ $S = +4 ; O = -2$

d. nitrogen (I) oxide   e. hypochlorite ion

$\overset{..}{\underset{..}{N}} = N = \overset{..}{\underset{..}{O}}$ $\left[:\overset{..}{\underset{..}{O}} - \overset{..}{\underset{..}{Cl}}: \right]^{-}$

$N = +1 ; O = -2$ $Cl = +1 ; O = -2$

58.

	Atom	Formal Charge	Oxidation Number
a. H_2O	H	$1 - 1/2(2) = 0$	$+1$
	O	$6 - 4 - 1/2(4) = 0$	-2
b. CH_4	C	$4 - 1/2(8) = 0$	-4
	H	$1 - 1/2(2) = 0$	$+1$
c. NO_2^+	N	$5 - 1/2(8) = +1$	$+5$
	O	$6 - 4 - 1/2(4) = 0$	-2
d. HOF	H	$1 - 1/2(2) = 0$	$+1$
	O	$6 - 4 - 1/2(4) = 0$	0
	F	$7 - 6 - 1/2(2) = 0$	-1

60. Resonance structures for NO_2^- :

$$\left[\overset{..}{\underset{..}{O}} = N - \overset{..}{\underset{..}{O}}: \right]^{-} \longleftrightarrow \left[:\overset{..}{\underset{..}{O}} - N = \overset{..}{\underset{..}{O}} \right]^{-}$$

        1   2   1   2

Formal charges:

O_1 $6 - 4 - 1/2(4) = 0$ O_1 $6 - 6 - 1/2(2) = -1$

N $5 - 2 - 1/2(6) = 0$ N $5 - 2 - 1/2(6) = 0$

O_2 $6 - 6 - 1/2(2) = -1$ O_2 $6 - 4 - 1/2(4) = 0$

110

62.   a. Resonance structures of N_2O :

$$:N\equiv N-\ddot{\text{O}}: \quad\longleftrightarrow\quad \ddot{N}=N=\ddot{\text{O}} \quad\longleftrightarrow\quad :\ddot{N}-N\equiv O:$$

  1   2   1   2   1   2

b. Formal Charges:

N_1	5 - 2 - 1/2(6) = 0	5 - 4 - 1/2(4) = -1	5 - 6 - 1/2(2) = -2
N_2	5 - 0 - 1/2(8) = +1	5 - 0 - 1/2(8) = +1	5 - 0 - 1/2(8) = +1
O	6 - 6 - 1/2(2) = -1	6 - 4 - 1/2(4) = 0	6 - 2 - 1/2(6) = +1

c. Of these three structures, the first is the most reasonable in that the most electronegative atom, O, bears a formal charge of -1.

Molecular Geometry

64. Using the Lewis structure describe the Electron-pair and molecular geometry:

a.

$$H-\ddot{N}-\ddot{\text{Cl}}: \\ \quad| \\ \quad H$$

Electron-pair: tetrahedral
Molecular : trigonal pyramidal

b.

$$:\ddot{\text{Cl}}-\ddot{\text{O}}-\ddot{\text{Cl}}:$$

Electron-pair: tetrahedral
Molecular: bent

c.

$$\left[\ddot{N}=C=\ddot{S}\right]^{-}$$

Electron-pair: linear
Molecular: linear

d.

$$H-\ddot{\text{O}}-\ddot{F}:$$

Electron-pair: tetrahedral
Molecular: bent

66. Using the Lewis structure describe the Electron-pair and molecular geometry:

a.

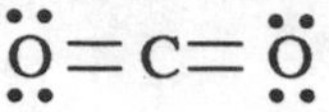

$$\ddot{O} = C = \ddot{O}$$

Electron-pair: linear
Molecular : linear

b.

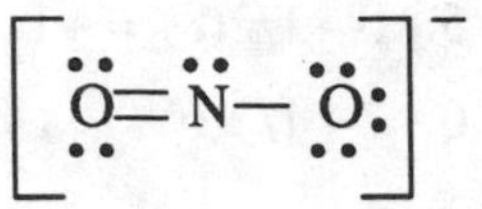

$$\left[\ddot{O} = N - \ddot{O}: \right]^{-}$$

Electron-pair: trigonal planar
Molecular: bent

c.

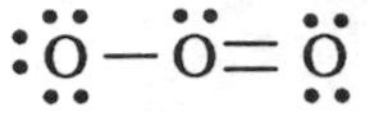

$$:\ddot{O} - \ddot{O} = \ddot{O}$$

Electron-pair: trigonal planar
Molecular: bent

d.

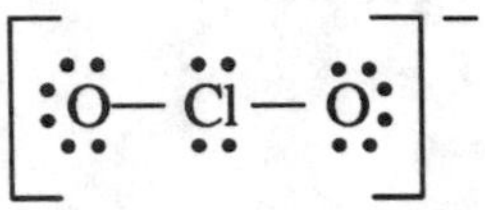

$$\left[:\ddot{O} - \ddot{C}l - \ddot{O}: \right]^{-}$$

Electron-pair: tetrahedral
Molecular: bent

e.

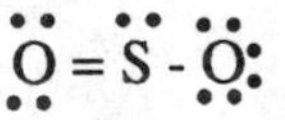

$$\ddot{O} = \ddot{S} - \ddot{O}:$$

Electron-pair: trigonal planar
Molecular: bent

68. Using the Lewis structure describe the electron-pair and molecular geometry.
[Lone pairs on F have been omitted for clarity.]

a.

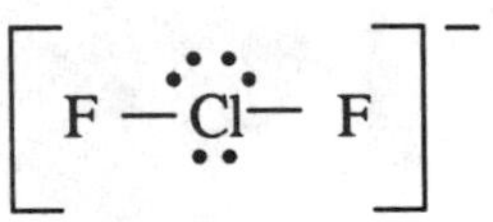

$$\left[F - \ddot{C}l - F \right]^{-}$$

Electron-pair: trigonal bipyramidal
Molecular : linear

b.

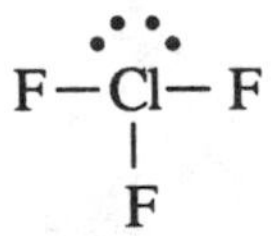

$$F - \ddot{C}l - F$$
$$|$$
$$F$$

Electron-pair: trigonal bipyramidal
Molecular: T-shaped

c.

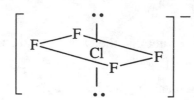

Electron-pair: octahedral
Molecular: square planar

d.

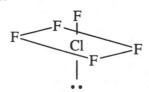

Electron-pair: octahedral
Molecular: square pyramidal

70.   a. O-S-O angle in SO_2 :   Slightly less than 120°; The lone pair of S should
                                                          reduce the predicted 120° angle slightly.

       b. F-B-F angle in BF_3 :   120°

       c. (1) H-C-H angle in CH_3CN :   109°
          (2) C-C ≡N angle in CH_3CN :   180°
          (3) H-N-H angle in CH_3CN 109°

72. For the molecule acetylacetone, the estimated bond angles are:

     Angle 1:   C-C-C   120°
     Angle 2:   C-C-H   109°
     Angle 3:   C-C-O   120°

74. Approximate values for the angles: (A = axial ; E= equatorial)

	Bond	Compound	Angles	
a.	F-Se-F	SeF_4	F(A)-Se-F(A)	180° see SF_4 p 438
			F(A)-Se-F(E)	90°
			F(E)-Se-F(E)	120°
b.	O-S-F	OSF_4	O-S-F(A)	90°
			O-S-F(E)	120°
			F(A)-S-F(E)	90°
			F(E)-S-F(E)	120°

113

| c. | F-Br-F | BrF$_5$ | F(E)-Br-F(E) | 90° |
| | | | F(A)-Br-F(E) | 90° |

76. NO_2^+ has two structural pairs around the N atom. We predict that the O-N-O bond angle would be approximately 180°. NO_2^- has three stuctural pairs (one lone pair). The geometry around this central atom would be trigonal planar with a bond angle of approximately 120°.

Molecular Polarity

78. For the molecules:

| H$_2$O | NH$_3$ | CO$_2$ | ClF | CCl$_4$ |

i. Using the electronegativities to determine bond polarity:

$$\frac{\Delta\chi}{\Sigma\chi} \qquad \frac{1.4}{5.6} \qquad \frac{0.9}{5.1} \qquad \frac{1.0}{6.0} \qquad \frac{10}{7.0} \qquad \frac{0.5}{5.5}$$

Reducing these fractions to a decimal form indicates that the H-O bonds in water are the most polar of these bonds.

ii. The nonpolar compounds are:

CO$_2$ The O-C-O bond angle is 180°, thereby cancelling the C-O dipoles.

CCl$_4$ The Cl-C-Cl bond angles are approximately 109°, with the Cl atoms directed at the corners of a tetrahedron. Such an arrangement results in a net dipole moment of zero.

iii. The F atom in ClF is more negatively charged.(Electronegativity of F = 4.0, Cl = 3.0)

80.   Molecular polarity of the following: BeCl$_2$, HBF$_2$, CH$_3$Cl, SO$_3$

BeCl$_2$ and SO$_3$ are nonpolar. For HBF$_2$, the hydrogen and fluorine atoms are arranged at the corners of a triangle. The "negative end" of the molecule lies on the plane between the fluorine atoms, and the H atom is the "positive end." For CH$_3$Cl, the chlorine atom is the negative end and the H atoms form the positive end.

82.

Group Number	Number of Valence Electrons
1A	1
3A	3
4A	4

84. The bond order for NO_2^+ is 2, while the bond order for NO_2^- is 1.5. The Lewis dot structure for the NO_2^+ ion indicates that both NO bonds are double, while in the nitrite ion, any resonance structure (there are two) shows one π bond and two σ bonds. Hence the NO bonds in nitrite will be longer (124 pm) than those in the NO_2^+ ion (110 pm).

86. a. Angle 1 = 120°; Angle 2 = 180°; Angle 3 = 120°

   b. The C=C double bond is shorter than the C-C single bond.

   c. The C=C double bond is stronger than the C-C single bond.

   d. The C≡N bond is the most polar bond, with the N atom being the negative end of the bond dipole.

88. Using the Lewis structure describe the Electron-pair and Molecular geometry:

   a.

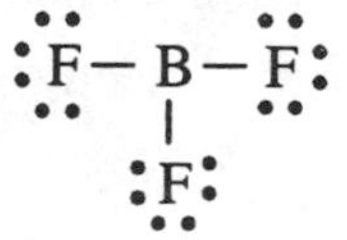

   :F— B— F:
        |
       :F:

   Electron-pair: trigonal planar
   Molecular : trigonal planar

   b.

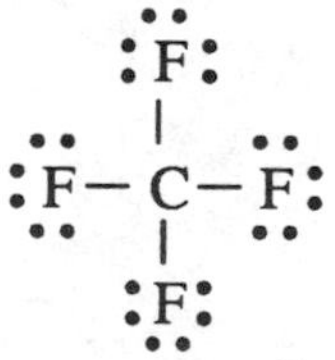

       :F:
        |
   :F— C— F:
        |
       :F:

   Electron-pair: tetrahedral
   Molecular: tetrahedral

   c.

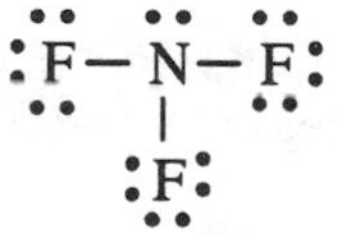

   :F— N— F:
        |
       :F:

   Electron-pair: tetrahedral
   Molecular: trigonal pyramidal

   d.

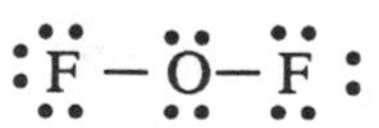

   :F — O— F :

   Electron-pair: tetrahedral
   Molecular: bent

115

e.

H—$\ddot{\underset{..}{F}}$:   Electron-pair: tetrahedral

Molecular: linear

90.   a. In XeF_2 the bonding pairs occupy the axial positions with the lone pairs located in the (preferred) equatorial plane.

b. In ClF_3 two of the three equatorial positions are occupied by the lone pairs of electrons on the Cl atom.

92. a. Approximate values for bond angles:

#	Angle	Value
1	O-C-C	120°
2	N-C-H	120°
3	N-C-N	120°
4	C-N-C	120° (probably a bit smaller owing to the lone pair of electrons on N)
5	H-N-H	109° (probably a bit smaller, once again, owing to the lone pair of electrons on N)

b. The C = O bond is the most polar bond.

Conceptual Questions

94.   a. Resonance structures for CNO^- with formal charges.

$$\left[\ddot{C}=N=\ddot{O} \right]^- \longleftrightarrow \left[:\ddot{C}-N\equiv O: \right]^- \longleftrightarrow \left[:C\equiv N-\ddot{O}: \right]^-$$

Formal Charges:

  -2   +1   0   -3   +1   +1   -1   +1   -1

b. The most reasonable structure is the one at the far right because it is the one in which the formal charges on the atoms are at a minimum, and oxygen has the negative formal charge.

c. The instability of the ion could be attributed to the fact that the least electronegative atom in the ion bears a negative charge in all resonance structures.

96. Estimate the bond dissociation energy for the B—F bond.

$$B(solid) + 3/2\ F_2(g) \longrightarrow BF_3(g) \qquad \Delta H^\circ_f\ [BF_3(g)] = -1137\ kJ/mol$$

$$\Delta H^\circ_f\ [BF_3(g)] = -1137\ kJ/mol = \Delta H^\circ_{vap}\ [B(s)] + 3/2\ D_{F\text{-}F} - 3\ D_{B\text{-}F}$$

$$-1137\ kJ/mol = 1\ mol\ (563\ kJ/mol) + 3/2\ mol\ (159\ kJ/mol) - 3\ D_{B\text{-}F}$$

$$D_{B\text{-}F} = 646\ kJ/mol$$

98.   a. The production of ClF_3 from the elements may be represented:

$$Cl_2(g) + 3\ F_2(g) \rightarrow 2\ ClF_3(g)$$

b. Mass of ClF_3 expected:

$$\text{Moles } Cl_2: 0.71\ g\ Cl_2 \cdot \frac{1\ mol\ Cl_2}{70.9\ g\ Cl_2} = 0.010\ mol\ Cl_2$$

$$\text{Moles } F_2: 1.00\ g\ F_2 \cdot \frac{1\ mol\ F_2}{38.00\ g\ F_2} = 0.0263\ mol\ F_2$$

Ratios: Moles required: $\dfrac{3\ mol\ F_2}{1\ mol\ Cl_2}$

Moles available: $\dfrac{0.0263\ mol\ F_2}{0.010\ mol\ Cl_2}$

F_2 is the **limiting reagent.**

$$0.0263\ mol\ F_2 \cdot \frac{2\ mol\ ClF_3}{3\ mol\ F_2} \cdot \frac{92.45\ g\ ClF_3}{1\ mol\ ClF_3} = 1.62\ g\ ClF_3$$

c. The electron dot structure for ClF_3 is:

[The three lone pairs on F's have
been omitted for clarity.]

$$F-\overset{\displaystyle \cdot\cdot\ \cdot}{Cl}-F$$
$$|$$
$$F$$

d. The electron-pair geometry for the molecule is trigonal bipyramidal.

e. The molecular geometry is T-shape. The polarity of this molecule implies that there are polar bonds that are asymmetrically arranged. This **does not** unambiguously establish a geometry. Variations in the location of F atoms in axial or equatorial positions are possible.

f. The enthalpy of formation of ClF_3 from bond energies:

Bonds broken:

3 mol F-F bond	= 3 mol • 159 kJ/mol =	477 kJ
1 mol Cl-Cl bond	= 1 mol • 243 kJ/mol =	243 kJ
	Energy input =	720 kJ

Bonds formed:

6 mol Cl-F bond	= 6 mol • 255 kJ/mol =	1530 kJ
	Energy released =	1530 kJ

$$\Delta H^{\circ}_{rxn} = 720 - 1530 = -810 \text{ kJ}$$

The enthalpy change per mole is then : $\dfrac{-810 \text{ kJ}}{2 \text{ mol ClF}_3}$ or -405 kJ/mol

Chapter 10:
Bonding and Molecular Structure: Orbital Hybridization, Molecular Orbitals, and Metallic Bonding

Hybrid Orbitals

15. OF_2 has **tetrahedral** electron-pair geometry and **bent** molecular geometries.

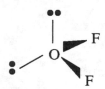

   With four electron-pairs attaached, the O is sp^3 hybridized. The lone pairs occupy two of the four **sp^3** hybrid orbitals with the remaining two orbitals overlapping the **p** orbitals on F, forming the O-F sigma bonds.

17. The Lewis electron dot structure of CH_2Cl_2 :

   Electron pair geometry = tetrahedral

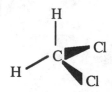

      Molecular geometry = tetrahedral

      The H-C bonds are a result of the overlap of the hydrogen **s** orbital with **sp^3** hybrid orbitals on carbon. The Cl-C bonds are formed by the overlap of the **sp^3** hybrid orbitals on carbon with the **p** orbitals on chlorine.

   19. Orbital sets used by the underlined atoms:

  a. $\underline{B}Cl_3$: sp^2

  b. $\underline{N}O_2^+$: sp

  c. $H\underline{C}Cl_3$: sp^3

  d. $H_2\underline{C}O$: sp^2

21. Hybrid orbital sets used by the underlined atoms:

  a. $\underline{C}$: sp^3 ; $\underline{O}$: sp^3

  b. $\underline{C}H_3$: sp^3 ; $\underline{C}H$ and $\underline{C}H_2$: sp^2

  c. $\underline{N}$: sp^3; $\underline{C}H_2$: sp^3, $\underline{C}=O$: sp^2

23. For the molecule XeF_2 :

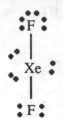

Electron pair geometry = trigonal bipyramidal

Molecular geometry = linear

Bonding is a result of overlap of sp^3d orbitals on xenon with **p** orbitals on fluorine.

25. Orbital sets used by the underlined atoms:

a. $\underline{S}F_6$: sp^3d^2 b. $\underline{S}F_4$: sp^3d c. $\underline{I}Cl_2^-$: sp^3d

27. For the molecule CO_2: Hybridization of $C = sp$

$$\left[\ddot{O} = C = \ddot{O} \right]$$

Bonding: 1 sigma bond between each oxygen and carbon

(**sp** hybrid orbitals)

1 pi bond between each oxygen and carbon (**p** orbitals)

29. For the molecule $COCl_2$: Hybridization of $C = sp^2$

Bonding: 1 sigma bond between each chlorine and carbon
(**sp** hybrid orbitals)

1 sigma bond between carbon and oxygen

(**sp** hybrid orbitals)

1 pi bond between carbon and oxygen

(**p** orbital)

Molecular Orbital Theory

31. Configuration for H_2^+: $(\sigma_{1s})^1$

Bond order for H_2^+: 1/2 (no. bonding e^- - no. antibonding e^-) = 1/2

The bond order for molecular hydrogen is <u>one</u> (1), and the H-H bond is stronger in the H_2 molecule than in the H_2^+ ion.

33. The molecular orbital diagram for C_2^{2-}, the acetylide ion:

σ^*2p ____   There are 2 net pi bonds and 1 net

π^*2p ____   ____   sigma bond in the ion, giving a

$\sigma2p$ ↑↓   bond order of 3. On adding two

$\pi2p$ ↑↓   ↑↓   electrons to C_2 (added to $\sigma2p$)to

σ^*2s ↑↓   obtain C_2^{2-}, the bond order

$\sigma2s$ ↑↓   increases by one.The ion is

diamagnetic.

35. Using Figure 10.22 as a model for heteronuclear diatomic molecules, the electron configuration (showing only the outer level electrons) for CO is:

σ^*2p ____

π^*2p ____   ____   There are no unpaired electrons,

$\sigma2p$ ↑↓   hence CO is diamagnetic.

$\pi2p$ ↑↓   ↑↓   There is one net sigma bond, and

σ^*2s ↑↓   two net pi bonds for an overall

$\sigma2s$ ↑↓   bond order of 3.

General Questions

37. <u>The hybrid orbitals used by sulfur</u>

    a. SO_2 3 electron-pair groups sp^2

    b. SO_3 3 electron-pair groups sp^2

    c. SO_3^{2-} 4 electron-pair groups sp^3

    d. SO_4^{2-} 4 electron-pair groups sp^3

39.   For the ion NO_3^- :

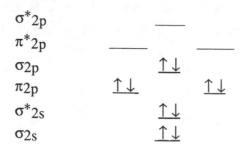

Hybridization of N: sp^2

The N=O bond is formed by overlap of an **sp^2** hybrid orbital on both N and O to form a σ bond,while the unhybridized **p** orbitals on N and O overlap side-to-side to form a π bond.

41. Both carbon 1 and 2 are **sp^2** hybridized (3 electron-pair groups). Since angles A, B, and C are associated with sp^2 hybridized carbons, each of the three angles is approximately 120°.

43.   a. Number of π and σ bonds in lactic acid:

    1 π bond (C=O)

    11 σ bonds (every atom attached to any other atom with a σ bond)

    b. Hybridizations:

Atom 1	C (4 electron-pair groups)	sp^3
Atom 2	C=O (3 electron-pair groups)	sp^2
Atom 3	O-H (4 electron-pair groups)	sp^3

    c. Shorter O-H bond

    The delocalization of the π bond associated with C(2) should shorten the C-O bond between C(2) and O(3) weakening (and lengthening) the O(3)-H bond over the OH bond associated with angle B.

    d. Bond angles:

    A & B   109°   tetrahedral   C   120°   trigonal planar

45.   a. Geometry around B in BF_3 : trigonal planar
                         in H_3N-BF_3 : tetrahedral

    b. Hybridization of B
                    in BF_3 : sp^2
                    in H_3N-BF_3 : sp^3

    c. Yes, the hybridization of B changes from sp^2 to sp^3

47. N_2^+ ion: [core electrons] $(\sigma 2s)^2(\sigma^* 2s)^2(\pi 2p)^4(\sigma 2p)^1(\pi^* 2p)^0$
    N_2^- ion: [core electrons] $(\sigma 2s)^2(\sigma^* 2s)^2(\pi 2p)^4(\sigma 2p)^2(\pi^* 2p)^1$

Property	N_2^+	N_2	N_2^-
a. magnetic	paramagnetic	diamagnetic	paramagnetic
b. net π bonds	2	2	1.5
c. bond order	2.5	3	2.5
d. bond length	Should be of similar length; and longer than that of N≡N		
e. bond strength	Should be of similar strength; and both should be weaker than N≡N		

49.   a. SbF_5, antimony pentafluoride, has 40 valence electrons and so has the dot structure below (where the three lone pairs on each F atom are not shown).

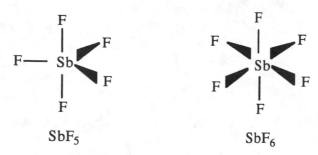

SbF_5 SbF_6

Since the electron pair geometry is trigonal bipyramidal, the Sb atom hybridization is **sp^3d**. In SbF_6^-, which has 48 valence electrons, the octahedral electron pair geometry implies that Sb has **sp^3d^2** hybridization.

b. The H_2F^+ ion has 8 valence electrons (like water), and so it has a bent geometry with an **sp^3** hybridized F atom.

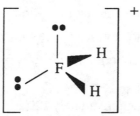

51.

XeO_3   XeO_4

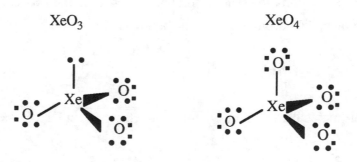

	XeO_3	XeO_4
Electron-pair geometry	tetrahedral	tetrahedral
Molecular geometry	trigonal pyramidal	tetrahedral
Xe hybrid orbitals	sp^3	sp^3

53. | Molecule or Ion | Number of Valence Electrons | Magnetic Behavior |
|---|---|---|
| NO | $5 + 6 = 11$ | paramagnetic |
| OF^- | $6 + 7 + 1 = 14$ | diamagnetic |
| O_2^- | $(2 \times 6) + 2 = 14$ | diamagnetic |
| Ne_2^+ | $(2 \times 8) - 1 = 15$ | paramagnetic |
| CN | $4 + 5 = 9$ | paramagnetic |

55. | Atom | Hybridization in Reactant | Hybridization in Product |
|---|---|---|
| a. C | sp^2 | sp^3 |
| b. P | sp^3 | sp^3d |
| c. Xe | sp^3d | sp^3d^2 |
| d. Sn | sp^3 | sp^3d^2 |

57. For the molecule: $CH_3-N=C=S$:

a. Orbital sets used by:

$C_1 : sp^3$ (bonds formed by C_1 to four atoms)

$C_2 : sp$ (bonds formed by C_2 to two atoms)

$N : sp^2$ (bonds formed by N to two atoms)

b. Bond angles:
   C-N=C angle: approximately 120°   N=C=S angle: approximately 180°

c. For the C_2 atom, the s and p_x orbitals would form the two sp hybrid orbitals, leaving the p_y and p_z orbitals to form the π bonds to N and S.

Conceptual Questions

58. Possible molecular structures: [Lone pair electrons on F have been omitted for clarity.]

Hybridization of B= sp^2
F-B-B bond angle: approximately 120°

Hybridization of C = sp^2
H-C-C bond angle: approximately 120°

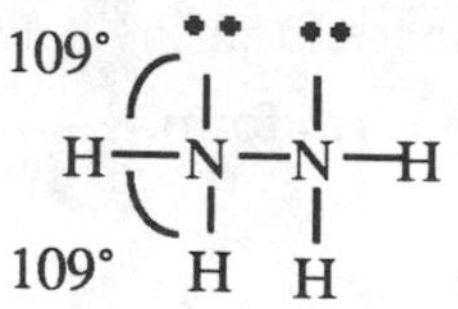

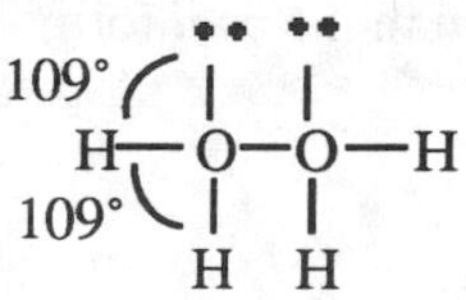

Hybridization of N = sp^3

H-N-N bond angle: approximately 109°

Hybridization of O = sp^3

H-O-O bond angle: approximately 109°

For both hydrazine (N_2H_4) and hydrogen peroxide (H_2O_2), the lone pairs on N and O respectively will reduce the bond angles to slightly less than 109°.

Only the carbon compound contains a double bond.

Summary Question

61.   a. The *enol* → *keto* transformation

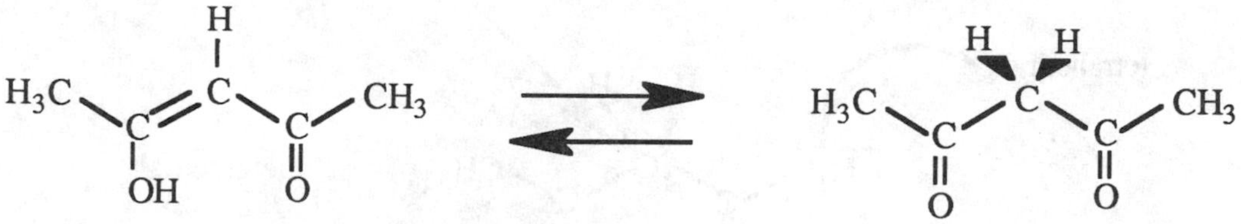

occurs by breaking a C—O single bond, an O—H bond, and a C=C double bond and by then making a C=O double bond, a C—C single bond, and a C—H bond. Therefore, we can estimate the reaction enthalpy change using only these bond energies:

$$\Delta H = D_{C-O} + D_{O-H} + D_{C=C} - D_{C=O} - D_{C-C} - D_{C-H}$$
$$= 351 \text{ kJ} + 464 \text{ kJ} + 611 \text{ kJ} - 745 \text{ kJ} - 347 \text{ kJ} - 414 \text{ kJ}$$
$$= -80 \text{ kJ}$$

This result indicates the reaction is expected to be **exothermic**.

b. Hybridization in the *enol* form   This C atom changes from sp^2 in the *enol*
                                              form to sp^3 in the *keto* form.

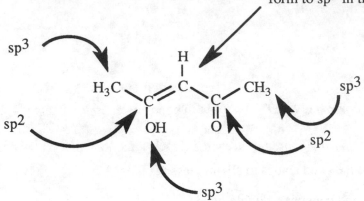

c. C atom geometries in *enol* and *keto* forms.   This C atom changes from a trigonal
                                                    planar geometry to tetrahedral
                                                    geometry in the *keto* form.

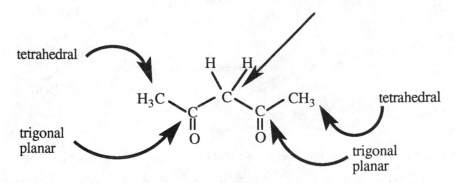

d. 15.0 g Cr compound $\cdot \dfrac{1 \text{ mol compound}}{349.33 \text{ g}} = 4.29 \times 10^{-2}$ mol Cr compound

   4.29×10^{-2} mol Cr compound $\cdot \dfrac{1 \text{ mol CrCl}_3}{1 \text{ mol Cr compound}} \cdot \dfrac{158.36 \text{ g}}{1 \text{ mol compound}}$

   $= 6.80$ g $CrCl_3$ required

   4.29×10^{-2} mol Cr compound $\cdot \dfrac{3 \text{ mol CH}_3\text{C(OH)CHCOCH}_3}{1 \text{ mol Cr compound}} \cdot \dfrac{100.12 \text{ g}}{\text{mol compound}}$

   $= 12.9$ g $CH_3C(OH)CHCOCH_3$ required

   4.29×10^{-2} mol Cr compound $\cdot$ 3 mol NaOH $\cdot \dfrac{40.0 \text{ g}}{\text{mole compound}} = 5.15$ g NaOH
   required

Chapter 11:
Bonding and Molecular Structure: Organic Chemistry

Alkanes

1. With eight carbons in the chain, the name is **octane**.

3. Alkanes have the formula C_nH_{2n+2}. A cycloalkane "loses" a H from each "end" of the chain to form a ring. The cycloalkane therefore has the formula C_nH_{2n}.
   (b) C_5H_{10} is one solution. Since C_2H_4 can't form a ring (c) is the only possible one of the four offered.

5. The compound is named : 2,3-dimethylbutane. There are 4 carbons in the longest chain (butane). The two (di) methyl groups are on carbons 2 and 3.

   A structural isomer for 2,3-dimethylbutane:

   $CH_3CH_2CH_2CH_2CH_2CH_3$ hexane

   $CH_3CHCH_2CH_2CH_3$ 2-methylpentane
   $\quad\quad |$
   $\quad\quad CH_3$

   $CH_3CH_2CHCH_2CH_3$ 3-methylpentane
   $\quad\quad\quad |$
   $\quad\quad\quad CH_3$

   $\quad\quad CH_3$
   $\quad\quad |$
   $CHCCH_2CH_3$ 2,2-dimethylbutane
   $\quad\quad |$
   $\quad\quad CH_3$

7.   a. 2,3-dimethylpentane
   $\quad\quad\quad CH_3\ \ CH_3$
   $\quad\quad\quad |\quad\quad |$
   $CH_3 - CH - CH - CH_2 - CH_3$

   b. 2,4-dimethyloctane
   $\quad\quad\quad CH_3\quad\quad\quad CH_3$
   $\quad\quad\quad |\quad\quad\quad\quad\quad |$
   $CH_3 - CH - CH_2 - CH - CH_2 - CH_2 - CH_2 - CH_3$

c. 3-ethylhexane

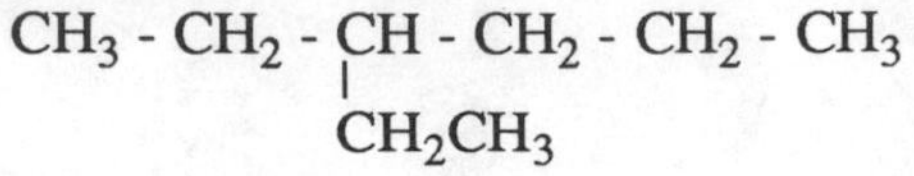

$$CH_3 - CH_2 - CH - CH_2 - CH_2 - CH_3$$
$$|$$
$$CH_2CH_3$$

d. 2-methyl-3-ethylhexane

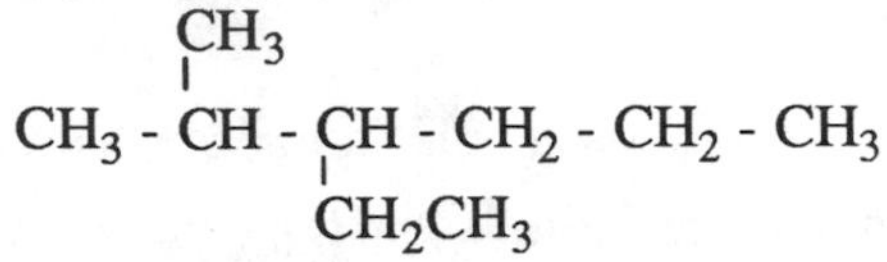

$$CH_3$$
$$|$$
$$CH_3 - CH - CH - CH_2 - CH_2 - CH_3$$
$$|$$
$$CH_2CH_3$$

Alkenes and Alkynes

9. cis- and trans- isomers of :

a. 1,2-dichloroethene

cis   trans

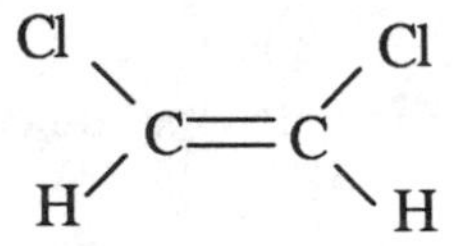

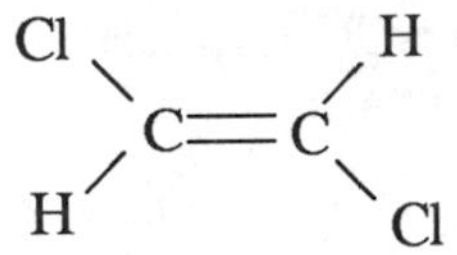

b. 3-methyl-3-hexene

cis   trans

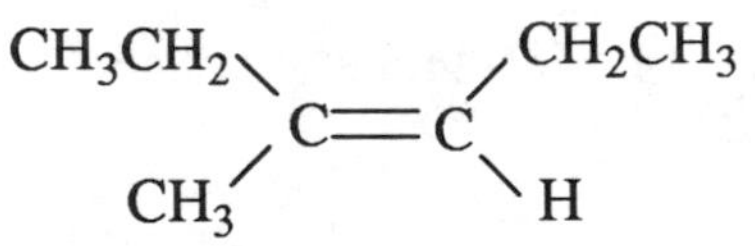

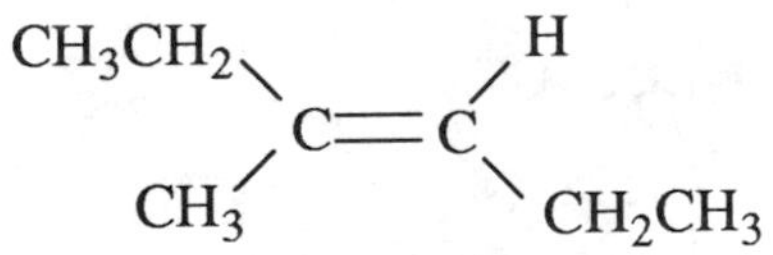

11.   a. Structures and names for alkenes with formula C_5H_{10}.

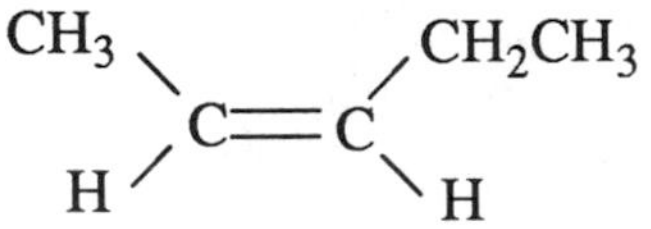

cis-2-pentene 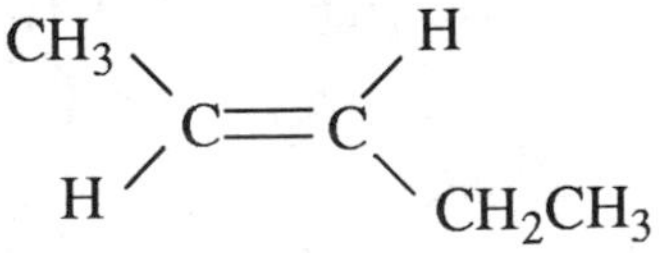   trans-2-pentene

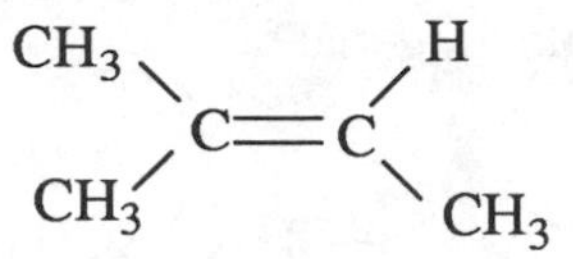

$$CH_3 \diagdown \diagup CH_3$$
$$C=C$$
$$CH_3 \diagup \diagdown H$$

2-methyl-2-butene

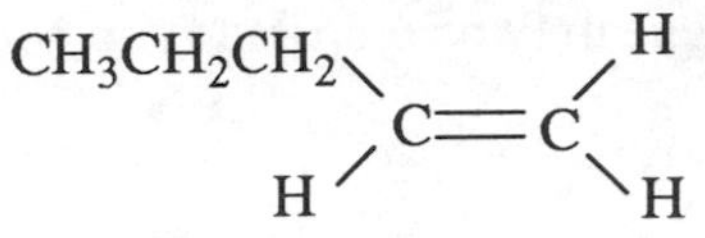

$$CH_3CH_2CH_2 \diagdown \diagup H$$
$$C=C$$
$$H \diagup \diagdown H$$

1-pentene

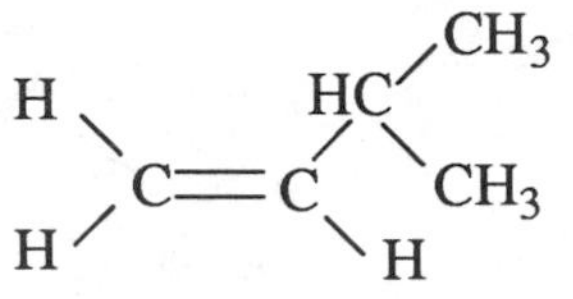

$$H \diagdown \quad HC \diagup CH_3$$
$$C=C \diagdown CH_3$$
$$H \diagup \quad \diagdown H$$

3-methyl-1butene

$$H \diagdown \quad \diagup CH_2CH_3$$
$$C=C$$
$$H \diagup \quad \diagdown CH_3$$

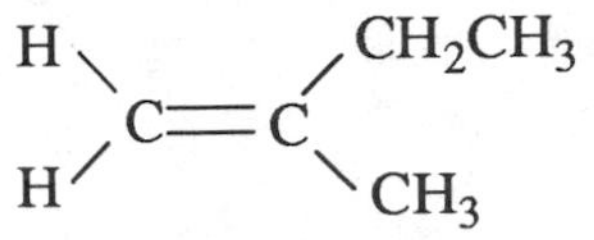

2-methyl-1-butene

b. The cycloalkane with the formula C_5H_{10}

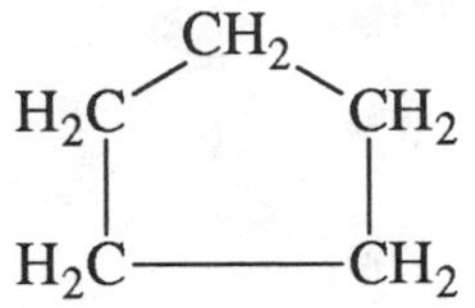

$$CH_2$$
$$H_2C \qquad CH_2$$
$$H_2C \text{——} CH_2$$

13. Structure and names for products of:

  a. $CH_3CH_2CH=CH_2$ + Br_2 $\longrightarrow$ $CH_3CH_2CHBrCH_2Br$

     1,2-dibromobutane

  b. $CH_3CH=CHCH_3$ + H_2 $\longrightarrow$ $CH_3CH_2CH_2CH_3$

     butane

Benzene and Aromatic Compounds

15.   a. p-Dichlorobenzene   b. m-Bromotoluene 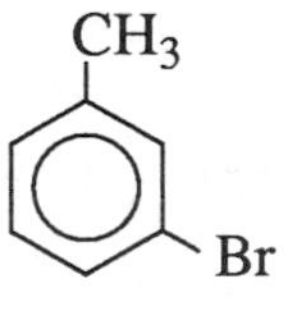   c. p-Diethylbenzene

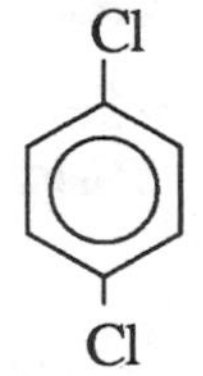

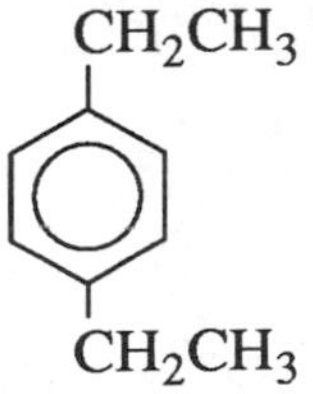

$$Cl$$
$$\bigcirc$$
$$Cl$$

$$CH_3$$
$$\bigcirc$$
$$Br$$

$$CH_2CH_3$$
$$\bigcirc$$
$$CH_2CH_3$$

17. Propylbenzene may be prepared from benzene in the following manner:

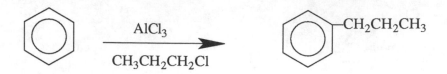

Alcohols

19. Systematic names of the alcohols:

    a. 1-propanol   primary alcohol

    b. 1-butanol   primary alcohol

    c. 2-methyl-2-propanol   tertiary alcohol

    d. 2-methyl-2-butanol   tertiary alcohol

21.   1-butanol $CH_3CH_2CH_2CH_2OH$

     2-butanol $CH_3CH_2\overset{\displaystyle |}{\underset{\displaystyle OH}{C}}HCH_3$

     2-methyl-1-propanol $CH_3\overset{\displaystyle |}{\underset{\displaystyle CH_3}{C}}HCH_2OH$

     2-methyl-2-propanol $CH_3\overset{\displaystyle OH}{\overset{\displaystyle |}{\underset{\displaystyle |}{\underset{\displaystyle CH_3}{C}}}}CH_3$

23.   a. $CH_3CH_2CH_2CH_2OH \xrightarrow{\;H_2SO_4,\;180°C\;} CH_3CH_2CH=CH_2$ and depending upon conditions an ether could be formed by condensation of two molecules of the alcohol: $(H_3CH_2CH_2CH_2)_2O$

    b. $CH_3CH_2CH_2OH + Na \longrightarrow CH_3CH_2CH_2O^-Na^+ + 1/2\,H_2$

    c. $CH_3CH_2CH_2CH_2OH + HBr \longrightarrow CH_3CH_2CH_2CH_2Br + H_2O$

Compounds with a Carbonyl Group

25. Structural formulas for :

    a. 2-hexanone   b. pentanal

$CH_3-CH_2-CH_2-CH_2-\underset{\underset{O}{\|}}{C}-CH_3$ $CH_3-CH_2-CH_2-CH_2-\underset{\underset{O}{\|}}{C}-H$

    c. hexanoic acid

    $CH_3CH_2CH_2CH_2CH_2CO_2H$ or $CH_3(CH_2)_4CO_2H$

27.   a. $CH_3CH_2\underset{\underset{CH_3}{|}}{C}HCH_2CO_2H$ 3-methylpentanoic acid

       b. $CH_3CH_2COCH_3$ methyl propanoate

       c. $CH_3COCH_2CH_2CH_2CH_3$ butyl ethanoate

       d. 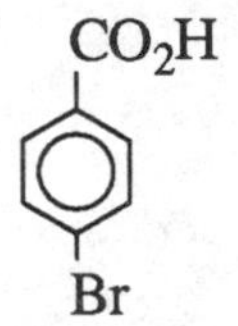 $\underset{Br}{\overset{CO_2H}{\bigcirc}}$ p-bromobenzoic acid

29.   a. The product of oxidation is propanoic acid:

           $CH_3CH_2\underset{}{\overset{\overset{O}{\|}}{C}}-OH$

       b. The reduction product is 1-propanol:

           $CH_3CH_2CH_2\,OH$

       c. The reduction yields 2-butanol

           $CH_3CH_2\underset{\overset{|}{}}{\overset{OH}{C}}HCH_3$

       d. The oxidation of 2-butanone with potassium permanganate gives <u>no reaction</u>.

131

31. $CH_3CH_2CO_2H$ $\xrightarrow{\text{LiAlH}_4}$ $CH_3CH_2CH_2OH$

$CH_3CH_2CO_2H$ + $HOCH_2CH_2CH_3$ $\xrightarrow{H^+}$ $CH_3CH_2\overset{\displaystyle O}{\overset{\|}{C}}OCH_2CH_2CH_3$

33.   The products of the hydrolysis of the ester are 1-butanol and acetic acid (and Na^+):

$$CH_3CH_2CH_2CH_2OH \qquad\qquad CH_2CO_2H$$

Polymers

35.   a. The structure for polyvinylacetate:

b. Prepare polyvinyl alcohol from polyvinyl acetate: Polyvinyl acetate is an ester. Hydrolysis of the ester with NaOH will produce the sodium salt, $NaC_2H_3O_2$, and polyvinyl alcohol. Acidification with a strong acid (e.g. HCl) will produce polyvinyl alcohol.

37.

1,1-dichloroethene   chloroethene

The reaction proceeds with the free radical addition of the copolymers:

132

General Questions

39.   a. Isomers for C_3H_8O:

Compound	Name	Class
$CH_3CH_2CH_2$-OH	1-propanol	primary alcohol
CH_3CHCH_3 OH	2-propanol	secondary alcohol
CH_3-O-CH_2CH_3	methyl ethyl ether	ether

b. Aldehyde with molecular formula C_4H_8O:

    $CH_3CH_2CH_2CHO$ butanal   aldehyde

Ketone with molecular formula C_4H_8O:

$$\overset{\overset{\textstyle O}{\|}}{CH_3CH_2CCH_3}$$

                butanone   ketone

41.

$$\begin{array}{c} CH_2\text{-O-}\overset{\overset{\textstyle O}{\|}}{C}\text{ - }(CH_2)_{10}CH_3 \\ | \qquad\quad \overset{\overset{\textstyle O}{\|}}{} \\ HC\text{- O - }\overset{\overset{\textstyle O}{\|}}{C}\text{ - }(CH_2)_{10}CH_3 \\ | \qquad\quad \overset{\overset{\textstyle O}{\|}}{} \\ CH_2\text{-O-}\overset{\overset{\textstyle O}{\|}}{C}\text{ - }(CH_2)_{10}CH_3 \end{array} \xrightarrow{\text{NaOH}} \begin{array}{c} CH_2OH \\ | \\ CHOH \\ | \\ CH_2OH \end{array} + \ 3\ Na^+\ O_2C(CH_2)_{10}CH_3$$

   glyceryl trilaurate   glycerol   sodium laurate

43.   a. Lewis structure for the BF_3 compound with dimethyl ether :

$$\begin{array}{ccc} & F & CH_3 \\ F & \ddot{\underset{\ddots}{B}} : \underset{\ddots}{\overset{\cdot\cdot}{O}} : & CH_3 \\ & F & \end{array}$$

b. The hybridization of the oxygen is sp^3. Two of the orbitals contain the lone pairs and two contain the bonding pairs that attach the carbon atoms to the oxygen. The hybridization of the O doesn't change during compound formation. The hybridization of B in BF_3 is **sp^2**. The unhybridized p orbital of the B atom in BF_3 accepts an electron pair, providing 4 groups around the B with a **change to sp^3 hybridization**.

c. The F—F—F bond angle with sp^2 hybridization is approximately 120° and with sp^3 hybridization — approximately 109°.

45.

In each case, every C is bonded directly to 3 other carbons.

47.   a. $CH_3CH_2CH=CH_2$ + Br_2 $\longrightarrow$ $CH_3CH_2CHBrCH_2Br$

  1-butene   1,2-dibromobutane

                            catalyst
b. $HC\equiv CCH_3$ + H_2 $\longrightarrow$ $CH_2=CHCH_3$

  propyne   propene

                            catalyst
c. $HC\equiv CCH_3$ + 2 H_2 $\longrightarrow$ $CH_3CH_2CH_3$

  propyne   propane

d. $CH_3CH=CHCH_3$ + H_2O $\longrightarrow$ $CH_3CH_2CHOHCH_3$

  2-butene   2-butanol

49.

Kevlar

51.

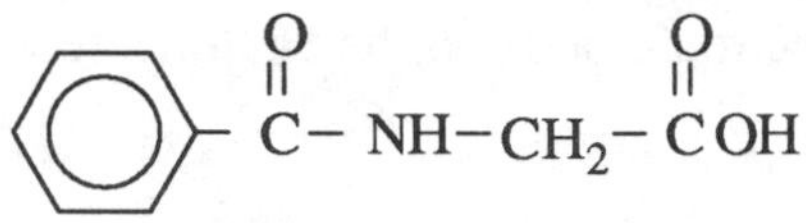

Hippuric Acid

Hippuric acid retains the carboxylic acid group of glycine, with the amide linkage formed between the carboxylate group of benzoic acid and the amine group of glycine.

Conceptual Questions

53. Balanced equations for the combustion of ethane and ethanol:

$$C_2H_6 + 7/2\ O_2 \longrightarrow 2\ CO_2 + 3\ H_2O$$
$$C_2H_5OH + 3\ O_2 \longrightarrow 2\ CO_2 + 3\ H_2O$$

Compound with the most negative enthalpy of combustion.

Combustion of ethane: (Using the ΔH_f from the Appendix)

ΔH_{rx} = [(2mol)(-393.509 kJ/mol) + (3 mol)(-285.830 kJ/mol)] -

[(1 mol)(-84.68 kJ/mol) + (7/2 mol)(O)]

= (-1644.508 kJ) - (-84.68 kJ)

ΔH_{rx} = -1559.83 kJ

For the combustion of ethanol, the difference lies in the ΔH_f of C_2H_5O.

The value is: -277.69 kJ/mol

ΔH_{rx} = (-1644.508 kJ) - (-277.69 kJ)

ΔH_{rx} = -1366.82 kJ

Ethane has the more negative enthalpy of combustion. The effect of partially oxidizing the ethane reduces the energy released upon combustion.

55. To discriminate between the two isomers, react the two with elemental bromine.
Cyclopentane will not react with bromine, while 1-pentene will react with bromine.

57.   a. Energy is required to rotate the C—C bond because the pi bond present in the alkene prohibits free rotation of the C—C bond.

b. Energy is required for the rotation since as the rotation occurs the hydrogen atoms on C_2 and C_3 are brought closer together, and the electron clouds on the H atoms repel each other as they are brought to their closest approach.

59.   a. The empirical formula of maleic acid:

Calculate the mass of C, H, and O respectively

$$0.190 \text{ g CO}_2 \cdot \frac{12.01 \text{ g C}}{44.01 \text{ g CO}_2} = 0.0518 \text{ g C}$$

$$0.0388 \text{ g H}_2\text{O} \cdot \frac{2.0158 \text{ g H}}{18.02 \text{ g H}_2\text{O}} = 0.00434 \text{ g H}$$

The mass of O is then 0.125 g - (0.0518 + 0.00434) = 0.0689 g O

The moles of atoms of C, H, and O are:

$$0.0518 \text{ g C} \cdot \frac{1 \text{ mol C}}{12.011 \text{ g C}} = 0.00431 \text{ mol C}$$

$$0.00434 \text{ g H} \cdot \frac{1 \text{ mol H}}{1.0079 \text{ g H}} = 0.00431 \text{ mol H}$$

$$0.0689 \text{ g O} \cdot \frac{1 \text{ mol O}}{16.00 \text{ g O}} = 0.00431 \text{ mol O}$$

The empirical formula is then CHO.

b. 0.261 g of acid requires 34.60 mL of 0.130 M NaOH

Moles of NaOH = 0.130 mol/L $\cdot$ 0.03460 L = 0.00450 mol NaOH

Since maleic is a diprotic acid, the moles of maleic acid present is half the number of moles of NaOH, and the molar mass

$$\frac{0.261 \text{ g acid}}{0.00225 \text{ mol acid}} = 116 \text{ g/mol}$$

The empirical formula CHO would have a mass of 29.0 g

$$\frac{116 \text{ g maleic acid}}{1 \text{ mol maleic acid}} \cdot \frac{1 \text{ empirical formula}}{29.0 \text{ g maleic acid}} = 4 \frac{\text{empirical formulas}}{\text{mole maleic acid}} \text{ or } C_4H_4O_4$$

c. Lewis structure for $C_4H_4O_4$

d. Hybridization used by the C atoms: Each carbon in the molecule has three electron-pair groups around it, making the C atoms sp^2 hybridized.

e. Bond angles: Each C atom would have three 120° angles separating the three groups attached.

Chapter 12:
Gases

Measuring Pressure

11. a. $725 \text{ mmHg} \cdot \dfrac{1 \text{ atm}}{760 \text{ mmHg}} = 0.954 \text{ atm}$

  b. $0.67 \text{ atm} \cdot \dfrac{760. \text{ mmHg}}{1 \text{ atm}} = 510 \text{ mmHg}$

  c. $740 \text{ mmHg} \cdot \dfrac{101.325 \text{ kPa}}{760 \text{ mmHg}} = 99 \text{ kPa}$

  d. $0.75 \text{ atm} \cdot \dfrac{101.3 \text{ kPa}}{1.0 \text{ atm}} = 76 \text{ kPa}$

  e. $745 \text{ mmHg} \cdot \dfrac{1.013 \text{ bar}}{760 \text{ mmHg}} = 0.993 \text{ bar}$

  f. $125 \text{ mmHg} \cdot \dfrac{760 \text{ torr}}{760 \text{ mmHg}} = 125 \text{ torr}$

13. The levels in the U-tube manometer indicate a gas pressure of 56.3 mmHg.

  $56.3 \text{ mmHg} \cdot \dfrac{1 \text{ atm}}{760 \text{ mmHg}} = 0.0741 \text{ atm}$

  $56.3 \text{ mmHg} \cdot \dfrac{1 \text{ torr}}{1 \text{mmHg}} = 56.3 \text{ torr}$

  $56.3 \text{ mmHg} \cdot \dfrac{101.3 \text{ k Pa}}{760.0 \text{ mmHg}} = 7.50 \text{ kPa}$

15. **Boyle's law** states that the pressure a gas exerts is inversely proportional to the volume it occupies, or for a given amount of gas--PV = constant . We can write this as:

$$P_1 V_1 = P_2 V_2$$

  So $(67.5 \text{ mmHg})(125 \text{ mL}) = (P_2)(500. \text{ mL})$

  and $\dfrac{(67.5 \text{ mmHg})(125 \text{ mL})}{500. \text{ mL}} = 16.9 \text{ mmHg}$

The Ideal Gas Law

29. The pressure of 1.25 g of gaseous carbon dioxide may be calculated with the ideal gas law:

$$1.25 \text{ g CO}_2 \cdot \frac{1 \text{ mol CO}_2}{44.01 \text{ g CO}_2} = 0.0284 \text{ mol CO}_2$$

Rearranging PV = nRT to solve for P, we obtain:

$$P = \frac{nRT}{V} = \frac{(0.0284 \text{ mol})(0.082057 \frac{L \cdot atm}{K \cdot mol})(296 \text{ K})}{0.850 \text{ L}} = 0.812 \text{ atm}$$

31. The volume of the flask may be calculated by realizing that the gas will expand to fill the flask.

$$4.4 \text{ g CO}_2 \cdot \frac{1 \text{ mol CO}_2}{44.0 \text{ g CO}_2} = 0.10 \text{ mol CO}_2$$

$$P = 635 \text{ mmHg} \cdot \frac{1 \text{ atm}}{760 \text{ mmHg}} = 0.84 \text{ atm}$$

$$V = \frac{(0.10 \text{ mol})(0.082057 \frac{L \cdot atm}{K \cdot mol})(295 \text{ K})}{0.84 \text{ atm}} = 2.9 \text{ L}$$

33. The number of moles of N_2 can be calculated with the ideal gas law:

$$n = \frac{41.8 \text{ mm Hg} \cdot \frac{1 \text{ atm}}{760 \text{ mmHg}} \cdot 150. \text{ L}}{0.082057 \frac{L \cdot atm}{mol \cdot K} \cdot 298.2 \text{ K}} = 0.337 \text{ mol N}_2$$

35. Calculate the number of moles of compound:

Note that $700. \text{ mmHg} \cdot \frac{1 \text{ atm}}{760 \text{ mmHg}} = 0.921 \text{ atm}$

$$n = \frac{PV}{RT} = \frac{(0.921 \text{ atm})(0.450 \text{ L})}{(0.082057 \frac{L \cdot atm}{K \cdot mol})(296 \text{ K})} = 0.0171 \text{ mol}$$

So 0.982 g = 0.0171 mol and $\frac{0.982 \text{ g}}{0.0171 \text{ mol}} = 57.5 \text{ g/mol}$

37. Using the ideal gas law, we can calculate the moles of gas represented by 0.218 g of the compound.

$$n = \frac{374 \text{ mm Hg} \cdot \dfrac{1 \text{ atm}}{760 \text{ mmHg}} \cdot 0.185 \text{ L}}{0.082057 \dfrac{\text{L} \cdot \text{atm}}{\text{mol} \cdot \text{K}} \cdot 296.2 \text{ K}} = 3.75 \times 10^{-3} \text{ mol}$$

So the molar mass of the compound is: $\dfrac{0.218 \text{ g}}{3.75 \times 10^{-3} \text{ mol}} = 58.1 \text{ g/mol}$

The compound is 82.66 % C and 17.34 % H. The empirical formula would then be:

Carbon: $82.66 \text{ g} \cdot \dfrac{1 \text{ mol C}}{12.011 \text{ g C}} = 6.88 \text{ mol C}$

Hydrogen: $17.34 \text{ g} \cdot \dfrac{1 \text{ mol H}}{1.008 \text{ g H}} = 17.2 \text{ mol H}$

The ratio of H : C is 2.5 : 1 or 5 H : 2 C

With a molar mass of about 58 grams and an empirical formula of C_2H_5 (formula mass ≈29), the molecular formula would be C_4H_{10} .

39. Calculate the empirical formula:

% F = 100.0% - (11.79 % C + 69.57 % Cl) = 18.64 % F

The moles of each element:

$18.64 \text{ g F} \cdot \dfrac{1 \text{ mol F}}{18.998 \text{ g F}} = 0.9812 \text{ mol F}$

$11.79 \text{ g C} \cdot \dfrac{1 \text{ mol C}}{12.011 \text{ g C}} = 0.9816 \text{ mol C}$

$69.57 \text{ g Cl} \cdot \dfrac{1 \text{ mol Cl}}{35.453 \text{ g Cl}} = 1.962 \text{ mol Cl}$

Giving an empirical formula of $FCCl_2$

Calculate the molar mass:

$21.3 \text{ mmHg} \cdot \dfrac{1 \text{ atm}}{760. \text{ mmHg}} = 0.0280 \text{ atm}$

$n = \dfrac{PV}{RT} = \dfrac{(0.0280 \text{ atm})(0.458 \text{ L})}{(0.082057 \dfrac{\text{L} \cdot \text{atm}}{\text{K} \cdot \text{mol}})(298 \text{ K})} = 5.25 \times 10^{-4} \text{ mol}$

Since 0.107 g corresponds to 5.25×10^{-4} mol , the molar mass is:

$$\frac{0.107 \text{ g}}{5.25 \times 10^{-4} \text{ mol}} = 204 \text{ g/mol}$$

With an empirical formula of CCl_2F (Empirical formula weight = 102), the molecular formula must be $C_2Cl_4F_2$.

41.   Write the ideal gas law as: Molar Mass $= \frac{dRT}{P}$ where d = density in grams per liter.

Solving for d, we obtain: $\frac{(\text{Molar Mass}) \cdot P}{R \cdot T} = d$

The average molecular weight for air is approximately 29 g.

$$\frac{(29 \text{ g/mol})(0.20 \text{ mmHg} \cdot 1 \text{ atm}/760 \text{ mmHg})}{(0.082057 \frac{L \cdot atm}{K \cdot mol})(250 \text{ K})} = 3.7 \times 10^{-4} \text{ g/L} = d$$

43.   Molar mass $= \dfrac{(0.259 \text{ g/L})(0.082057 \frac{L \cdot atm}{K \cdot mol})(400. \text{ K})}{(190. \text{ mmHg} \cdot 1 \text{ atm}/760 \text{ mmHg})} = 34.0$

The molar mass of methyl fluoride, CH_3F, to three significant figures is 34.0.

45.   Solve the ideal gas law for volume to obtain

(1) $V_1 = \dfrac{n_1RT_1}{P_1}$

We are interested in the mass (and therefore moles) of oxygen needed to fill this balloon to the same volume, measured at the same pressure but a different temperature. We could express this as:

(2) $V_1 = \dfrac{n_2RT_2}{P_1}$

In equations (1) and (2) we have the common term V_1 so we may write:

(3) $\dfrac{n_1RT_1}{P_1} = \dfrac{n_2RT_2}{P_1}$

Since both R and P_1 are constant in these conditions, equation (3) may be simplified:

$$n_1T_1 = n_2T_2$$

Calculating the moles of oxygen represented by 12.0 g gives:

$$n_1 = 12.0 \text{ g } O_2 \cdot \frac{1 \text{ mol } O_2}{32.00 \text{ g } O_2} = 0.375 \text{ mol } O_2$$

Substituting for T_1 (300. K) and T_2 (354 K) and solving for n_2 we obtain:

$$\frac{(0.375 \text{ mol } O_2) \cdot (300. \text{ K})}{354 \text{ K}} = 0.318 \text{ mol } O_2 \text{ or } 10.2 \text{ g } O_2$$

47. To calculate the temperature at which P = 7.25 atm, rearrange the Ideal Gas Law.

$$\frac{PV}{nR} = T$$

$$\frac{(7.25 \text{ atm})(1.52 \text{ L})}{(0.406 \text{ mol})(0.082057 \frac{L \cdot atm}{K \cdot mol})} = 331 \text{ K or } 58 \text{ °C}$$

Gas Laws and Stoichiometry

49. Determine the amount of H_2 generated when 1.0 g Fe reacts:

$$1.0 \text{ g Fe} \cdot \frac{1 \text{ mol Fe}}{55.85 \text{ g Fe}} \cdot \frac{1 \text{ mol } H_2}{1 \text{ mol Fe}} = 0.018 \text{ mol } H_2$$

The pressure of this amount of H_2 is :

$$P = \frac{nRT}{V} = \frac{(0.018 \text{ mol } H_2)(62.4 \frac{L \cdot torr}{K \cdot mol})(298 \text{ K})}{15.0 \text{ L}}$$

$$= 22.2 \text{ torr or } 22.2 \text{ mmHg}$$

51. Calculate the moles of N_2 needed:

$$n = \frac{PV}{RT} = \frac{(1.3 \text{ atm})(25.0 \text{ L})}{(0.082057 \frac{L \cdot atm}{K \cdot mol})(298 \text{ K})} = 1.33 \text{ mol } N_2$$

The mass of NaN_3 needed to produce this is obtained from the stoichiometry of the equation:

$$1.33 \text{ mol } N_2 \cdot \frac{2 \text{ mol } NaN_3}{3 \text{ mol } N_2} \cdot \frac{65.0 \text{ g } NaN_3}{1 \text{ mol } NaN_3} = 58 \text{ g } NaN_3 \text{ (to 2 sf)}$$

53. $N_2H_4 (g) + O_2 (g) \rightarrow N_2 (g) + 2 H_2O (g)$

1.00 kg N_2H_4 $\cdot \dfrac{1.0 \times 10^3 \text{ g } N_2H_4}{1.0 \text{ kg } N_2H_4} \cdot \dfrac{1 \text{ mol } N_2H_4}{32.0 \text{ g } N_2H_4} \cdot \dfrac{1 \text{ mol } O_2}{1 \text{ mol } N_2H_4}$

$$= 3.13 \times 10^1 \text{ mole } O_2$$

$$P(O_2) = \frac{n(O_2) \cdot R \cdot T}{V} = \frac{(3.13 \times 10^1 \text{mol})(0.082057 \text{ L} \cdot \text{atm/K} \cdot \text{mol})(296 \text{ K})}{450 \text{ L}}$$

$P(O_2) = 1.69$ atm or 1.7 atm to 2 sf

55. Calculate the number of moles of F_2 that the 50.0 L tank contains:

$$n = \frac{PV}{RT} = \frac{(8.0 \text{ atm})(50.0 \text{ L})}{(0.082057\frac{\text{L} \cdot \text{atm}}{\text{K} \cdot \text{mol}})(298 \text{ K})} = 16.4 \text{ mol } F_2$$

The moles of F_2 needed to consume the uranium:

$$1.0 \times 10^3 \text{ g U} \cdot \frac{1 \text{ mol U}}{238 \text{ g U}} \cdot \frac{1 \text{ mol } UF_6}{1 \text{ mol U}} \cdot \frac{3 \text{ mol } F_2}{1 \text{ mol } UF_6} = 12.6 \text{ mol } F_2$$

The amount of F_2 remaining is (16.4 - 12.6) or 3.8 mol F_2 (to 2 sf)

The pressure that this amount of F_2 would exert:

$$P = \frac{nRT}{V} = \frac{(3.8 \text{ mol})(0.082057 \frac{\text{L} \cdot \text{atm}}{\text{K} \cdot \text{mol}})(298 \text{ K})}{50.0 \text{ L}} = 1.8 \text{ atm}$$

Rounding the amount of F_2 in the first two steps (16 and 13 respectively) results in 1 sf for moles of F_2 remaining, and a final pressure of 2 atm.

57. The reaction can be illustrated

$Fe_x(CO)_y + O_2 \longrightarrow Fe_2O_3 + CO_2$

Since each mole of CO in the compound produces 1 mol CO_2, we can calculate the moles of CO by calculating the number of moles of CO_2.

$$n_{CO_2} = \frac{PV}{RT} = \frac{(44.9 \text{ torr})(1.50 \text{ L})}{(62.4 \frac{\text{L} \cdot \text{torr}}{\text{K} \cdot \text{mol}})(298 \text{ K})} = 3.62 \times 10^{-3} \text{ mol } CO_2 \text{ (and mol CO)}$$

This amount of CO will have a mass of :

$$3.62 \times 10^{-3} \text{ mol CO} \cdot \frac{28.01 \text{ g CO}}{1 \text{ mol CO}} = 0.101 \text{ g CO}$$

The mass of Fe in the 0.142 g sample is : 0.142 g sample - 0.101 g CO = 0.0405 g Fe

The amount of Fe is: $0.0405 \text{ g Fe} \cdot \dfrac{1 \text{ mol Fe}}{55.847 \text{ g Fe}} = 7.26 \times 10^{-4} \text{ mol Fe}$

The ratio of mole CO : mol Fe is:

$$\dfrac{3.62 \times 10^{-3} \text{ mol CO}}{7.26 \times 10^{-4} \text{ mol Fe}} = \dfrac{5 \text{ mol CO}}{\text{mol Fe}}$$

The formula for the compound is $Fe_1(CO)_5$.

59. The moles of KO_2 present : $6.0 \text{ g } KO_2 \cdot \dfrac{1 \text{ mol } KO_2}{71.10 \text{ g } KO_2} = 0.225 \text{ mol } KO_2$

the amount of CO_2 present:
$$n = \dfrac{PV}{RT} = \dfrac{(1.24 \text{ atm})(4.00 \text{ L})}{(0.082057\dfrac{\text{L} \cdot \text{atm}}{\text{K} \cdot \text{mol}})(296 \text{ K})} = 0.204 \text{ mol } CO_2$$

The reaction requires 2 mol of KO_2 per mol of CO_2 , so 0.225 mol of **KO_2 will be completely consumed.** The amount of O_2 produced is:
$$0.225 \text{ mol } KO_2 \cdot \dfrac{3 \text{ mol } O_2}{4 \text{ mol } KO_2} = 0.169 \text{ mol } O_2$$

The pressure exerted would be:

$$P = \dfrac{nRT}{V} = \dfrac{(0.169 \text{ mol})(0.082057 \dfrac{\text{L} \cdot \text{atm}}{\text{K} \cdot \text{mol}})(298 \text{ K})}{2.50 \text{ L}} = 1.65 \text{ atm}$$

Gas Mixtures

61. Given that the partial pressure of gas is proportional to the number of moles of gas present:
The relative pressure of He : H_2 is 150 mmHg : 25 mmHg or 6 : 1, so the relative # of moles is 6 mol He : 1 mol H_2 .

Given the 0.56 g of He is $0.56 \text{ g} \cdot \dfrac{1 \text{ mol He}}{4.00 \text{ g He}} = 0.14 \text{ mol He}$

The amount of H_2 present is : $0.14 \text{ mol He} \cdot \dfrac{1 \text{ mol } H_2}{6 \text{ mol He}} \cdot \dfrac{2.00 \text{ g } H_2}{1 \text{ mol } H_2} = 0.047 \text{ g } H_2$

63. First the % N is: $100.0 - (4.5 \% \; H_2S + 3.0 \% \; CO_2) = 92.5 \% \; N$

   If the percentages given are mol percentages, then the partial pressures of each gas are directly proportional to the percentages.

   $P_{N_2} = (46 \text{ atm})(92.5 \% \; N) = 42.5 \text{ atm or } 43 \text{ atm (to 2 sf)}$

   $P_{H_2S} = (46 \text{ atm})(4.5 \% \; N) = 2.1 \text{ atm}$

   $P_{CO_2} = (46 \text{ atm})(3.0 \% \; N) = 1.4 \text{ atm}$

65.   a. Grams of He

   $$n = \frac{PV}{RT} = \frac{(1.0 \text{ atm})(12 \text{ L})}{(0.082057 \frac{L \cdot atm}{K \cdot mol})(293 \text{ K})} = 0.50 \text{ mol He}$$

   $0.50 \text{ mol He} \cdot \frac{4.003 \text{ g He}}{1 \text{ mol He}} = 2.0 \text{ g He}$

   b. Final partial pressure of He

   $$P = \frac{nRT}{V} = \frac{(0.50 \text{ mol})(0.082057 \frac{L \cdot atm}{K \cdot mol})(293 \text{ K})}{26 \text{ L}} = 0.46 \text{ atm}$$

   c. Partial pressure of O_2:

   $P_{O_2} = P_t - P_{He} = 1.0 \text{ atm} - 0.46 \text{ atm} = 0.54 \text{ or } 0.5 \text{ atm} \; (1 \text{ sf})$

   d. Mole fraction of each gas : Since the presure each gas exerts is proportional to the mole fraction of that gas, we can write :

   $P_{O_2} = P_T \cdot mf_{O_2}$ and $P_{He} = P_T \cdot mf_{He}$

   Since the total pressure is 1.0 atm, the mole fraction of each gas will equal the partial pressure calculated in parts b and c.

67. The molar ratio of 2 mol Cl_2 per 1 mol SO_2 means that the partial pressure of Cl_2 would have to be twice the partial pressure of SO_2, or 250. mmHg.

69. Dalton's Law tells us that in a mixture of gases the total pressure is the sum of the pressures of the individual gases (partial pressures)

$$P_T = P_{N_2} + P_{H_2O} \quad \text{so}$$

$$P_{N_2} = P_T - P_{H_2O} = 747 \text{ mmHg} - 15.5 \text{ mmHg}$$

$$= 732 \text{ mmHg}$$

71. a. Since equal masses of O_2 and N_2 are placed in flasks of equal volume at the same temperature the pressures will be proportional to the number of moles of each gas present. Oxygen has 32 g/mole while N_2 has 28 g/mole. There will be greater number of moles of N_2 — and a greater pressure. **Statement a is true.**

   b. Given that there are more moles of N_2 in one flask than moles of O_2 in the other, there are more molecules of N_2 in the flask than molecules of O_2 . **Statement b is false.**

Kinetic Molecular Theory

73. a. Kinetic energy depends only on the temperature so the average kinetic energies of these two gases are equal.

   b. Since the kinetic energies are equal, we can state:

$$KE(H_2) = KE(CO_2)$$

$$1/2 \; m(H_2) \cdot \overline{V}^2 (H_2) = 1/2 \; m(CO_2) \cdot \overline{V}^2(CO_2)$$

Where m = mass of a molecule and $\overline{V}$ = average velocity of a molecule

So $\qquad m(H_2) \cdot \overline{V}^2(H_2) = m(CO_2) \cdot \overline{V}^2(CO_2)$

and $\qquad \dfrac{\overline{V}^2(H_2)}{\overline{V}^2(CO_2)} = \dfrac{m(CO_2)}{m(H_2)}$

Now the molar mass of $H_2 = 2.0$ g and the molar mass of $CO_2 = 44$ g

$$\dfrac{\overline{V}_{H_2}}{\overline{V}_{CO_2}} = \sqrt{\dfrac{m_{CO_2}}{m_{H_2}}} = \sqrt{\dfrac{44}{2.0}} = 4.7$$

146

The hydrogen molecules have an average velocity which is 4.7 times the average velocity of the CO_2 molecules.

c. Since the temperatures and the volumes are equal for these two gas samples, the pressure is proportional to the amount of gas present.

$$V_A = \frac{n_A R T_A}{P_A} \quad \text{and} \quad V_B = \frac{n_B R T_B}{P_B} \quad \text{now } T_A = T_B \text{ and } V_A = V_B \text{ so}$$

$$\frac{n_A R}{P_A} = \frac{n_B R}{P_B} \quad \text{or} \quad \frac{n_A}{P_A} = \frac{n_B}{P_B}$$

Since the pressure in Flask B (2 atm) is twice that of Flask A (1 atm), there are two times as many moles (and molecules) of gas in Flask B (CO_2) as there are in Flask A (H_2).

d. Since Flask B contains twice as many moles of CO_2 as Flask A contains of H_2, the ratio of masses of gas present are:

$$\frac{\text{Mass (Flask B)}}{\text{Mass (Flask A)}} = \frac{(2 \text{ mole } CO_2)(44 \text{ g } CO_2/\text{mol } CO_2)}{(1 \text{ mol } H_2)(2 \text{ g } H_2/\text{mol} H_2)} = \frac{44}{1}$$

Note that any number of moles of CO_2 and H_2 (in the ratio of 2 : 1) would provide the same answer.

75. Since two gases at the same temperature have the same kinetic energy

$$KE_{O_2} = KE_{CO_2}$$

and since the average $\quad KE = 1/2 \, m\bar{u}^2$

where $\bar{u}$ is the average speed of a molecule, we can write.

$$1/2 \, M_{O_2}\bar{U}_{O_2}^2 = 1/2 \, M_{CO_2}\bar{U}_{CO_2}^2 \quad \text{or} \quad M_{O_2}\bar{U}_{O_2}^2 = M_{CO_2}\bar{U}_{CO_2}^2$$

$$\text{and} \quad \frac{M_{O_2}}{M_{CO_2}} = \frac{\bar{U}_{CO_2}^2}{\bar{U}_{O_2}^2}$$

and solving for the average velocity of CO_2 :

$$\bar{U}_{CO_2}^2 = \frac{M_{O_2}}{M_{CO_2}} \cdot \bar{U}_{O_2}^2$$

Taking the square root of both sides

$$\bar{U}_{CO_2} = \sqrt{\frac{M_{O_2}}{M_{CO_2}}} \cdot \bar{U}_{O_2} = \sqrt{\frac{32.0 \text{ g } O_2 \text{ / mol } O_2}{44.0 \text{ g } CO_2 \text{ / mol } CO_2}} \cdot 4.28 \times 10^4 \text{cm/s}$$

$$= 3.65 \times 10^4 \text{ cm/s}$$

77. The species will have average molecular speeds which are inversely proportional to their molar masses.

$$CH_2F_2 \quad < \quad Ar \quad < \quad N_2 \quad < \quad CH_4$$
$$54 \qquad\qquad 40 \qquad 28 \qquad\quad 16 \qquad \text{(to integral values)}$$

Diffusion and Effusion

79. The rates at which argon and helium gas effuse can be calculated by Graham's Law:

$$\frac{\text{Rate of effusion of He}}{\text{Rate of effusion of Ar}} = \sqrt{\frac{M \text{ of Ar}}{M \text{ of He}}}$$

Since $D = \frac{M}{V}$, and Ar is 10 times as dense as He at the same T and P, the molar mass of Ar is 10 times that of He, and we can write:

$$\frac{\text{Rate of effusion of He}}{\text{Rate of effusion of Ar}} = \sqrt{\frac{10}{1}} = 3.2$$

He effuses 3.2 times faster than Ar

81. Determine the molar mass of a gas which effuses at a rate 1/3 that of He:

$$\frac{\text{Rate of effusion of He}}{\text{Rate of effusion of unknown}} = \sqrt{\frac{M \text{ of unknown}}{M \text{ of He}}}$$

$$\frac{3}{1} = \sqrt{\frac{M \text{ of unknown}}{4.0 \text{ g/mol}}}$$

Squaring both sides gives: $9 = \frac{M}{4.0}$ or M = 36 g/mol

Non-Ideal Gases

83.   The van der Waals Equation may be written

$$\left[P_{obs} + a(\tfrac{n}{V})^2\right]\left[V-bn\right] \;=\; nRT \text{ where a and b are the van der Waals constants for a}$$

specific substance Values for Cl_2 are found in Table 12.2.

so $P_{obs} = \dfrac{nRT}{(V-bn)} - a(\tfrac{n}{V})^2$

$$P_{obs} \;=\; \dfrac{(8.00 \text{ mol})(0.082057 \ \text{L} \cdot \text{atm/K} \cdot \text{mol})(300 \text{ K})}{(4.00 \text{ L} - (0.0562 \text{ L/mol} \cdot 8.00 \text{ mol}))}$$

$$- \dfrac{6.49 \text{ atm} \cdot \text{L}^2}{\text{mol}^2} \cdot (\dfrac{8.00 \text{ mol}}{4.00 \text{ L}})^2$$

$$P_{obs} \;=\; \dfrac{199 \text{ L} \cdot \text{atm}}{3.55 \text{ L}} - 26.0 \text{ atm} \;=\; 55.5 \text{ atm} - 26.0 \text{ atm} = 29.5 \text{ atm}$$

From the ideal gas law: $P = \dfrac{nRT}{V} = \dfrac{(8.00 \text{ mol})(0.082057 \ \text{L} \cdot \text{atm/K} \cdot \text{mol})(300 \text{ K})}{4.00 \text{ L}}$

$$P \;=\; 49.3 \text{ atm}$$

General Questions

85.

	atm	mmHg	kPa	bar
Standard atmosphere:	**1**	$1 \text{ atm} \cdot \dfrac{760. \text{ mmHg}}{1 \text{ atm}}$ $= 760. \text{ mmHg}$	$1 \text{ atm} \cdot \dfrac{101.325 \text{ kPa}}{1 \text{ atm}}$ $= 101.325 \text{ kPa}$	$1 \text{ atm} \cdot \dfrac{1.013 \text{ bar}}{1 \text{ atm}}$ $= 1.013 \text{ bar}$
Partial pressure of N_2 in the atmosphere	$593 \text{ mmHg} \cdot \dfrac{1 \text{ atm}}{760 \text{ mmHg}}$ $= 0.780 \text{ atm}$	**593**	$0.780 \text{ atm} \cdot \dfrac{101.3 \text{ kPa}}{1 \text{ atm}}$ $= 79.0 \text{ kPa}$	$0.780 \text{ atm} \cdot \dfrac{1.013 \text{ bar}}{1 \text{ atm}}$ $=0.790 \text{ bar}$
Tank of compressed H_2	$133 \text{ bar} \cdot \dfrac{1 \text{ atm}}{1.013 \text{ bar}}$ $= 131 \text{ atm}$	$131 \text{ atm} \cdot \dfrac{760. \text{ mmHg}}{1 \text{ atm}}$ $= 99800 \text{ mmHg}$	$131 \text{ atm} \cdot \dfrac{101.3 \text{ kPa}}{1 \text{ atm}}$ $= 13300 \text{ kPa}$	**133**
Atmospheric pressure at top of Mt. Everest	$33.7 \text{ kPa} \cdot \dfrac{1 \text{ atm}}{101.3 \text{ kPa}}$ $= 0.333 \text{ atm}$	$0.333 \text{ atm} \cdot \dfrac{760 \text{ mmHg}}{1 \text{ atm}}$ $= 253 \text{ mmHg}$	**33.7**	$0.333 \text{ atm} \cdot \dfrac{1.013 \text{ bar}}{1 \text{ atm}}$ $= 0.337 \text{ bar}$

87.   Rewriting the ideal gas law we obtain:

$$P \cdot V = n \cdot R \cdot T$$

$$P \cdot V = \frac{mass}{Molar\ mass} \cdot R \cdot T$$

Rearranging this equation gives: $Molar\ mass \cdot P = \frac{mass}{V} \cdot R \cdot T$

and noting that $D = \frac{mass}{V}$ we write: $Molar\ mass = \frac{D \cdot R \cdot T}{P}$

Converting 331 mmHg to atm yields:

$$331\ mmHg \cdot \frac{1\ atm}{760\ mmHg} = 0.436\ atm$$

$$Molar\ mass = \frac{0.855\ \frac{g}{L} \cdot 0.082057\ \frac{L \cdot atm}{K \cdot mol} \cdot 273.2\ K}{0.436\ atm} = 44.0\ \frac{g}{mol}$$

89. Since P and the amount of gas are fixed, the ideal gas law in these situations can be written

$$\frac{P_1 V_1}{T_1} = \frac{P_2 V_2}{T_2} \quad\quad and \quad\quad since\ P1 = P2 \quad\quad \frac{V_1}{T_1} = \frac{V2}{T2}$$

$$\frac{25.5\ mL}{363\ K} = \frac{21.5\ mL}{T_2} \quad and \quad T_2 = \frac{(21.5\ mL)(363\ K)}{25.5\ mL}$$

$$T_2 = 306\ K \quad or \quad 33\ ^\circ C$$

91. The average (or root mean square) speed at -33 °C can be calculated:

$$\bar{U} = \sqrt{\frac{3\ RT}{M}} = \sqrt{\frac{(3)(8.314\ J/K\bullet mol)(240\ K)}{4.00 \times 10^{-3}\ kg/mol}}$$

$$\bar{U} = \sqrt{1.50 \times 10^6\ m^2/s^2} = 1.22 \times 10^3\ m/s$$

If we increase the average speed by 10.0 % — to 1.35×10^3 m/s, we can calculate T

$$\frac{\bar{U}^2 \bullet M}{3\ R} = T = \frac{(1.35 \times 10^3\ m/s)^2 (4.00 \times 10^{-3}\ kg/mol)}{(3)(8.314\ J/K \bullet mol)} = 290\ K$$

or $290 - 273 = 17\ ^\circ C$

93. The partial pressure of each gas can be calculated from

$$P_2 = P_1 \cdot \frac{V_1}{V_2}$$ where V_2 in each case is 5.0 L

Partial pressure of He $= 145 \text{ mmHg} \cdot \frac{3.0 \text{ L}}{5.0 \text{ L}} = 87 \text{ mmHg}$

Partial pressure of Ar $= 355 \text{ mmHg} \cdot \frac{2.0 \text{ L}}{5.0 \text{ L}} = 140 \text{ mmHg}$

$P_{total} = P_{He} + P_{Ar} = 87 \text{ mmHg} + 140 \text{ mmHg} = 227 \text{ mmHg}$

95.   a. $1.0 \text{ L H}_2 \cdot \frac{1 \text{ mol H}_2}{22.4 \text{ L H}_2} = 0.045 \text{ mol H}_2$

b. $1.0 \text{ L Ar} \cdot \frac{1 \text{ mol Ar}}{22.4 \text{ L Ar}} = 0.045 \text{ mol Ar}$

c. $n = \frac{PV}{RT} = \frac{(1 \text{ atm})(1.0 \text{ L})}{(0.082057 \frac{\text{L} \cdot \text{atm}}{\text{K} \cdot \text{mol}})(300 \text{ K})} = 0.041 \text{ mol H}_2$

d. $n = \frac{PV}{RT} = \frac{(900 \text{ mmHg} \cdot \frac{1 \text{ atm}}{760 \text{mmHg}})(1.0 \text{ L})}{(0.082057 \frac{\text{L} \cdot \text{atm}}{\text{K} \cdot \text{mol}})(273 \text{ K})} = 0.053 \text{ mol He}$

The number of molecules may be calculated by multiplying the number of moles by Avogadro's number. However, the number of moles is proportional to the number of molecules, so it is sufficient to note that 1.0 L of **He at 0 °C and 800 mmHg contains the largest number of molecules (of this group of gases)** and **1.0 L of H$_2$ at 27 °C and 760 mmHg contains the smallest number of molecules.** The largest mass can be calculated by multiplying the number of moles of each gas by its molar mass. This is largest for Argon: $(0.045 \text{ mol Ar})(39.9 \text{ g Ar/mol Ar}) = 1.8 \text{ g Ar}$

97. The amount of N$_2$ can be calculated.

$P(\text{Total}) = P(H_2O) + P(N_2)$

$736.0 \text{ mmHg} = 18.7 \text{ mmHg} + P(N_2)$

$717.3 \text{ mmHg} = P(N_2) = 0.944 \text{ atm}$

$$n(N_2) = \frac{(0.944 \text{ atm})(0.295 \text{ L})}{(0.082057 \frac{\text{L} \cdot \text{atm}}{\text{K} \cdot \text{mol}})(294.2 \text{ K})} = 1.15 \times 10^{-2} \text{ mol N}_2$$

According to the equation in which sodium nitrite reacts with sulfamic acid, one mole of $NaNO_2$ produces one mole of N_2.

$$1.15 \times 10^{-2} \text{ mol } N_2 \cdot \frac{1 \text{ mol } NaNO_2}{1 \text{ mol } N_2} \cdot \frac{69.00 \text{ g } NaNO_2}{1 \text{ mol } NaNO_2} = 0.796 \text{ g } NaNO_2$$

$$\text{Weight percentage of } NaNO_2 = \frac{0.796 \text{ g } NaNO_2}{1.232 \text{ g sample}} \times 100 = 64.6\% \ NaNO_2$$

99.   a. Mass of NiO that reacts with ClF_3

Calculate the moles of ClF_3 :

$$n = \frac{PV}{RT} = \frac{(250 \text{ torr})(2.5 \text{ L})}{(62.4 \frac{L \cdot torr}{K \cdot mol})(293 \text{ K})} = 0.034 \text{ mol } ClF_3$$

The mol of NiO that react : $0.034 \text{ mol } ClF_3 \cdot \frac{6 \text{ mol NiO}}{4 \text{ mol } ClF_3} = 0.051 \text{ mol NiO}$

The mass of NiO would be : $0.051 \text{ mol NiO} \cdot \frac{74.7 \text{ g NiO}}{1 \text{ mol NiO}} = 3.8 \text{ g NiO}$

b. Partial pressure of $O_2 + Cl_2$

$$0.051 \text{ mol NiO} \cdot \frac{3 \text{ mol } O_2}{6 \text{ mol NiO}} = 0.026 \text{ mol } O_2$$

$$0.051 \text{ mol NiO} \cdot \frac{2 \text{ mol } Cl_2}{6 \text{ mol NiO}} = 0.017 \text{ mol } Cl_2$$

The pressure can be found by substituting into the Ideal Gas Law:

$$P_{O_2} = \frac{nRT}{V} = \frac{(0.026 \text{ mol } O_2)(62.4 \frac{L \cdot torr}{K \cdot mol})(293 \text{ K})}{2.5 \text{ L}}$$

$$= 187.5 \text{ torr or } 190 \text{ mmHg (to 2 sf)}$$

The same procedure with Cl_2 gives 124 mmHg or 120 mmHg   (to 2 sf)

The total pressure is then $190 + 120 = 310 \text{ mmHg}$

101. To calculate molecules of water in 1 cm^3, let's calculate the number of moles of water in 1 cm^3 $(1 \times 10^{-3} \text{ L})$:

$$n = \frac{PV}{RT} = \frac{(23.8 \text{ torr})(1 \times 10^{-3} \text{ L})}{(62.4 \frac{L \cdot torr}{K \cdot mol})(298 \text{ K})} = 1.28 \times 10^{-6} \text{ mol } H_2O$$

and multiplying by Avogadro's number

$$(1.28 \times 10^{-6} \text{ mol})(\frac{6.022 \times 10^{23} \text{ molecules H}_2\text{O}}{1 \text{ mol}}) = 7.71 \times 10^{17} \text{ molecules}$$

103. Use the Ideal Gas Law to calculate the molar mass of the new gas.

$$PV = \frac{\text{mass}}{M} RT \qquad \text{or} \qquad M = \frac{\text{mass } RT}{PV}$$

$$M = \frac{(0.150 \text{ g})(62.4 \frac{L \cdot torr}{K \cdot mol})(294 \text{ K})}{(17.2 \text{ torr})(1.850 \text{ L})} = 86.5 \text{ g/mol}$$

Since we know the compound **must** have at least one atom of Cl, F, and O, let's subtract the mass of 1 mol of each of these atoms from the molar mass of 86.5.

$$86.5 \text{ g cpd} - (35.5 \text{ g Cl} + 16.0 \text{ g O} + 19.0 \text{ g F}) = 16 \text{ g}$$

So the compound must have an additional oxygen atom — ClO_2F.

105.   a. O_2 (32 g/mol) < B_2H_6 (27.7 g/mol) < H_2O (18g/mol)

—increasing speed⟶

b. Both reaction gases are contained in tanks of equal volume at the same temperature. Therefore, their pressures are proportional to the quantity of gas present. Stoichiometry demands that there must be 1/3 as many moles of B_2H_6 as there are of O_2. Alternatively, the pressure of B_2H_6 must be 1/3 as large as that of O_2. Since the pressure of O_2 is 45 atm, this means that the pressure of B_2H_6 should be 15 atm.

107. The number of moles of He in the balloon can be calculated with the Ideal Gas Law. First calculate the P of He in the balloon:

gauge = total - barometric or

gauge + barometric = total pressure = 22 mmHg + 755 mmHg = 777 mmHg

or 777 torr

$$n = \frac{PV}{RT} = \frac{(777 \text{ torr})(0.300 \text{ L})}{(62.4 \frac{L \cdot torr}{K \cdot mol})(298 \text{ K})} = 0.0125 \text{ mol He}$$

109. The rates of diffusion:

$$\frac{\text{Rate of diffusion of NH}_3}{\text{Rate of diffusion of HCl}} = \frac{\sqrt{M \text{ (HCl)}}}{\sqrt{M \text{ (NH}_3)}} = \frac{\sqrt{36.46 \text{ g/mol}}}{\sqrt{17.03 \text{ g/mol}}} = 1.463$$

The calculation tells us that, if a HCl molecule travels a distance x in a given time, then a NH$_3$ molecule travels a distance 1.463 x. Therefore

$$x + 1.463\, x = 50.\ cm$$

$$x = 20.3 \text{ cm} = \text{ distance traveled by HCl in a given time}$$

$$1.463\, x = 29.7 \text{ cm} = \text{ distance traveled by NH}_3 \text{ in a given time}$$

(or 30. cm to 2 sf)

111.   a. Empirical Formula

1. Find the moles of B and H in the sample of B$_x$H$_y$:

$$0.540 \text{ g H}_2\text{O} \cdot \frac{1 \text{ mol H}_2\text{O}}{18.02 \text{ g}} \cdot \frac{2 \text{ mol H}}{1 \text{ mol H}_2\text{O}} = 0.0599 \text{ mol H}$$

$$0.0599 \text{ mol H} \cdot \frac{1.008 \text{ g H}}{1 \text{ mol H}} = 0.0604 \text{ g H}$$

The amount of B would be: 0.492 g sample - 0.0604 g H = 0.432 g B

$$0.432 \text{ g B} \cdot \frac{1 \text{ mol}}{10.81 \text{ g}} = 0.0400 \text{ mol B}$$

2. Find the mole ratio of H to B:

$$\frac{0.0599 \text{ mol H}}{0.0400 \text{ mol B}} = \frac{3 \text{ H}}{2 \text{ B}} \qquad \text{indicating an empirical formula } = \text{ B}_2\text{H}_3$$

b. Molecular Formula

1. Find the molar mass:

$$n = \frac{PV}{RT} = \frac{(0.130 \text{ atm})(0.120 \text{ L})}{(0.082057 \frac{\text{L} \cdot \text{atm}}{\text{K} \cdot \text{mol}})(296 \text{ K})} = 6.41 \times 10^{-4} \text{ mol}$$

$$\frac{0.0631 \text{ g}}{6.41 \times 10^{-4} \text{ mol}} = 98.4 \text{ g/mol}$$

2. The empirical formula would have a mass of 24.65 g.

$$\text{So } \frac{98.4 \text{ g/mol}}{24.65 \text{ g/formula unit}} = 3.99 \text{ formula units per mole}$$

Therefore, the molecular formula is (B$_2$H$_3$)$_4$ = B$_8$H$_{12}$

Summary Questions

119.   a. Valence electrons for ClO_2 : $1(7) + 2(6) = 19$ electrons

b. Electron dot structure for ClO_2^- :   (20 electrons)

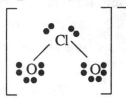

c. The hybridization for Cl is **sp^3** (4 electron pairs attached). The ion has a **bent shape**.

d. The molecule ozone has a "central" oxygen atom with 3 electron groups attached. (1 lone pair; 1 double bond; 1 single bond — with the double bond being delocalized between the three O atoms.) The predicted geometry would give a bond angle of about 120 ° (measured is approximately 117 °). The bond angle for ClO_2^- would have a smaller angle (approximately 109°).

e. To determine the mass of ClO_2, determine the limiting reagent (if there is one).

Moles of ClO_2:

$$15.6 \text{ g } NaClO_2 \cdot \frac{1 \text{ mol } NaClO_2}{54.99 \text{ g } NaClO_2} = 0.284 \text{ mol } NaClO_2$$

Moles of Cl_2:

$$n = \frac{(1050 \text{ torr})(1.45 \text{ L})}{(62.4 \frac{L \cdot torr}{K \cdot mol})(295 \text{ K})} = 0.0827 \text{ mol } Cl_2$$

Note (from the balanced equation) that each mol of Cl_2 requires 2 mol of $NaClO_2$ (which we have in excess). So Cl_2 will limit the amounts of product obtainable.

The mass of ClO_2 obtainable is:

$$0.0827 \text{ mol } Cl_2 \cdot \frac{2 \text{ mol } ClO_2}{1 \text{ mol } Cl_2} = 0.165 \text{ mol } ClO_2$$

$$0.165 \text{ mol } ClO_2 \cdot \frac{67.45 \text{ g } ClO_2}{1 \text{ mol } ClO_2} = 11.2 \text{ g } ClO_2$$

The pressure the ClO_2 will exert is:

$$P = \frac{nRT}{V} = \frac{(0.165 \text{ mol } ClO_2)(62.4 \frac{L \cdot torr}{K \cdot mol})(298 \text{ K})}{(1.25 \text{ L})}$$

$$= 2460 \text{ torr or } 2460 \text{ mmHg (or } 3.24 \text{ atm)}$$

Chapter 13:
Intermolecular Forces, Liquids, and Solids

Intermolecular Forces

16. Must overcome both **dipole-dipole forces** and **hydrogen bonds**. To separate the KCl ions, **ion-ion** forces between K^+ and Cl^- must be overcome. The K^+ and Cl^- ions are attracted to the water molecules by **ion-dipole** forces.

18. **Ion - dipole** forces bind the waters of hydration (dipole) to the $NiSO_4$ species (ions).

20. To convert <u>species</u> from a liquid to a gas one must overcome <u>intermolecular</u> forces.

<u>species</u>	<u>intermolecular</u>
a. liquid O_2	induced dipole-induced dipole
b. mercury	induced dipole-induced dipole
c. methyl iodide	dipole-dipole
d. ethanol	hydrogen bonding and dipole-dipole

22. Increasing strength of intermolecular forces:

    $Ne \; < \; CH_4 \; < \; CO \; < \; CCl_4$

    Neon and methane are nonpolar species and possess only induced dipole-induced dipole interactions. Neon has a smaller molar mass than CH_4, and therefore weaker London (dispersion) forces. Carbon monoxide is a polar molecule. Molecules of CO would be attracted to each other by dipole-dipole interactions, but the CO molecule is not a very strong dipole. The CCl_4 molecule is a non-polar molecule, but very heavy (when compared to the other three). Hence the greater London forces that accompany larger molecules would result in the strongest attractions of this set of molecules.

    The lower molecular weight molecules with weaker interparticle forces should be gases at 25 °C and 1 atmosphere: Ne, CH_4, CO.

24. H_2S molecules interact through dipole-dipole forces. Water forms the stronger hydrogen bonds.

26. The higher boiling point for iodomethane indicates stronger intermolecular forces between CH3I molecules than for the CH3Cl analogs. While the C-I bond isn't as polar as the C-Cl bond, the more polarizable I atom creates stronger induced dipole-induced dipole forces.

28. Compounds which are capable of forming hydrogen bonds with water are those containing polar O-H bonds and lone pairs of electrons on N,O, or F.
    (a) CH_3-O-CH_3 no; no "polar H's" and the C-O bond is not very polar
    (b) CH_4 no
    (c) HF   yes: lone pairs of electrons on F and a "polar hydrogen".
    (d) CH_3COOH yes: lone pairs of electrons on O atoms, and a "polar hydrogen" attached to one of the oxygen atoms
    (e) Br_2 no
    (f) CH_3OH yes: "polar H" and lone pairs of electrons on O

30.   a. LiCl would be more strongly hydrated than CsCl, since the smaller Li^+ would be more strongly attracted to water than Cs^+.

     b. $Mg(NO_3)_2$ will be more likely hydrated since Mg^{2+} will be more strongly attracted to water than Na^+ ($+2 > +1$).

     c. $NiCl_2$ — for the same reason as Mg^{2+} in part b.

Liquids

32. Heat required is:
$$0.500 \text{ mL} \cdot \frac{13.6 \text{ g Hg}}{1 \text{ mL}} \cdot \frac{1 \text{ mol Hg}}{200.59 \text{ g Hg}} \cdot \frac{59.11 \text{ kJ}}{1 \text{ mol Hg}} = 2.00 \text{ kJ}$$

34. From Figure 13.21:

     a. The equilibrium vapor pressure of diethyl ether at room temperature is slightly greater than 400 mmHg.

     b. In order of increasing intermolecular forces (weakest to strongest)
         ether < alcohol < water

c. At 40 °C and 400 mmHg ether is a gas, while water and alcohol are predominantly in the liquid state.

36. The vapor pressure of $(C_2H_5)_2O$ at 30. °C is **600 mmHg**.

Calculate the amount of $(C_2H_5)_2O$ to furnish this vapor pressure at 30.°C.

$$n = \frac{PV}{RT} = \frac{600 \text{ mm} \cdot \dfrac{1 \text{ atm}}{760 \text{ mm}} \cdot 0.1 \text{ L}}{0.082057 \dfrac{L \cdot atm}{K \cdot mol} \cdot 303 \text{ K}} = 3.2 \times 10^{-3} \text{ mol}$$

The total mass of $(C_2H_5)_2O$ [FW = 74.1 g] needed to create this pressure is about 0.23 g. As the flask is cooled from 30.°C to 0 °C, **some of the gaseous ether will condense** to form liquid ether.

38.   a. From the figure, we can read the vapor pressure of CS_2 as approximately 630 mmHg and for nitromethane as approximately 80 mmHg.

b. The principle intermolecular forces for CS_2 (a non-polar molecule) are **induced dipole-induced dipole**; for nitromethane they are **dipole-dipole**.

c. The normal boiling point from the figure for CS_2 is 46 °C and for CH_3NO_2, 101 °C.

d. The temperature at which the vapor pressure of CS_2 is 600 mmHg is about 39 °C.

e. The vapor pressure of CH_3NO_2 is 60 mmHg at approximately 35 °C.

40. The critical temperature for propane (96.7 °C) is well above room temperature, so propane can be liquefied. Liquefied propane is used for heating and cooking. One can frequently see vessels (tractor-trailers) by which LPG is transported.

42. This compound would have the formula AB since each black square (A) has one corresponding white square (B).

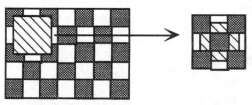

The diagonally crossed square is the "outline" of the unit cell. Note that it contains 1 solid square and quarters of four solid squares.

The diagonally crossed square also contains quarters of four "white" squares, making the overall ratio 4 "A" and 4 "B" squares, or a unit cell of AB.

44. To calculate the radius of a calcum atom we must find the diagonal distance of the unit cell—a distance which contains 4 Ca radii (1 atom + 2 "half-atoms"). To calculate the volume of the cell we need to determine the mass of the uit cell — which contains 4 Ca atoms The **mass of one Ca atom** is:

$$\frac{40.078 \text{ g}}{1 \text{mol}} \cdot \frac{1 \text{ mol}}{6.022 \times 10^{23} \text{ atoms}} = 6.655 \times 10^{-23} \text{ g/atom}$$

So **one unit cell has a mass** of:

$$\frac{4 \text{ Ca atoms}}{\text{unit cell}} \cdot 6.655 \times 10^{-23} \text{ g/atom} = 2.66 \times 10^{-22} \text{ g/unit cell}$$

With a density of $1.54 \frac{\text{g Ca}}{\text{cm}^3}$, the **volume of the cell** would be:

$$2.662 \times 10^{-22} \text{ g/unit cell} \cdot \frac{1 \text{ cm}^3}{1.54 \text{ g}} = 1.73 \times 10^{-22} \text{ cm}^3$$

The **edge of the cell** would be $(\text{volume})^{1/3}$ or $(1.73 \times 10^{-22} \text{ cm}^3)^{1/3}$

$$= 5.57 \times 10^{-8} \text{ cm} = \ell$$

The body diagonal of the unit cell would be $\sqrt{2} \cdot \ell$ or 7.89×10^{-8} cm. Since this diagonal distance contains 4 radii, the **atomic radius would be**

$$1.97 \times 10^{-8} \text{ cm or } \textbf{197 pm}.$$

46. Let's calculate the density for simple cubic, body-centered cubic, and face-centered cubic lattice. The number of vanadium atoms in the unit cell for each lattice is 1 (simple) 1 (body-centered) or 4 (face-centered).

First, let's calculate the volume of a body-centered cell. The body diagonal (d) of such a cell would be $d^2 = 3\ell^2$

And since d would contain 4 radii:

$$\left(4 \times 132 \text{ pm/radius} \times \frac{1 \times 10^2 \text{ cm}}{1 \times 10^{12} \text{ pm}}\right)2 = 3\ell^2$$

$$2.79 \times 10^{-15} \text{ cm}^2 = 3\ell^2$$

$$3.05 \times 10^{-8} = \ell \qquad \text{The volume would be } \ell^3 \text{ or } 2.8 \times 10^{-23} \text{ cm}^3$$

The mass of one Vanadium ion is:

$$\frac{50.9415 \text{ g V}}{1 \text{ mol V}} \cdot \frac{1 \text{ mol V}}{6.022 \times 10^{23} \text{ atoms}} = 8.459 \times 10^{-23} \frac{\text{g V}}{\text{atom}}$$

So the density of a body-centered cell would be $\dfrac{(2 \text{ atom})(8.459 \times 10^{-23} \frac{\text{g V}}{\text{atom}})}{2.8 \times 10^{-23} \text{ cm}^3}$

$$= 5.97 \text{ g/cm}^3$$

For a face-centered lattice the edge (ℓ) wouldbe related to the face diagonal (f)

$$f^2 = \ell^2 + \ell^2$$

Since f contains 4 radii as calculated earlier

$$f^2 = 2.79 \times 10^{-15} \text{ cm}^2 = 2\ell^2$$
$$= 3.73 \times 10^{-8} \text{ cm} = \ell$$

And a corresponding volume of ℓ^3 or 5.20×10^{-23} cm^3

and a density of $\dfrac{(4 \text{ atoms})(8.459 \times 10^{-23} \frac{\text{g V}}{\text{atom}})}{5.20 \times 10^{-23} \text{ cm}^3} = 6.50 \text{ g/cm}^3$

For a simple cubic lattice, the vanadium atoms at the corners of the unit cell are in contact. The length of the side would then be equal to two radii. The volume of the cube would be:

$$[(2 \text{ radii})(132 \frac{\text{pm}}{\text{radii}})(\frac{1 \times 10^2 \text{ cm}}{1 \times 10^{12} \text{ pm}})]^3 = 1.84 \times 10^{-23} \text{ cm}^3$$

With 1 atom of V per unit cell, the density would be

$$\frac{(8 \text{ atoms})(\frac{1 \text{ atom}}{8 \text{ unit cells}})(8.459 \times 10^{-23} \frac{\text{g V}}{\text{atom}})}{1.84 \times 10^{-23} \text{ cm}^3} = 4.60 \text{ g/cm}^3$$

So the V atom appears to use the **body-centered cubic lattice**.

48. Solid's density $= 0.77$ g/cm^3 Edge of unit cell $= 4.086 \times 10^{-8}$ cm

Volume of unit cell $= (4.086 \times 10^{-8} \text{ cm})^3 = 6.822 \times 10^{-22}$ cm^3

Mass $=$ Density $\cdot$ Volume $= 0.77$ g/cm^3 $\cdot$ 6.822×10^{23} cm$^3 = 5.3 \times 10^{-23}$ g

1 LiH ion pair has a mass of:

$$\frac{7.94 \text{ LiH}}{1 \text{ mol LiH}} \cdot \frac{1 \text{ mol LiH}}{6.022 \times 10^{23} \text{ LiH ion pairs}} = 1.32 \times 10^{-23} \text{ g/LiH ion pair}$$

$$5.3 \times 10^{-23} \text{ g/unit cell} \cdot \frac{1 \text{ LiH ion pair}}{1.32 \times 10^{-23} \text{ g}} = 4 \text{ ion pairs/ unit cell}$$

The **face-centered cubic lattice is appropriate** for LiH.

50. To determine the perovskite formula, determine the number of each atom belonging **uniquely** to the unit cell shown. The Ca atom is wholly contained within the unit cell. There are Ti atoms at each of the eight corners. Since each of these atoms belong to eight unit cells, the portion of each Ti atom belonging to the pictured unit cell is 1/8 so 8 Ti atoms x 1/8 = 1 Ti atom. The O atoms on an edge belong to 4 unit cells, so the fraction contained within the pictured cell is 1/4. There are twelve such O atoms, leading to 12 x 1/4 = 3 O atoms.

52. Note that each corner is occupied by a Zn atom. Also each face is occupied by a Zn atom. The tetrahedral holes (four of them) are occupied by A atoms. The net formula would be

    8 Zn (corner) x 1/8 = 1
    6 Zn (faces) x 1/2 = 3/4
    4 S (tetrahedral holes) x 1 = 4 $\Rightarrow$ ZnS

Molecular and Network Solids

54. For the unit cell shown there are 8 corner atoms (1/8 in the cell), 6 face atoms (1/2 in the cell), and 4 atoms wholly within the cell, for a total of **8 carbon atoms**.

    a. The volume of the unit cell may be calculated if we first calculate the masses of the atoms involved:

    $$\frac{12.01 \text{ g C atom}}{1 \text{ mol C atom}} \cdot \frac{1 \text{ mol C atom}}{6.022 \times 10^{23} \text{ C atom}} \cdot \frac{8 \text{ atoms}}{1 \text{ unit cell}} = 1.60 \times 10^{-22} \frac{\text{g C atom}}{\text{unit cell}}$$

    $$D = \frac{M}{V} \quad \text{and} \quad V = \frac{M}{D} = \frac{1.60 \times 10^{-22} \text{ g/unit cell}}{3.51 \text{ g/cm}^3} = 4.55 \times 10^{-23} \text{ cm}^3/\text{unit cell}$$

    b. The length of an edge: Volume $= l^3$
    $$4.55 \times 10^{-23} \text{ cm}^3 = l^3$$
    $$3.57 \times 10^{-8} \text{ cm} = l \text{ or } 3.57 \times 10^{-8} \text{ cm} \cdot \frac{1.00 \times 10^{12} \text{ pm}}{1 \times 10^2 \text{ cm}} = 357 \text{ pm}$$

Phase Changes

56.   a. The positive slope of the solid/liquid equilibrium line means the liquid CO_2 is **less dense** than solid CO_2.

b. At 5 atm and 0 °C, CO_2 is in the **gaseous phase**.

c. The phase diagram for CO_2 shows the critical pressure for CO_2 to be 73 atm, and the critical temperature to be +31 °C.

58. (1) Heat evolved when 20.0 g of freon is cooled from +40 °C to - 29.8 °C :

$$q = 20.0 \text{ g } CCl_2F_2 \cdot \frac{1 \text{ mol } CCl_2F_2}{120.915 \text{ g } CCl_2F_2} \cdot 117.2 \frac{J}{mol \cdot K} \cdot (-69.8 \text{ K}) = -1350 \text{ J}$$

cooling gas to boiling point

(2) Heat evolved when freon condenses from gaseous to liquid state at the boiling point:

$$q = 20.0 \text{ g } CCl_2F_2 \cdot \frac{1 \text{ mol } CCl_2F_2}{120.915 \text{ g } CCl_2F_2} \cdot \frac{-20.11 \text{ kJ}}{1 \text{ mol } CCl_2F_2} = -3.326 \text{ kJ } (-3.33 \times 10^3 \text{ J})$$

(3) Heat evolved as the liquid freon cools from - 29.8°C to - 40°C:

$$\text{Heat evolved} = 20.0 \text{ g } CCl_2F_2 \cdot \frac{1 \text{ mol } CCl_2F_2}{120.915 \text{ g } CCl_2F_2} \cdot \frac{72.3 \text{ J}}{mol \cdot K} \cdot (-10.2 \text{ K}) = -122 \text{ J}$$

The quantity of heat evolved is then: $1350 \text{ J} + 3.33 \times 10^3 \text{ J} + 122 \text{ J}$ or 4802 J (4.80 kJ)

Physical Properties of Solids

60. Given that the attraction between cations and anions varies inversely with the square of the distance between them, the smaller Mg^{2+} ion would be closer to the Cl^- ions than the larger Ba^{2+} ion, so we anticipate that **$MgCl_2$ would have the greater lattice energy**.

62. Since the iodide ion is larger than the fluoride ion, the ion distances would be greater in CsI, and we expect **CsF to have a higher melting point than CsI**.

64. The heat evolved when 15.5 g of benzene freezes at 5.5 °C :

$$15.5 \text{ g benzene} \cdot \frac{1 \text{ mol benzene}}{78.1 \text{ g benzene}} \cdot \frac{9.95 \text{ kJ}}{1 \text{ mol benzene}} = -1.97 \text{ kJ}$$

Note once again the negative sign indicates that heat is evolved.

The quantity of heat needed to remelt this 15.5 g sample of benzene would be +1.97 kJ.

General Questions

66. Increasing strength of intermolecular forces :

$$Ar \ < \ CO_2 \ < \ CH_3OH \ < \ CaO$$

Argon and CO_2 are nonpolar species and possess only induced dipole-induced dipole (London) interactions. Ar has a smaller mass than CO_2, so London forces are expected to be less. The polar molecular CH_3OH is capable of forming the stronger hydrogen-bonds. Lastly, CaO is an ionic compound, and the ion-ion interactions are stronger than the hydrogen bonds of CH_3OH.

68.

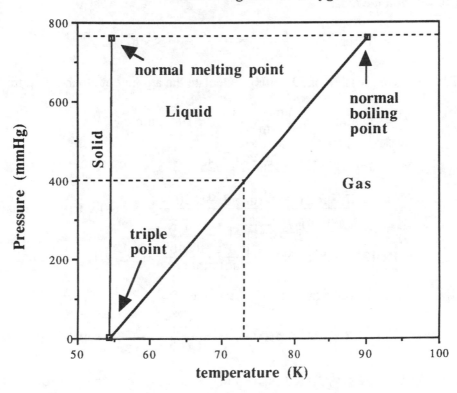

Phase Diagram of Oxygen

The estimated vapor pressure at 77 K is approximately 400 mmHg. The very slight positive slope of the solid/liquid equilibrium line indicates the solid is more dense than the liquid.

70. Acetone readily absorbs water owing to hydrogen bonding between the C = O oxygen atom and the O—H bonds of water.

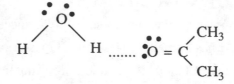

72. $BeSO_4$ will be more likely hydrated, owing to the larger charge density of Be^{2+} over Ba^{2+}.

74. Volume of room $= 3.0 \times 10^2$ cm $\cdot$ 2.5×10^2 cm $\cdot$ 2.5×10^2 cm $= 1.9 \times 10^7$ cm^3
    Convert volume to L:

$$1.9 \times 10^7 \text{ cm}^3 \cdot \frac{1 \text{ L}}{1.0 \times 10^3 \text{ cm}^3} = 1.9 \times 10^4 \text{ L}$$

To produce a pressure of 59 mmHg, calculate the amount of ethanol required.

$$P = 59 \text{ mm} \cdot \frac{1 \text{ atm}}{760 \text{ mm}} = 7.8 \times 10^{-2} \text{ atm}$$

$$V = 1.9 \times 10^4 \text{ L} \qquad \text{and } T = 25 + 273 = 298 \text{ K}$$

$$\frac{PV}{RT} = \frac{7.8 \times 10^{-2} \text{ atm} \cdot 1.9 \times 10^4 \text{ L}}{0.082057 \frac{\text{L} \cdot \text{atm}}{\text{K} \cdot \text{mol}} \cdot 298 \text{ K}} = 6.0 \times 10^1 \text{ mol ethanol}$$

$$6.0 \times 10^1 \text{ mol ethanol} \cdot \frac{46.1 \text{ g ethanol}}{1 \text{ mol ethanol}} = 2.8 \times 10^3 \text{ g ethanol}$$

This mass of ethanol would occupy a volume of:

$$2.8 \times 10^3 \text{ g C}_2\text{H}_5\text{OH} \cdot \frac{1 \text{ cm}^3}{0.785 \text{ g}} = 3.5 \times 10^3 \text{ cm}^3$$

As only 1 L of C_2H_5OH (1.0×10^3 cm^3) was introduced into the room, **all the ethanol would evaporate**.

76. The **viscosity of ethylene glycol would be predicted to be greater** than that of ethanol since the glycol possesses two O-H groups per molecule while ethanol possesses one.

78.

Higher boiling Substance		Intermolecular Forces
a.	ICl	dipole-dipole stronger than induced dipole (Br_2)
b.	Kr	induced dipole-induced dipole greater than larger atoms.
c.	CH_3CH_2OH	hydrogen bonding (stronger than dipole-dipole)
d.	piperidine	hydrogen bonding (not possible for N-methylpyrrolidine)

80. For a simple cubic unit cell, each corner is occupied by an atom or ion. Each of these is contained within EIGHT unit cells contributing 1/8 to each. Within one unit cell, therefore, there is ($8 \times \frac{1}{8}$) 1 atom or ion. The volume occupied by that one net atom would be equal to $4/3 \pi r^3$ with r representing the radius of the spherical atom or ion. The volume of the unit cell may be calculated by noting that the length of one side of the cell (an edge) corresponds to two radii (2r)-- since the spheres touch. The volume of this cube is therefore $(2r)^3$. The empty space within the cell is therefore:

$(2r)^3 - 4/3 \pi r^3$ and the fraction of space unoccupied is:

$$\frac{(8 - 4/3 \pi)r^3}{8r^3} = \frac{8 - 4/3 \pi}{8} = 0.476 \text{ or approximately } 48 \%$$

82.   a.

water   ethanol

Water has **two polar hydrogen atoms** and **two lone pairs of electrons** on the O, whereas **ethanol has only one polar hydrogen atom**, leading to less extensive hydrogen bonding in ethanol, and a lower boiling point.

b. The extensive hydrogen bonding in HF makes possible (F—H—F) ions, and therefore salts of the HF_2^- ion.

c. The water molecules interact extensively with the ethanol molecules — through hydrogen bonding — and can therefore occupy less than the anticipated 100 mL.

Conceptual Questions

85. $CaCl_2$ cannot have the NaCl structure. As shown in Figure 13.31 of your text, the cubic structure possesses 4 net lattice ions (occupied by anions) per face-centered lattice and 4 octahedral holes (occupied by cations). This is suitable for salts of a 1:1 composition.

87. CO_2 will be a larger molecule than O_2 or N_2, and therefore be more polarizable. This polarizability leads to a reaction with water, forming the HCO_3^- ion:
$$CO_2(g) + H_2O(\ell) \longrightarrow HCO_3^-(aq) + H^+(aq)$$
The formation of bicarbonate ion increases the amount of CO_2 that dissolves in the sea water.

89. The fluoride ion is considerably smaller than the other halide ions ($F^- = 119$ pm, $Cl^- = 167$ pm). The smaller radius results in a smaller distance between the fluoride ion and the alkali metal cation, and an increased attraction —hence the greater energy needed to disrupt the attraction.

91.   a. The normal boiling point is the temperature at which the vapor pressure is equal to atmospheric pressure (1 atm) or **approximately -28 °C.**

          b. At a temperature of 25 °C, the substance should exert a pressure of ~ 6.5 atmospheres.

          c. The rapid initial flow is due to the great pressure inside the cylinder at 25 °C (about 6.5 atm). As the gas escapes, additional CCl_2F_2 vaporizes —and with each gram extracts 165 kJ of heat from the bulk liquid, resulting in the rapid cooling of the liquid inside the tank. As water vapor from the surrounding air contacts the colder surface, condensation of the water (g) to water (ℓ) and (s) occur.

          d. Knocking off the top of a cylinder is never recommended. The dry ice would reduce the vapor pressure of the CCl_2F_2 and slow the emptying of the cylinder.

92. Alternative unit cells:

 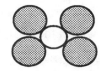
In each case, there is a net ratio of one cation to one anion — with a formula MX, e.g. NaCl.

93.   a. The electron dot structure for SO_2:

(i) The OSO angle is approximately 120°. (Slightly less due to the lone pair on S).

(ii) The electron-pair geometry is trigonal planar and the molecular geometry is bent.

(iii) The S atom is using sp^2 hybrid orbitals.

b. The forces binding SO_2 molecules to each other are **dipole-dipole** forces since SO_2 molecules are polar.

c. Listed in order of increasing intermolecular forces:

$CH_4 < NH_3 < SO_2 < H_2O$

For H_2O hydrogen bonding is possible. SO_2 is a relatively heavy molecule that will exhibit dipole-dipole forces. The lighter NH_3 molecule will have the hydrogen bonding forces that would be absent in CH_4.

d. Enthalpy change for SO_2 (g) $\rightarrow$ SO_3 (g):

$$\Delta H_{rxn} = [\Delta H_f\ SO_3\ (g)] - [\Delta H_f\ SO_2\ (g) + \Delta H_f\ O_2\ (g)]$$
$$= (-395.72\ kJ/mol) - (-296.83\ kJ/mol)$$
$$= -98.89\ kJ/mol$$

Enthalpy change for H_2SO_4 formation:

$$\Delta H_{rxn} = [\Delta H_f\ H_2SO_4\ (aq)] - [\Delta H_f\ SO_3\ (g) + \Delta H_f\ H_2O\ (l)]$$
$$= -909.27\ kJ/mol - [-395.72\ kJ/mol + (-285.83\ kJ/mol)]$$
$$= -227.72\ kJ/mol$$

Chapter 14:
Solutions and Their Behavior

Concentration Units

13. For a solution containing 2.56 g of $C_4H_6O_5$ in 500.0 g of water, the molarity is:

$$2.56 \text{ g } C_4H_6O_5 \cdot \frac{1 \text{ mol } C_4H_6O_5}{134.1 \text{ g } C_4H_6O_5} \cdot \frac{1}{0.500 \text{ L}} = 0.0382 \text{ M}$$

The molality is:

$$2.56 \text{ g } C_4H_6O_5 \cdot \frac{1 \text{ mol } C_4H_6O_5}{134.1 \text{ g } C_4H_6O_5} \cdot \frac{1}{0.500 \text{ kg}} = 0.0382 \text{ molal}$$

The mole fraction of malic acid is:

$$500.0 \text{ g water} \cdot \frac{1 \text{ mol water}}{18.015 \text{ g water}} = 27.75 \text{ mol } H_2O$$

and the moles of malic acid calculated above are: 0.0191 mol $C_4H_6O_5$

giving a mf malic acid of $\dfrac{0.0191 \text{ mol } C_4H_6O_5}{0.0191 \text{ mol } C_4H_6O_5 + 27.75 \text{ mol } H_2O} = 6.88 \times 10^{-4}$

and a mf water of (1 - 0.000680) or 0.9993

The weight percent of malic acid is:

$$\frac{2.54 \text{ g } C_4H_6O_5}{2.54 \text{ g } C_4H_6O_5 + 500.0 \text{ g } H_2O} \cdot 100 = 0.509 \text{ \% malic acid}$$

15. Complete the following transformations for

KI:

Weight percent:

$$\frac{0.15 \text{ mol KI}}{1 \text{ kg solvent}} \cdot \frac{166.0 \text{ g KI}}{1 \text{ mol KI}} = \frac{24.9 \text{ g KI}}{1 \text{ kg solvent}}$$

$$\frac{24.9 \text{ g KI}}{1000 \text{ g solvent} + 24.9 \text{ g KI}} \cdot 100 = 2.4 \text{ \% KI}$$

Mole fraction:

$$1000 \text{ g } H_2O = 55.51 \text{ mol } H_2O$$

$$X_{KI} = \frac{0.15 \text{ mol KI}}{55.51 \text{ mol } H_2O + 0.15 \text{ mol KI}} = 2.7 \times 10^{-3}$$

C_2H_5OH:

Molality:

$$\frac{3.0 \text{ g } C_2H_5OH}{100 \text{ g solution}} \cdot \frac{1 \text{ mol } C_2H_5OH}{46.07 \text{ g } C_2H_5OH} \cdot \frac{100 \text{ g solution}}{97 \text{ g solvent}} \cdot \frac{1000 \text{ g solvent}}{1 \text{ kg solvent}}$$

$$= 0.67 \text{ molal}$$

Mole fraction:

$$\frac{3.0 \text{ g } C_2H_5OH}{1} \cdot \frac{1 \text{ mol } C_2H_5OH}{46.07 \text{ g } C_2H_5OH} = 0.065 \text{ mol } C_2H_5OH$$

and for water : $\dfrac{97 \text{ g } H_2O}{1} \cdot \dfrac{1 \text{ mol } H_2O}{18.02 \text{ g } H_2O} = 5.38 \text{ mol } H_2O$

$$X_{C_2H_5OH} = \frac{0.065 \text{ mol } C_2H_5OH}{5.38 \text{ mol } H_2O + 0.065 \text{ mol } C_2H_5OH} = 0.012$$

$C_{12}H_{22}O_{11}$:

Weight percent:

$$\frac{0.10 \text{ mol } C_{12}H_{22}O_{11}}{1 \text{ kg solvent}} \cdot \frac{342.3 \text{ g } C_{12}H_{22}O_{11}}{1 \text{ mol } C_{12}H_{22}O_{11}} = \frac{34.2 \text{ g } C_{12}H_{22}O_{11}}{1 \text{ kg solvent}}$$

$$\frac{34.2 \text{ g } C_{12}H_{22}O_{11}}{1000 \text{ g } H_2O + 34.2 \text{ g } C_{12}H_{22}O_{11}} \times 100 = 3.3 \% \ C_{12}H_{22}O_{11}$$

Mole fraction:

$$X_{C_{12}H_{22}O_{11}} = \frac{0.10 \text{ mol } C_{12}H_{22}O_{11}}{55.51 \text{ mol } H_2O + 0.10 \text{ mol } C_{12}H_{22}O_{11}} = 1.8 \times 10^{-3}$$

17.   a. To prepare a solution that is 0.200 m $NaNO_3$:

$$\frac{0.200 \text{ mol } NaNO_3}{1 \text{ kg } H_2O} \cdot \frac{0.250 \text{ kg } H_2O}{1} \cdot \frac{84.99 \text{ g } NaNO_3}{1 \text{ mol } NaNO_3} = 4.25 \text{ g } NaNO_3$$

The mole fraction of $NaNO_3$ in the resulting solution:

$$\frac{250. \text{ g } H_2O}{1} \cdot \frac{1 \text{ mol } H_2O}{18.02 \text{ g } H_2O} = 13.9 \text{ mol } H_2O$$

$$X_{NaNO_3} = \frac{0.0500 \text{ mol } NaNO_3}{0.0500 \text{ mol } NaNO_3 + 13.9 \text{ mol } H_2O} = 0.00359$$

19. The mole fraction (mf) of glycol is 0.125 which means the mf (water) is 0.875.

The number of moles of water contained in 950. g H_2O

$$950. \text{ g } H_2O \cdot \frac{1 \text{ mol } H_2O}{18.02 \text{ g } H_2O} = 52.7 \text{ mol } H_2O$$

Since 27.8 mol H_2O corresponds to a mf of 0.875, we can write the ratio:

$$\frac{52.7 \text{ mol}}{0.875 \text{ mf}} = \frac{x}{0.125 \text{ mf}}$$

$$7.53 = x \text{ mol glycol}$$

which would correspond to

$$7.53 \text{ mol } C_2H_4(OH)_2 \cdot \frac{62.07 \text{ g } C_2H_4(OH)_2}{1 \text{ mol } C_2H_4(OH)_2} = 468 \text{ g } C_2H_4(OH)_2$$

The molality of $C_2H_4(OH)_2$ in the solution is then

$$\frac{7.53 \text{ mol } C_2H_4(OH)_2}{0.950 \text{ kg } H_2O} = 7.93 \text{ molal}$$

21.   For the compound K_2CO_3:

$$0.0125 \text{ molal solution} = \frac{x \text{ mol } K_2CO_3}{0.250 \text{ kg } H_2O} \text{ and}$$

$$x = 3.13 \times 10^{-3} \text{ mol } K_2CO_3 \text{ or } 0.432 \text{ g } K_2CO_3$$

Since 250. g H_2O corresponds to 13.9 mol H_2O,

$$\text{the mf}(K_2CO_3) = \frac{3.13 \times 10^{-3} \text{ mol } K_2CO_3}{3.13 \times 10^{-3} \text{ mol } K_2CO_3 + 13.9 \text{ mol } H_2O} = 2.25 \times 10^{-4}$$

For the compound C_2H_5OH:

13.5 g C_2H_5OH = 0.293 mol and 150. g H_2O = 8.32 mol

The molality of the solution is: $\dfrac{0.293 \text{ mol}}{0.150 \text{ kg}} = 1.95 \text{ molal}$

The mf (ethanol) of the solution is: $\dfrac{0.293}{0.421 + 8.32} = 0.0340$

For the compound $NaNO_3$:

The mf (water) = 1 - 0.0934 = 0.9066

and 555 g water = $555 \text{ g } H_2O \cdot \dfrac{1 \text{ mol } H_2O}{18.02 \text{ g } H_2O} = 30.8 \text{ mol } H_2O$

We can write: $\dfrac{\text{mol } H_2O}{\text{mol } H_2O + \text{mol } NaNO_3}$ = 0.9066 and substituting

$\dfrac{30.8}{30.8 + x}$ = 0.9066 so x = 3.17 and 3.17 mol of $NaNO_3$ = 270. g $NaNO_3$

For the molality:

$$\dfrac{3.17 \text{ mol of } NaNO_3}{0.555 \text{ kg } H_2O} = 5.72 \text{ molal}$$

23.   a. Calculate the molality of a solution of 12.0 M HCl with a density of 1.18 g/cm^3:

$$\dfrac{1.18 \text{ g HCl}}{1 \text{ mL}} \cdot \dfrac{1000 \text{ mL}}{1 \text{ L}} = 1180 \text{ g/L}$$

The mass of HCl in 12.0 M HCl: $\dfrac{12.0 \text{ mol HCl}}{1 \text{ L}} \cdot \dfrac{36.5 \text{ g HCl}}{1 \text{ mol HCl}} = 438 \text{ g HCl/L}$

The mass of water in the solution is: 1180 g solution - 438 g HCl = 742 g H_2O

The molality of HCl = $\dfrac{12.0 \text{ mol HCl}}{0.742 \text{ kg}} = 16.2 \text{ molal}$

b. Calculate the weight percent of HCl:

$$\dfrac{438 \text{ g HCl}}{1180 \text{ g solution}} \times 100 = 37.1 \text{ \% HCl}$$

25.   a. Mole fraction of NaOH:

$$\dfrac{10.7 \text{ mol NaOH}}{1 \text{ kg solvent}} \cdot \dfrac{1\text{kg solvent}}{1000 \text{ g solvent}} \cdot \dfrac{18.02 \text{ g solvent}}{1 \text{ mol solvent}} = \dfrac{10.7\text{mol NaOH}}{55.5 \text{ mol } H_2O}$$

$$\dfrac{10.7 \text{ mol NaOH}}{55.5 \text{ mol } H_2O + 10.7 \text{ mol NaOH}} = 0.162 \text{ mf NaOH}$$

b. Weight percentage of NaOH:

$$\dfrac{10.7 \text{ mol NaOH}}{1000 \text{ g solvent}} \cdot \dfrac{40.0 \text{ g NaOH}}{1 \text{ mol NaOH}} = \dfrac{428 \text{ g NaOH}}{1000 \text{ g solvent}}$$

The mass of solution would be (428 g + 1000. g) 1428 g.

$$\dfrac{428 \text{ g NaOH}}{1428 \text{ g solution}} \times 100 = 30.0\% \text{ NaOH}$$

171

c. Molarity of the solution:

$$\frac{10.7 \text{ mol NaOH}}{1428 \text{ g solution}} \cdot \frac{1.33 \text{ g NaOH}}{1 \text{ cm}^3 \text{ solution}} \cdot \frac{1000 \text{ cm}^3}{1 \text{ L solution}} = 9.97 \text{ M NaOH}$$

27.   The molality of the $Ca(NO_3)_2$ solution:

$$\frac{2.00 \text{ g } Ca(NO_3)_2}{0.750 \text{ kg solvent}} \cdot \frac{1 \text{ mol } Ca(NO_3)_2}{164.1 \text{ g } Ca(NO_3)_2} = 0.0163 \text{ molal } Ca(NO_3)_2$$

One mol $Ca(NO_3)_2$ provides 3 mol of ions (1 Ca^{2+} and 2 NO_3). The total molality would be (3 x 0.0163) or 0.0489 molal.

29.   The concentration of ppm expressed in grams is:

$$0.18 \text{ ppm} = \frac{0.18 \text{ g solute}}{1.0 \times 10^6 \text{ g solvent}} = \frac{0.18 \text{ g solute}}{1.0 \times 10^3 \text{ kg solvent}} \text{ or } \frac{0.00018 \text{ g solute}}{1 \text{ kg water}}$$

$$\frac{0.00018 \text{ g } Li^+}{1 \text{ kg water}} \cdot \frac{1 \text{ mol } Li^+}{6.939 \text{ g } Li^+} = 2.6 \times 10^{-5} \text{ molal } Li^+$$

31.   Raising the temperature of the solution will increase the solubility of NaCl in water. Hence to increase the amount of dissolved NaCl in solution one must **(c) raise the temperature of the solution and add some NaCl.**

33.   The enthalpy of formation (ΔH_f) for NH_4Cl:

$$\Delta H^{\circ}\text{solution} = \Delta H_f^{\circ} \ NH_4Cl \ (1m) - \Delta H_f^{\circ} \ NH_4Cl \ (s)$$
$$+ \ 14.8 \text{ kJ/mol} = \Delta H_f^{\circ} \ NH_4Cl \ (l \ m) - (- 314.4 \text{ kJ/mol})$$
$$- \ 299.6 \text{ kJ/mol} = \Delta H_f^{\circ} \ NH_4Cl \ (l \ m)$$

The compound should become more soluble as the temperature increases. As heat is added (T increases), the energy for NH_4Cl is provided.

35.   The greatest attraction between Na^+ and solvent molecules will occur if the solvent is very polar. Of the four solvents given, H_2O fits that description.

Henry's Law

37. Molality of O_2 = $k \cdot P_{O_2}$

$$= (1.66 \times 10^{-6} \frac{molal}{mmHg}) \cdot 40. \, mmHg = 6.6 \times 10^{-5} \, molal$$

39.   Solubility = $k \cdot P_{CO_2}$

$$0.0506 \, M = (4.48 \times 10^{-5} \frac{M}{mmHg}) \cdot P_{CO_2}$$

$$1130 \, mmHg = P_{CO_2} \quad or \quad 1130 \, mmHg \cdot \frac{1 \, atm}{760 \, mmHg} = 1.49 \, atm$$

Vapor Pressure Changes

41. To calculate the vapor pressure of the mixture; calculate the mf of water

$$5.000 \times 10^2 \, g \, H_2O \longrightarrow 27.75 \, mol \, H_2O$$

$$35.0 \, g \, glycol \longrightarrow 0.564 \, mol \, glycol$$

and the mf of water is: $\dfrac{27.75 \, mol \, H_2O}{(27.75 + 0.564) \, mol} = 0.980$

The vapor pressure is then

$$P_{water} = X_{water} \, P°_{water}$$

$$= 0.980 \cdot 35.7 \, mmHg = 35.0 \, mmHg \, (to \, 3 \, sf)$$

43.   Using Raoult's Law, we know that the vapor pressure of pure water (P°) multiplied by the mole fraction(X) of the solute gives the vapor pressure of the solvent above the solution (P).

$$P_{water} = X_{water} \, P°_{water}$$

The vapor pressure of pure water at 90 °C is 525.8 mmHg (from Appendix D).
Since the P_{water} is given as 457 mmHg, the mole fraction of the water is:

$$\frac{457 \, mmHg}{525.8 \, mmHg} = 0.869$$

The 2.00 kg of water correspond to a mf of 0.869. This mass of water corresponds to:

$$2.00 \times 10^3 \text{ g H}_2\text{O} \cdot \frac{1 \text{ mol H}_2\text{O}}{18.02 \text{ g H}_2\text{O}} = 111 \text{ mol water.}$$

Representing moles of ethylene glycol as x we can write:

$$\frac{\text{mol H}_2\text{O}}{\text{mol H}_2\text{O} + \text{mol C}_2\text{H}_4\text{(OH)}_2} = \frac{111}{111 + x} = 0.869$$

$$\frac{111}{0.869} = 111 + x \; ; \quad 16.7 = x \text{ (mol of ethylene glycol)}$$

$$16.7 \text{ mol C}_2\text{H}_4\text{(OH)}_2 \cdot \frac{62.07 \text{ g C}_2\text{H}_4\text{(OH)}_2}{1 \text{ mol C}_2\text{H}_4\text{(OH)}_2} = 1.04 \times 10^3 \text{ g C}_2\text{H}_4\text{(OH)}_2$$

45.   The molar mass of the solute may be calculated by using Raoult's Law:

$P_{benzene} = 121.8$ mmHg at 30 °C $P_{solution} = 113.0$ mmHg at 30 °C

$$P_{solution} = P_{benzene} \cdot X_{benzene}$$

$$113.0 \text{ mm} = 121.8 \text{ mm} \cdot X_{benzene}$$
$$0.9278 = X_{benzene}$$

$$0.9278 = \frac{\text{mol C}_6\text{H}_6}{\text{mol C}_6\text{H}_6 + \text{mol unknown}}$$

Calculating the amount of C_6H_6:

$$100. \text{ g C}_6\text{H}_6 \cdot \frac{1 \text{ mol C}_6\text{H}_6}{78.11 \text{ g C}_6\text{H}_6} = 1.28 \text{ mol C}_6\text{H}_6$$

Substituting into the mf expression above
$$0.9278 = \frac{1.28 \text{ mol C}_6\text{H}_6}{1.28 + x} \qquad \text{and solving gives: } x = 0.0996 \text{ mol}$$

Since 10.0 g corresponds to 0.0996 mol, we can calculate the molar mass:

10.0 g = 0.0996 mol   and 100. g = 1 mol

Boiling Point Elevation

47. The boiling point is a function of (1) the solvent and (2) the molality of the solute

Calculate the molality :

$$15.0 \text{ g urea} \cdot \frac{1 \text{ mol urea}}{60.06 \text{ g urea}} = 0.250 \text{ mol urea and } \frac{0.250 \text{ mol urea}}{0.500 \text{ kg water}} = 0.500 \text{ molal urea}$$

The boiling point elevation is then

$$\Delta t = m \cdot K_{bp} = 0.500 \cdot \frac{0.512 \text{ }^\circ C}{\text{molal}} = 0.256 \text{ }^\circ C$$

So the boiling point is $100.00 + 0.256$ or $100.26 \text{ }^\circ C$.

49. As in 47, calculate the molality of $C_{12}H_{10}$ in the solution.

$$0.515 \text{ g } C_{12}H_{10} \cdot \frac{1 \text{ mol } C_{12}H_{10}}{154.2 \text{ g } C_{12}H_{10}} = 3.34 \times 10^{-3} \text{ mol } C_{12}H_{10}$$

and the molality is : $\dfrac{3.34 \times 10^{-3} \text{ mol acenaphthalene}}{0.0150 \text{ kg CHCl}_3} = 0.223 \text{ molal}$

the boiling point elevation is:

$$\Delta t = m \cdot K_{bp} = 0.223 \cdot \frac{+3.63 \text{ }^\circ C}{\text{molal}} = 0.808 \text{ }^\circ C$$

and the boiling point will be $61.70 + 0.808 = 62.51 \text{ }^\circ C$

51. The boiling point has been elevated by $4.3 \text{ }^\circ C$.

The molality of the solution is : $\Delta T_{bp} = K_{bp} \cdot m$

$4.3 \text{ }^\circ C = +0.512 \text{ }^\circ C/\text{molal} \cdot m$ and $8.4 = m$

Since molality is defined as:

$$m = \frac{\text{moles solute}}{\text{kg solvent}} \quad \text{then} \quad 8.4 = \frac{\text{moles glycerol}}{0.750 \text{ kg water}}$$

$$6.3 = \text{mol glycerol } (C_3H_5(OH)_3)$$

$$6.3 \text{ mol } C_3H_5(OH)_3 \cdot \frac{62.07 \text{ g } C_3H_5(OH)_3}{1 \text{ mol } C_3H_5(OH)_3} = 580 \text{ g } C_3H_5(OH)_3$$

The mf(glycerol) is: $750. \text{ g H}_2O \cdot \dfrac{1 \text{ mol H}_2O}{18.02 \text{ g H}_2O} = 41.6 \text{ mol H}_2O$

$$\frac{6.3 \text{ mol glycerol}}{6.3 \text{ mol glycerol H}_2O} = 0.13$$

175

53. The **solution with the highest boiling point** will have the **greatest number of particles** in solution.

If we assume total dissociation of the solutes given the molality of particles for the solution will be:

$$0.10 \text{ m KCl} \rightarrow 0.10 \text{ m K}^+ + 0.10 \text{ m Cl}^- = 0.20 \text{ m}$$

$$0.10 \text{ m sugar} \rightarrow \text{(covalently bonded specie)} = 0.10 \text{ m}$$

$$0.080 \text{ m MgCl}_2 \rightarrow 0.080 \text{ m Mg}^{2+} + 0.16 \text{ m Cl}^- = 0.24 \text{ m}$$

In order of increasing boiling point: $0.10 \text{ m sugar} < 0.10 \text{ m KCl} < 0.080 \text{ m MgCl}_2$

55. The change in boiling point of chloroform is :

$$62.22 \,°C - 61.70 \,°C = +0.52 \,°C$$

and the molality of the solution

$$+0.52 \,°C = m \cdot +3.63\frac{°C}{\text{molal}} \quad \text{or} \quad 0.14 \text{ molal}$$

$$0.14 \text{ molal} = \frac{\frac{0.640g \text{ BHA}}{MM}}{0.25 \text{ kg solvent}}$$

and solving for MM : 178.7 or 180 (to 2 sf)

57. Calculate the molality of the solution from $\Delta t = m \cdot K_{fp} \cdot i$ (where $i = 1$, since the compound is not ionic).

$$80.34 \,°C - 80.10 \,°C = (+2.53 \,°C/ \text{ molal}) \cdot m$$

$$m = 0.095 \text{ mol/kg}$$

Calculate the moles of anthracene dissolved

$$0.095 \text{ mol/kg} \cdot 0.0300 \text{ kg} = 0.0028 \text{ mol}$$

Calculate the molar mass of anthracene

$$\frac{0.500 \text{ g}}{0.0028 \text{ mol}} = 180 \text{ g/mol} \quad \text{(actual molar mass is 178.2 g)}$$

With an empirical formula of C_7H_5 (empirical weight of ~ 89), the molecular formula must be $C_{14}H_{10}$.

Freezing Point Depression

59. The solution freezes 16.0 °C lower than pure water. We can calculate the molality of the ethanol

$$\Delta t = mK_{fp}$$
$$-16.0 \text{ °C} = m \, (-1.86 \text{ °C/molal})$$
$$8.60 = \text{molality of the alcohol}$$

If the molality is 8.60 then there are 8.60 moles of C_2H_5OH (8.60 x 46.07 g/mol) in the 1000 kg of H_2O.

The weight percent of alcohol is $\dfrac{396 \text{ g}}{1396 \text{ g}}$ x 100 = 28.4 % ethanol

61. The number of moles of LiF is : 52.5 g LiF • $\dfrac{1 \text{ mol LiF}}{25.94 \text{ g LiF}}$ = 2.02 mol LiF

So Δt_{fp} = $\dfrac{2.02 \text{ mol LiF}}{0.300 \text{ kg H2O}}$ • -1.86 °C/molal • 2 = -25.1 °C

The anticipated freezing point is then -25.1 °C.

63. Determine the molality of the solution

$$-0.040 \text{ °C} = m \bullet -1.86 \text{ °C/molal} = 0.215 \text{ molal (or 0.22 to 2 sf)}$$

and 0.22 molal = $\dfrac{\dfrac{0.180 \text{ g solute}}{MM}}{0.0500 \text{ kg water}}$

$$MM = 167 \text{ or } 170 \text{ (to 2 sf)}$$

65. The molality of the solution is:

$$(5.41 - 5.50) = m \bullet -5.12 \text{ °C/molal}$$
$$0.018 \text{ molal} = m$$

and 0.018 molal = $\dfrac{\dfrac{0.125 \text{ g compound}}{MM}}{0.01565 \text{ kg benzene}}$ MM = 444 or 450 (to 2 sf)

The empirical formula, $(C_2H_5)_2AlF$ has a fromula weight of approximately 104. So it would appear that there are **four** units, or a molecular formula of $[(C_2H_5)_2AlF]_4$.

67. Solutions given in order of increasing melting point:

    The solution with the greatest **number** of particles will have the lowest melting point.

    The total molality of solutions is then:
    (a) 0.1 m sugar   x 1 particle/formula unit = 0.1 m
    (b) 0.1 m NaCl   x 2 particles/formula unit = 0.2 m   [Na^+ , Cl^-]
    (c) 0.08 m $CaCl_2$ x 3 particles/formula unit = 0.24 m   [Ca^{2+}, 2 Cl^-]
    (d) 0.04 m Na_2SO_4 x 3 particles/formula unit = 0.12 m   [2 Na^+, SO_4^{2-}]

    The melting points would increase in the order: $CaCl_2$ < NaCl < Na_2SO_4 < sugar

Osmosis

69. Assume we have 100 g of this solution, the number of moles of phenylalanine is

$$3.00 \text{ g phenylalanine} \cdot \frac{1 \text{ mol phenylalanine}}{165.2 \text{ g phenylalanine}} = 0.0182 \text{ mol phenylalanine}$$

The molality of the solution is $\dfrac{0.0182 \text{ mol phenylalanine}}{0.09700 \text{ kg water}}$ = 0.187 molal

a. The freezing point :

    Δt = 0.187 molal • -1.86 °C/molal = -0.348 °C

    The new freezing point is 0.0 - 0.348 °C = -0.348 °C.

b. The boiling point of the solution

    Δt = m K_{bp} = 0.187 molal + 0.5121 = +0.0959 °C

    The new boiling point is then 100.00 + 0.0959 = +100.10 °C

c. The osmotic pressure of the solution:

    If we assume that the **Molarity** of the solution is equal to the **molality**, then
    the osmotic pressure should be

    Π = (0.187 mol/L)(0.0821 $\frac{L \cdot atm}{K \cdot mol}$)(298 K) = 4.58 atm

The freezing point will be most easily measured.

71.   3.1 mmHg • $\dfrac{1 \text{ atm}}{760 \text{ mmHg}}$ = (M)(0.08205 $\frac{L \cdot atm}{K \cdot mol}$)(298 K)

       1.67×10^{-4} = Molarity or 1.7×10^{-4} (to 2 sf)

$$1.7 \times 10^{-4} \frac{\text{mol bovine insulin}}{\text{L}} = \frac{\frac{1.00 \text{ g bovine insulin}}{\text{MM}}}{1 \text{ L}}$$

Solving for MM = 6.0×10^3 g/mol

Colloids

73.   a. $BaCl_2(aq) + Na_2SO_4(aq) \longrightarrow BaSO_4(s) + 2\,NaCl(aq)$

b. The $BaSO_4$ formed is of a colloidal size — not large enough to precipitate fully.

c. The particles of $BaSO_4$ grow with time, owing to a gradual loss of charge and become large enough to have gravity affect them —and settle to the bottom.

General Questions

75.   a. The increased boiling point is calculated :

$$\Delta t = m \cdot K_{bp} \cdot 2$$

Since the solvent (and K_{bp}) is constant for both solutions, we should look at the **molality** and **van't Hoff factors**. Sugar will remain as molecular entities (i = 1), but Na_2SO_4 will dissociate into ions (2 Na^+ and 1 SO_4^{2-} : i = 3). The product, m • i, will be (0.10)(3) for Na_2SO_4 and (0.15)(1) for sugar. So the **Na_2SO_4 solution will have the higher boiling point**.

b. The reaction can be written:

$$NaOH(s) \rightleftharpoons NaOH(aq) + \text{heat}$$

LeChatelier's principle tells us that as heat is added (temperature increased) the equilibrium will shift to the left (less NaOH will dissolve).

c. Lowering of vapor pressure is proportional to the mole fraction (the relative number of particles) of solute. Note that the mf of NH_4NO_3 (0.30 molal) will be greater than the mf of Na_2SO_4 (0.15 molal). **The vapor pressure of water will be higher for the 0.15 molal Na_2SO_4 solution.**

77. For DMG, $(CH_3CNOH)_2$, the MM is 116.1 g/mol

So 53.0 g is : $53.0\text{g} \cdot \dfrac{1 \text{ mol DMG}}{116.1 \text{ g DMG}} = 0.456$ mol DMG

179

500. g of C_2H_5OH is : $500.\ g \cdot \dfrac{1\ mol\ C_2H_5OH}{46.07\ g\ C_2H_5OH}\ =\ 10.9\ mol\ C_2H_5OH$

a. the mole fraction of DMG: $\dfrac{0.456\ mol}{(10.9 + 0.456)\ mol}\ =\ 0.0404\ mf\ DMG$

b. The molality of the solution: $\dfrac{0.456\ mol\ DMG}{0.500\ kg}\ =\ 0.913\ molal\ DMG$

c. $P_{alcohol}\ =\ P°_{alcohol} \cdot X_{alcohol}$

$=\ (760.\ mmHg)(1 - 0.0402)\ =\ 729\ mmHg$

d. The boiling point of the solution:

$\Delta t\ =\ m \cdot K_{bp} \cdot i\ =\ (0.912)(+1.22\ °C/molal)(1)$

$=\ 1.11\ °C$

The new boiling point is $78.4\ °C + 1.11\ °C\ =\ 79.51\ °C$ or $79.5\ °C$

79. The density of water is $0.997\ g/cm^3$, so 1000. mL of water ($1000.\ cm^3$) will have a mass of 997 g.

$997\ g\ H_2O \cdot \dfrac{1\ mol\ H_2O}{18.02\ g\ H_2O}\ =\ 55.3\ mol\ H_2O$

So $55.3\ mol\ H_2O/1\ L\ =\ 55.3\ Molar$

The molality will be $\dfrac{55.3\ mol}{0.997\ kg}\ =\ 55.5\ molal$

81. The molality of aluminon is: $\Delta t\ =\ m \cdot K_{bp} \cdot i$

$\dfrac{-0.197\ °C}{(-1.86\ °C/molal)(1)}\ =\ m\ =\ 0.106\ molal$

And $0.106\ molal\ =\ \dfrac{\dfrac{2.50\ g\ aluminon}{MM}}{0.0500\ kg\ water}$ $MM\ =\ 472\ g/mol$

83.   a. Given that the solubility of KNO_3 increases with increased temperature, the enthalpy of solution of KNO_3 is expected to be positive (endothermic).

b. The higher boiling point will be with the 0.20 molal KBr, since the product of the molality and van't Hoff factor (2 for KBr; 1 for sugar) will be greater for KBr than for sugar.

c. The lower freezing point will be that of the Na_2CO_3 solution, since the product of the molality and van't Hoff factor (2 for NH_4NO_3 ; 3 for Na_2CO_3) will be greater for Na_2CO_3 than for the 0.12 molal NH_4NO_3.

85. Calculate the molar mass of hexachlorophene if 0.640 g of the solid dissolved in 25.0 g chloroform gives a solution boiling at 61.93 °C.

$$\Delta T_{bp} = K_{bp} \bullet m \quad \text{and} \quad m = \frac{\Delta T_{bp}}{K_{bp}}$$

$$\frac{(61.93 \text{ °C} - 61.70 \text{ °C})}{3.63 \text{ °C/molal}} = 0.0634 \text{ molal}$$

The number of moles of hexachlorophene is: 0.0634 molal $\bullet$ 0.0250 kg $= 1.58 \times 10^{-3}$ mol

The molar mass is then $\dfrac{0.640 \text{ g}}{1.58 \times 10^{-3} \text{ mol}} = 404$ g/mol

87.   The solubility of NaCl at 100 °C is approximately 39.1 g/100. g H_2O.

$$\text{molality} = \frac{39.1 \text{ g NaCl}}{100 \text{ g } H_2O} \bullet \frac{1 \text{ mol NaCl}}{58.44 \text{ g NaCl}} \bullet \frac{1000 \text{ g } H_2O}{1 \text{ kg } H_2O} = 6.69 \text{ molal NaCl}$$

If the van't Hoff factor for NaCl is 1.85, the effective molality is (1.85×6.69) 12.4.

The change in the boiling point of this solution is:

$$\Delta T_{bp} = K_{bp} \bullet m = +0.512 \text{ °C/molal} \bullet 12.4 \text{ molal} = +6.34 \text{ °C}.$$

The solution will begin to boil at $(100.00 + 6.34)$ or 106.34 °C.

89. $P_T = 800. \text{ mmHg} = P_{Cl_2} + P_{H_2O} = P_{Cl_2} + 23.8 \text{ torr}$

$= 776.2 \text{ mmHg} = P_{Cl_2}$

Solubility of Cl_2 $= 8.2 \times 10^{-5} \dfrac{M}{\text{mmHg}} \bullet P_{Cl_2} = 8.2 \times 10^{-5} \dfrac{M}{\text{mmHg}} \bullet 776.2 \text{ mmHg}$

$= 0.064$ Molar (or 0.064 molal in dilute solutions)

$$0.064 \text{ molal} = \frac{\dfrac{x \text{ g Cl}_2}{70.9 \text{ g/mol}}}{0.100 \text{ kg water}} \qquad 0.45 \text{ g Cl}_2 = x$$

91. a. Mole fraction of C_2H_5OH and H_2O

$$50.0 \text{ mL } C_2H_5OH \cdot \frac{0.785 \text{ g } C_2H_5OH}{1 \text{ mL } C_2H_5OH} \cdot \frac{1 \text{ mol } C_2H_5OH}{46.07 \text{ g } C_2H_5OH} = 0.852 \text{ mol } C_2H_5OH$$

$$50.0 \text{ mL } H_2O \cdot \frac{1.00 \text{ g } H_2O}{1 \text{ mL } H_2O} \cdot \frac{1 \text{ mol } H_2O}{18.02 \text{ g } H_2O} = 2.77 \text{ } H_2O$$

$$X_{C_2H_5OH} = \frac{0.852}{0.852 + 2.77} = 0.235$$

$$X_{H_2O} = 1 - 0.235 = 0.765$$

b. $P_{H_2O} = P^{\circ}_{H_2O} \cdot X_{H_2O} = (17.5 \text{ mmHg})(0.765) = 13.4 \text{ mmHg}$

$P_{C_2H_5OH} = P^{\circ}_{C_2H_5OH} \cdot X_{C_2H_5OH} = (43.6 \text{ mmHg})(0.235) = 10.2 \text{ mmHg}$

$P_T = P_{H_2O} + P_{C_2H_5OH} = 13.4 \text{ mmHg} + 10.2 \text{ mmHg} = 23.6 \text{ mmHg}$

93. a. The molality of the solution is:

$-0.229 \text{ °C} = m \cdot -1.86 \text{ °C/molal} \cdot 1$

$0.123 = \text{molality of maltose}$

$$0.123 = \frac{\dfrac{4.00 \text{ g maltose}}{MM}}{0.09600 \text{ kg water}} \qquad MM = 338$$

The literature value for $C_{12}H_{22}O_{11}$ is 342.

b. Given the density of the solution as 1.014 g/mL, the volume of 100.00 g of solution will be:

$$100.00 \text{ g} \cdot \frac{1 \text{ mL solution}}{1.014 \text{ g solution}} = 98.62 \text{ mL solution}$$

The molarity of the solution would be:

$$4.00 \text{ g maltose} \cdot \frac{1 \text{ mol maltose}}{338 \text{ g maltose}} \cdot \frac{1}{0.09862 \text{ L}} = 0.120 \text{ M (to 3 sf)}$$

and the osmotic pressure (at 25 °C)

$$\Pi = 0.120 \text{ mol/L} \cdot 0.0821 \frac{L \cdot atm}{K \cdot mol} \cdot 273 \text{ K} = 2.93 \text{ atm}$$

95. At 80 °C, 1092 g of ammonium formate will dissolve in 200 g of water. At 0 °C, only 204 g of ammonium formate will dissolve in this mass of water, so (1092 - 204 g) **888 g of ammonium formate** will precipitate.

97. The vapor pressure of water at 18 °C is 15.5 torr. So the desired humidity is 55 % of 15.5 torr or 8.53 torr

$$8.53 \text{ torr} = 15.5 \text{ torr} \cdot X_{H_2O}$$

$$0.55 = X_{H_2O} \qquad \text{and } 0.45 = X_{glycerol}$$

Assume that there is a total of 1.00 mol of water plus glycerol. Then there would be 0.55 mol of water and 0.45 mol of glycerol.

The masses of this number of moles would be

$$0.55 \text{ mol } H_2O \cdot \frac{18.0 \text{ g } H_2O}{1 \text{ mol } H_2O} = 9.9 \text{ g } H_2O$$

$$0.45 \text{ mol glycerol} \cdot \frac{92.1 \text{ g glycerol}}{1 \text{ mol glycerol}} = 41 \text{ g glycerol (to 2 sf)}$$

The % of glycerol is then: $\dfrac{41 \text{ g glycerol}}{50.9 \text{ g total}} = 81 \text{ % (to 2 sf)}$

99. Calculate the molar mass by freezing point depression:

$$m = \frac{\Delta T_{fp}}{K_{fp}} = \frac{(0.00 \text{ °C} - 1.28 \text{ °C})}{-1.86 \text{ °C/molal}} = 0.688 \text{ molal}$$

Since KX provides two particles/formula unit, the number of moles of KX is half this number or 0.344.

$$\text{moles KX} = \frac{0.344 \text{ mol KX}}{1 \text{ kg water}} \cdot 0.100 \text{ kg water} = 3.44 \times 10^{-2} \text{ mol}$$

Therefore the molar mass for KX is: $\dfrac{4.00 \text{ g}}{3.44 \times 10^{-2} \text{ mol}} = 116 \text{ g/mol}$

Subtracting the mass of potassium from this molar mass yields:

$$116 - 39.1 = 77 \text{ g/mol X} \qquad \text{Thus the element X must be } \mathbf{Br}.$$

101.   a. The 10 m tree is equal to a column of water 10 x 1000 or 10000 mm tall. The equivalent column of mercury would be $1.0 \times 10^4 \text{ mm } H_2O \cdot \dfrac{1.0 \text{ mmHg}}{13.6 \text{ mm } H_2O}$

735 mmHg or 0.967 atm (or 0.97 to 2 sf)

The molarity of solutes to produce this pressure at 25 °C:

$$\Pi \; = \; \text{molarity} \cdot R \cdot T$$

$$0.97 \text{ atm} \; = \; \text{Molarity} \cdot 0.0821 \, \frac{L \cdot atm}{K \cdot mol} \cdot 298 \text{ K}$$

$$0.040 \text{ mol/L} \; = \; \text{Molarity}$$

b. $\dfrac{0.040 \text{ mol sucrose}}{1L \text{ solution}} \cdot \dfrac{342 \text{ g sucrose}}{1 \text{ mol sucrose}} = \dfrac{13.68 \text{ g sucrose}}{L \text{ solution}}$ 14 g/L (to 2 sf)

$$\frac{13.68 \text{ g sucrose}}{1 \text{ L solution}} \cdot \frac{1 \text{ L solution}}{1.0 \times 10^3 \text{ g solution}} = \frac{13.68 \text{ g sucrose}}{1000 \text{ g solution}}$$

$$\text{or} \; \frac{1.368 \text{ g sucrose}}{100 \text{ g solution}} \quad \text{or} \quad 1.4 \text{ \% sucrose}$$

Conceptual Questions

103. When the egg is placed in water, the solute concentration is greater inside the membrane. Water flows into the membrane, reducing the concentration of solutes inside (the egg swells).

If the egg is placed in a solution with a higher solute concentration, water flows out of the egg into the external solution (the egg shrivels).

Similarly a cucumber placed into brine shrivels as water leaves the cucumber and enters the brine, reducing the solute concentration of the brine. Spices are organic in nature and absorb in the cellular structure of the cucumber's skin.

105. The apparent molecular weight of acetic acid in benzene, determined by the depression of benzene's freezing point.

$$\Delta t \; = \; m \cdot K_{fp} \cdot 2$$

$$3.37 \, ^\circ C - 5.50 \, ^\circ C \; = \; m(-5.12 \, ^\circ C/molal) \, i$$

$$\frac{-2.13 \, ^\circ C}{-5.12 \, ^\circ C/molal} \; = \; m_i$$

$$0.416 \text{ molal} \; = \; m_i \quad \text{(momentarily assume } i = 1\text{)}$$

and the apparent molecular weight is:

$$0.416 \text{ molal} \; = \; \frac{\dfrac{5.00 \text{ g acetic acid}}{MM}}{0.100 \text{ kg}}$$

$$120 \text{ g/mol} \; = \; MM$$

The apparent molecular weight of acetic acid in water

$$\Delta t = m \cdot K_{fp} \cdot i$$

$$-1.49 \,°C - 0.00 \,°C = m(-1.86 \,°C/molal) \, i$$

$$\frac{-1.49 \,°C}{-1.86 \,°C/molal} = m_i$$

$$0.801 \text{ molal} = m_i \quad \text{(once again, momentarily } i = 1)$$

and the apparent molecular weight is:

$$0.801 \text{ molal} = \frac{\dfrac{5.00 \text{ g acetic acid}}{MM}}{0.100 \text{ kg}}$$

$$62.4 \text{ g/mol} = MM$$

The accepted value for acetic acid's molecular weight is approximately 60 g/mol. Hence the value for i isn't much larger than 1, indicating that the degree of dissociation of acetic acid molecules in water is not great—a finding consistent with the designation of acetic acid as a weak acid. The apparently doubled molecular weight of acetic acid in benzene indicates that the acid must exist primarily as a dimer.

107. Note that all 3 alcohols—methanol, ethanol, and octanol—contain a polar "end"—the OH, and a non-polar "end"—the carbons. As the carbon chain lengthens, the non-polar carbon chain reduces the solubility of the alcohol in water.

109. The C—C and C—H bonds in hydrocarbons being non-polar would tend to make such dispersions hydrophobic (water hating). The C—O and O—H bonds in starch present opportunities for hydrogen bonding with water, and hence such dispersions are expected to be hydrophilic (water loving).

111.   a. Compound contains 73.87% C, 8.21% H, and 17.92% Cr

Calculate the empirical formula.

$$73.87 \text{ g C} \cdot \frac{1 \text{ mol C}}{12.011 \text{ g C}} = 6.150 \text{ mol C}$$

$$8.21 \text{ g H} \cdot \frac{1 \text{ mol H}}{1.008 \text{ g H}} = 8.14 \text{ mol H}$$

$$17.92 \text{ g Cr} \cdot \frac{1 \text{ mol Cr}}{52.00 \text{ g Cr}} = 0.3446 \text{ mol Cr}$$

The ratio of atoms is: $\frac{6.150 \text{ mol C}}{0.3446 \text{ mol Cr}} = 17.85$ $\frac{8.14 \text{ mol H}}{0.3446 \text{ mol Cr}} = 23.63$

or to the nearest integral values: $CrC_{18}H_{24}$.

b. $\Pi = 3.17$ mmHg or 4.17×10^{-3} atm

T = 298 K   and   V = 0.100 L   mass $= 5.00 \times 10^{-3}$ g

$\Pi = M \cdot R \cdot T$ and since $M = \frac{n}{V}$

$\Pi = \frac{n}{V} \cdot R \cdot T$

$n = \frac{\Pi \cdot V}{R \cdot T} = \frac{4.17 \times 10^{-3} \text{atm} \cdot 0.100 \text{ L}}{0.082057 \frac{\text{L} \cdot \text{atm}}{\text{K} \cdot \text{mol}} \cdot 298 \text{ K}} = 1.71 \times 10^{-5}$ mol

So 5.00 mg of the sample corresponds to 1.71×10^{-5} mol.

The molar mass is:

$$5.00 \times 10^{-3} \text{ g} = 1.71 \times 10^{-5} \text{ mol}$$

$$\frac{5.00 \times 10^{-3} \text{ g}}{1.71 \times 10^{-5} \text{mol}} = 293 \text{ g/mol}$$

Using the empirical formula from part (a), we calculate an empirical mass of 292.
The molecular formula must be $CrC_{18}H_{24}$.

Chapter 15:
Principles of Reactivity: Chemical Kinetics

Concentration Units

18.   a. $2\,O_3\,(g) \;\rightarrow\; 3\,O_2\,(g)$

$$\text{Reaction Rate} = -\frac{1}{2}\cdot\frac{\Delta[O_3]}{\Delta t} = +\frac{1}{3}\cdot\frac{\Delta[O_2]}{\Delta t}$$

   b. $2\,HOF\,(g) \;\rightarrow\; 2\,HF\,(g) + O_2\,(g)$

$$\text{Reaction Rate} = -\frac{1}{2}\cdot\frac{\Delta[HOF]}{\Delta t} = +\frac{1}{2}\cdot\frac{\Delta[HF]}{\Delta t} = +\frac{\Delta[O_2]}{\Delta t}$$

   c. $2\,NO(g) + Br_2(g) \;\rightarrow\; 2\,BrNO(g)$

$$\text{Reaction Rate} = -\frac{1}{2}\cdot\frac{\Delta[NO]}{\Delta t} = -\frac{\Delta[Br_2]}{\Delta t} = +\frac{1}{2}\cdot\frac{\Delta[BrNO]}{\Delta t}$$

20. Plot the data for the hypothetical reaction $A \rightarrow 2\,B$

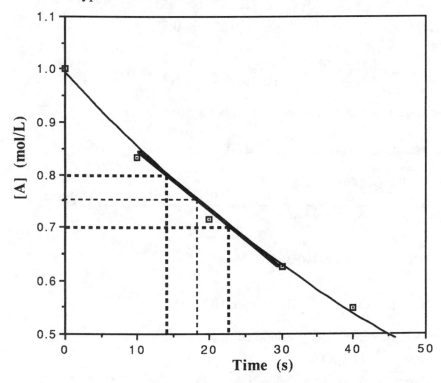

a. Rate $= -\dfrac{\Delta[A]}{\Delta t} = -\dfrac{(0.833 - 1.000)}{10.0 - 0.00} = +\dfrac{0.167}{10.0} = +0.0167 \dfrac{mol}{L \cdot s}$

$= -\dfrac{(0.714 - 0.833)}{20.0 - 10.00} = +\dfrac{0.119}{10.0} = +0.0119 \dfrac{mol}{L \cdot s}$

$= -\dfrac{(0.625 - 0.714)}{30.0 - 20.00} = +\dfrac{0.089}{10.0} = +0.0089 \dfrac{mol}{L \cdot s}$

$= -\dfrac{(0.555 - 0.625)}{40.0 - 30.00} = +\dfrac{0.070}{10.0} = +0.0070 \dfrac{mol}{L \cdot s}$

The rate of change decreases from one time interval to the next due to a continuing decrease in the amount of reacting material (A).

b. Since each A molecule forms 2 molecules of B

T	[A]	$[B] = 2([A]_0 - [A])$
10.0 s	0.833	$2(0.167) = 0.334$
20.0 s	0.714	$2(0.286) = 0.572$

Rate $= \dfrac{\Delta[B]}{\Delta t} = \dfrac{(0.572 - 0.334)}{20.0 - 10.00} = \dfrac{0.238}{10.0}$ or $0.0238 \dfrac{mol}{L \cdot s}$

The appearance of B is twice as great as the disappearance of A.

c. The instantaneous rate when [A] $= 0.75$ mol/L:

$-\dfrac{\Delta[A]}{\Delta t} = \dfrac{-(0.700\frac{mol}{L} - 0.800 \frac{mol}{L})}{22.5\ s - 14.0\ s} = \dfrac{0.100\ mol/L}{8.5\ s} = 0.012 \dfrac{mol}{L \cdot s}$

Concentration and Rate Equations

22. a. The rate equation :   Rate $= k[NO_2][O_3]$

b. Since k is constant, if $[O_3]$ is held constant, the rate would be tripled:

Rate$_1$ $= k[C][O_3]$

Rate$_2$ $= k[3\ C][O_3] = 3 \cdot k[C][O_3]$ or $3 \cdot$ Rate$_1$

c. Halving the concentration of O_3—assuming $[NO_2]$ is constant, would halve the rate.

Rate$_1$ $= k[NO_2][C]$

Rate$_2$ $= k[NO_2][1/2\ C] = 1/2[NO_2][C]$ or $1/2 \cdot$ Rate$_1$

24.   a. If we designate the three experiments (data sets in the table as i, ii, and iii respectively,

Experiment	[NO]	[O$_2$]	$-\dfrac{\Delta[NO]}{\Delta t}\left(\dfrac{mol}{L \cdot s}\right)$
i	0.020	0.010	1.0×10^{-4}
ii	0.040	0.010	4.0×10^{-4}
iii	0.020	0.040	4.0×10^{-4}

Note that experiment ii proceeds at a rate four times that of experiment i.

$$\frac{\text{experiment ii rate}}{\text{experiment i rate}} = \frac{4.0 \times 10^{-4} \dfrac{mol}{L \cdot s}}{1.0 \times 10^{-4} \dfrac{mol}{L \cdot s}} = 4$$

This rate change was the result of doubling the concentration of NO. The **order of dependence of NO must be second order**. Comparing experiments i and iii, we see that changing the concentration of O$_2$ by a factor of four, also affects the rate by a factor of four. **The order of dependence of O$_2$ must be first order**.

$$\text{Rate} = k[NO]^2[O_2]^1$$

b. To calculate the rate constant we have to have a rate. Note the data provided gives the rate of disappearance of NO. The relation of this concentration to the rate is:

$$\text{Rate} = -1/2 \cdot \frac{\Delta[NO]}{\Delta t}$$

Using experiment ii, the rate is $2.0 \times 10^{-4} \dfrac{mol}{L \cdot s}$

Substituting into the rate law

$$2.0 \times 10^{-4} \frac{mol}{L \cdot s} = k[0.040 \frac{mol}{L}]^2[0.010 \frac{mol}{L}]$$

$$12.5 \frac{L^2}{mol^2 \cdot s} = k$$

c. Rate when [NO] = 0.045 M and [O$_2$] = 0.025 M

$$\text{Rate} = k[NO]^2[O_2]$$

$$= 12.5 \frac{L^2}{mol^2 \cdot s} (0.045 \frac{mol}{L})^2 (0.025 \frac{mol}{L})$$

$$= 6.3 \times 10^{-4} \frac{mol}{L \cdot s}$$

d. The relation between reaction rate and concentration changes

$$\text{Rate} = -1/2 \cdot \frac{\Delta[NO]}{\Delta t} = -\frac{\Delta[O_2]}{\Delta t} = +1/2 \cdot \frac{\Delta[NO_2]}{\Delta t}$$

So when O_2 is reacting at $5.0 \times 10^{-4} \frac{mol}{L \cdot s}$ then NO will be reacting at $10.0 \times 10^{-4} \frac{mol}{L \cdot s}$ and NO_2 will be forming at $10.0 \times 10^{-4} \frac{mol}{L \cdot s}$

Concentration—Time Equations

26. Since the decomposition of N_2O_5 is first-order, the applicable integrated rate equation is

$$\ln \frac{[N_2O_5]}{[N_2O_5]_0} = -kt$$

$$\ln \left(\frac{2.50}{2.56}\right) = -k \cdot 4.26 \text{ min}$$

$$5.57 \times 10^{-3} \text{ min}^{-1} = k$$

Converting min^{-1} to s^{-1} (60 s = 1 min) : $9.28 \times 10^{-5} \text{ s}^{-1}$

28. This transformation is first-order (Example 15.3)

$$\ln \frac{[C_3H_6]}{[C_3H_6]_0} = -kt$$

$$\ln \frac{[0.025]}{[0.050]} = -(5.4 \times 10^{-2} \text{ hr}^{-1})t$$

$$12.8 \text{ hr} = t \qquad \text{or 13 hr (to 2 sf)}$$

30. The units of the rate constant, $\frac{L}{mol \cdot min}$, indicate that this decomposition is second order, so the integrated rate law is

$$\frac{1}{[NO_2]} - \frac{1}{[NO_2]_0} = kt$$

$$\frac{1}{1.50 \frac{mol}{L}} - \frac{1}{2.00 \frac{mol}{L}} = 3.40 \frac{L}{mol \cdot min} \cdot t$$

$$0.166 \frac{mol}{L} = 3.40 \frac{L}{mol \cdot min} \cdot t$$

$$\frac{0.166 \frac{mol}{L}}{3.40 \frac{L}{mol \cdot min}} = t \qquad \text{so } 0.0490 \text{ min} = t \text{ or 2.94 seconds}$$

Half—Life

32. The reaction is first order: $t_{1/2} = \dfrac{0.693}{k} = \dfrac{0.693}{3.33 \times 10^{-6} \text{ hr}^{-1}} = 2.08 \times 10^5 \text{ hr}$

Time for concentration to drop from 1.0 M to 0.20 M:

$$\ln\left(\frac{0.20}{1.00}\right) = -3.33 \times 10^{-6} \text{ hr}^{-1} \cdot t \quad \text{and solving: } t = 4.8 \times 10^5 \text{ hr}$$

34. If $t_{1/2} = 37.9$ sec, then $k = 1.83 \times 10^{-2}$ sec^{-1}. If 3/4 of PH_3 decomposes, the fraction remaining is 1/4.

$$\ln\left[\frac{[PH_3]_t}{[PH_3]_0}\right] = \ln\left(\frac{1}{4}\right) = -(1.83 \times 10^{-2})\, t$$
$$t = 7.58 \times 10^1 \text{ sec}$$

Alternately, for $[PH_3]$ to decrease to $\frac{1}{4}[PH_3]_0$ requires 2 half-lives

$[(\frac{1}{2})^2 = \frac{1}{4}]$ so t $= 2 \cdot t_{1/2}$ or 2(37.9 s) = 75.8 sec

36. Since the decomposition is first order: $\ln \dfrac{[\text{azomethane}]}{[\text{azomethane}]_0} = -kt$

Converting 2.00 g of azomethane to **moles** :

$$2.00 \text{ g azomethane} \cdot \frac{1 \text{ mol azomethane}}{58.08 \text{ g azomethane}} = 0.0344 \text{ mol}$$

$$\ln\frac{[\text{azomethane}]}{[0.0344 \text{ mol}]_0} = -(40.8 \text{ min}^{-1})(0.0500 \text{ min})$$

$$\ln\frac{[\text{azomethane}]}{[0.0344 \text{ mol}]_0} = -2.04$$
$$\frac{[\text{azomethane}]}{[0.0344 \text{ mol}]_0} = e^{-2.04} = 0.130$$

$$[\text{azomethane}] = 4.48 \times 10^{-3} \text{ mol}$$

Since 1 mol N_2 is produced when 1 mol of azomethane decomposes, the amount of N_2 is
0.0344 mol - 0.00448 mol = 0.0300 mol N_2

38. For the reaction HOF (g) $\rightarrow$ HF (g) + 1/2 O_2 (g)

the half-life [for HOF (g)] at room temperature is 30. minutes. If the partial pressure of HOF is initially 100. mm, after half of the HOF has decomposed, **its partial pressure will be 50. mmHg**. The HF formed will contribute 50. mm and the O_2 an additional 25 mm for a **total pressure of 125 mmHg**.

The first order kinetics for this decomposition allow us to calculate the rate constant, k.

$$\ln \left(\frac{[HOF]}{[HOF]_0} \right) = -kt$$

The [HOF] will be 1/2 of the original value, and substituting yields:

$$\ln (0.5) = -k (30. \text{ min}) \text{ and solving for k yields } k = 0.0231 \text{ min}^{-1}$$

Substitution of the rate constant into the concentration/time equation permits us to calculate the fraction of HOF remaining (X) after 45 minutes.

$$\ln X = -(0.0231 \text{min}^{-1})(45 \text{ min}) \qquad \text{and} \qquad X = 0.35$$

After 45 minutes the partial pressure of HOF is then 35% of its original value or 35 mmHg. The HF pressure will be 65 mmHg (100 - 35) and the O_2 pressure approximately 33 mmHg (1/2 O_2) for a total pressure (HOF + HF + O_2) of 133 mmHg.

40. Since this is a first-order process $\qquad \ln \frac{[Cu^{2+}]}{[Cu^{2+}]_0} = -kt \quad$ and $k = -\dfrac{0.693}{12.7 \text{ hr}}$

t is 2 days and 16 hr (64 hr)

$$\text{so } \ln \frac{[Cu^{2+}]}{[Cu^{2+}]_0} = -\frac{0.693}{12.7 \text{ hr}} \cdot 64 \text{ hr}$$

$$\ln \frac{[Cu^{2+}]}{[Cu^{2+}]_0} = -3.49 \quad \text{and} \quad \frac{[Cu^{2+}]}{[Cu^{2+}]_0} = 0.030 \quad \text{or 3.0 \% remains (2 sf)}$$

After 5 days (120 hrs)

$$\ln \frac{[Cu^{2+}]}{[Cu^{2+}]_0} = -\frac{0.693}{12.7 \text{ hr}} \cdot 120 \text{ hr} = -6.55$$

and $\qquad \dfrac{[Cu^{2+}]}{[Cu^{2+}]_0} = 0.0014 \quad$ or 0.14 % remains (2 sf)

Graphical Analysis of Rate Equations and k

42.   a. Plot of ln[sucrose] and $\dfrac{1}{[\text{sucrose}]}$ versus time.

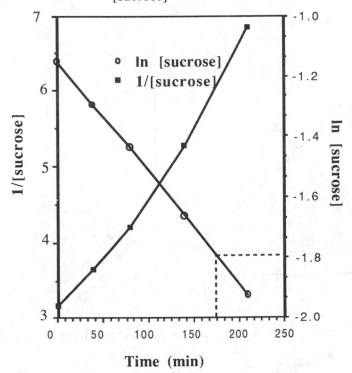

Time (min)

b. Since the reaction is first order with respect to sucrose, the rate expression may be written: Rate = k [sucrose]. The rate constant can be calculated using two data points: Using the first two points yields:

$$\ln\left(\frac{[A]}{[A]_0}\right) = -kt \quad \text{and substituting}: \quad \ln\left(\frac{0.274}{0.316}\right) = -k(39 \text{ min})$$

$$\frac{\ln(0.867)}{39 \text{ min}} = -k \quad \text{and } 3.7 \times 10^{-3} \text{ min}^{-1} = k$$

c. Using the graph of ln[sucrose] vs time, an estimate at 175 minutes yields:

ln[sucrose] = -1.8 corresponding to [sucrose] = 0.166 M

193

44. For the decomposition of N_2O:

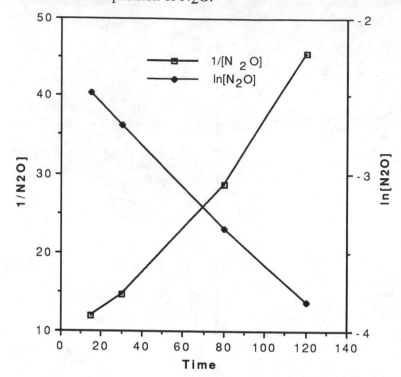

Since $\ln[N_2O]$ vs t gives a straight line, we know that the reaction is first order, and the line has a slope $= -k$.

$$\text{slope} = -k = \frac{(-3.8167) - (-2.4829)}{(120.0 - 15.0)\text{min}} = \frac{1.3338}{105.0 \text{ min}} = 0.0127 \text{ min}^{-1}$$

The rate equation is: Rate $=$ k [N$_2$O] and the rate of decomposition when $[N_2O] = 0.035$ mol/L :

$$\text{Rate} = (0.0127 \text{ min}^{-1})(0.035 \frac{\text{mol}}{\text{L}}) = 4.4 \times 10^{-4} \frac{\text{mol}}{\text{L} \cdot \text{min}}$$

Kinetics and Energy

46.   The reaction coordinate for the reaction $A + B \rightarrow C + D$

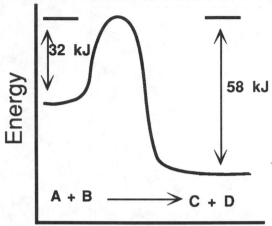

The reaction $A + B \rightarrow C + D$ is exothermic.

Note that the energy level of the products $C + D$ is lower than that of the reactants, $A + B$. As the reaction proceeds, the energy corresponding to the difference between these is (32- 58) kJ. So as the reaction proceeds, about 26 kJ of energy are liberated.

48.   The E* for the reaction N_2O_5 (g) $\rightarrow$ 2 NO_2 (g) + 1/2 O_2 (g)

Given k at 25 °C = 3.46 x 10^{-5} s^{-1} and k at 55 °C = 1.5 x 10^{-3} s^{-1}

The rearrangement of the Arrhenius equation shown in your text as equation 15.7 is helpful here.

$$\ln \frac{k_2}{k_1} = - \frac{E^*}{R} \left(\frac{1}{t_2} - \frac{1}{t_1} \right)$$

$$\ln \frac{1.5 \text{ x } 10^{-3} \text{ } s^{-1}}{3.46 \text{ x } 10^{-5} \text{ } s^{-1}} = - \frac{E^*}{8.31 \text{ x } 10^{-3} \text{ kJ/mol} \cdot \text{K}} \left(\frac{1}{328} - \frac{1}{298} \right)$$

and solving for E* yields a value of **102 kJ/mol for E*.**

50.   a. A plot of ln k vs $\frac{1}{T}$ is shown here

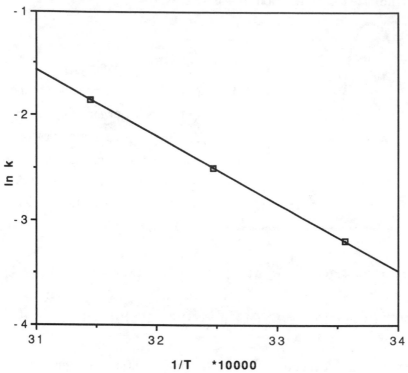

1/T   *10000

The straight has the equation $\ln k = \ln A - \frac{E_a}{R}(\frac{1}{T})$

So the slope is $-\frac{E_a}{R} = \frac{(-3.197) - (-1.852)}{(33.557 - 31.447) \times 10^{-4}} = \frac{-1.345}{2.110 \times 10^{-4}}$

$-\frac{E_a}{R} = -6.374 \times 10^3$ K

and since $-\dfrac{E_a}{8.31 \times 10^{-3} \frac{kJ}{k \cdot mol}} = -6.347 \times 10^3$ K

$E_a = 52.97$ or 53.0 kJ/mol (to 3 sf)

b. Calculate A: rearranging $\ln A = \ln K + \frac{E_a}{R}(\frac{1}{T})$

$= -3.197 + (\dfrac{53.0\frac{kJ}{mol}}{8.31 \times 10^{-3} \frac{kJ}{K \cdot mol}})(33.557 \times 10^{-4}$ K$^{-1})$

$= -3.197 + 21.391 = 18.194$

and $A = 7.97 \times 10^7$ or 8.0×10^7 (2 sf)

Determine k at 311 K

Substituting into $\ln k = \ln A - \frac{E_a}{R}(\frac{1}{T})$ where $T = 311$

$$\ln k = \ln(8.0 \times 10^7) - \left(\frac{53.0 \frac{kJ}{mol}}{8.31 \times 10^{-3} \frac{kJ}{K \cdot mol}} \right)(3.21 \times 10^{-3}\ K^{-1})$$

$$\ln k = 18.194 - 20.508$$

$$= -2.314$$

$$\text{and } k = 0.0985$$

Mechanisms

52.   Elementary Step   Rate law

a. $NO\ (g) + NO_3\ (g) \rightarrow 2\ NO_2\ (g)$ Rate $= k[NO][NO_3]$
Reaction is bimolecular

b. $Cl\ (g) + H_2\ (g) \rightarrow HCl\ (g) + H\ (g)$ Rate $= k[Cl][H_2]$
Reaction is bimolecular

c. $(CH_3)_3CBr\ (aq) \rightarrow (CH_3)_3C^+\ (aq) + Br^-\ (aq)$ Rate $= k[(CH_3)_3CBr]$
Reaction is unimolecular

54.   a. The rate determining step in a mechanism is the slowest step. For the steps given,
Step 2 is the rate determining step.

b. Rate $= k[O_3][O]$ is the rate expression for Step 2.

c. **Step 1 is unimolecular** and **Step 2 is bimolecular**.

Catalysis

56.   a. True—The concentration of a catalyst may appear in the rate expression. The function
may be to provide an alternative pathway or to assist in optimizing molecular orientations.

b. False—While a catalyst is involved in one or more steps of a process, the catalyst is
regenerated in subsequent steps, being available for further reaction.

c. False—Homogeneous catalysts are always in the same phase as the reactants, but heterogeneous catalysts are in different phases. As an example, the Pt gauze used in the production of nitric acid is in the solid state, while the reactants are not.

d. True—Nitric oxide (NO) catalyzes the conversion of O_2 to O_3 in the lower atmosphere, but in the upper atmosphere nitric oxide catalyzes the decomposition of O_3 to O_2.

e. True—by providing alternate mechanisms, catalysts can affect the step in a mechanism that is rate-determining.

58. One enzyme molecule hydrates 10^6 molecules of CO_2 / second.

so 1 L of 5×10^{-6}M enzyme would hydrate

$$5 \times 10^{-6} \frac{\text{mol enzyme}}{\text{L}} \cdot 1 \text{ L} \cdot \frac{6.022 \times 10^{23} \text{ molecules enzyme}}{1 \text{ mol enzyme}} \cdot \frac{10^6 \text{ molecules } CO_2}{1 \text{ molecule enzyme}}$$

$$= 3 \times 10^{24} \text{ molecules } CO_2 \text{ in 1 second}$$

and in 1 hr,

$$\frac{3 \times 10^{24} \text{ molecules } CO_2}{1 \text{ s}} \cdot \frac{3600 \text{ s}}{1 \text{ hr}} \cdot \frac{7.308 \times 10^{-23} \text{ g } CO_2}{1 \text{ moecule } CO_2} \cdot \frac{1 \text{ kg } CO_2}{1 \times 10^3 \text{ g } CO_2}$$

$$= 792 \text{ kg } CO_2 \text{ or } 800 \text{ kg } CO_2 \text{ (to 1 sf)}$$

Note the mass of CO_2 per molecule can be determined by dividing the molar mass (44.01) by Avogadro's number.

General Questions

60.   a. After you do part b, the rate expression may then be written as: Rate = k [CO][NO2]

b. Using the kinetic data, note that the rate of the second experiment is two times that of the first experiment. Note that this corresponds to a doubling in [NO2]—hence the reaction is **first order in [NO2]**. If you compare the fourth and fifth experiments, note that the [CO] increases by a factor of 1.5 , and the rate increases by a factor of $\frac{10.2}{6.8} = 1.5$. The reaction is therefore **first order with respect to CO**, and second order overall.

c. Using the Rate expression from part a, substitute some of the data from the table.

$$Rate = k \, [CO][NO_2]$$

$$3.4 \times 10^{-8} \frac{mol}{L \cdot hr} = k \cdot 5.1 \times 10^{-4} \frac{mol}{L} \cdot 0.35 \times 10^{-4} \frac{mol}{L} \quad and$$

$$\frac{3.4 \times 10^{-8} \frac{mol}{L \cdot hr}}{1.785 \times 10^{-8} \frac{mol^2}{L^2}} = 1.9 \frac{L}{mol \cdot hr}$$

62.   a. What quantity of NO_x remains after 5.25 hr?

Average half-life is 3.9 hr.   The rate constant $k = \dfrac{0.693}{3.9 \ hr}$

So $\ln \dfrac{[NO_x]}{[NO_x]_0} = -\dfrac{0.693}{3.9 \ hr} \cdot 5.25 \ hr = -0.933$

and $\dfrac{[NO_x]}{[NO_x]_0} = 0.393$

So 39.3 % of the original remains or : $0.393 \cdot 1.50 \ mg = 0.59 \ mg$

b. Time to decrease 1.50 mg of NOx to 2.50×10^{-6} mg:

$\ln \dfrac{[2.50 \times 10^{-6}]}{[1.50]} = -\dfrac{0.693}{3.9 \ hr} \cdot t \quad and \quad t = 75 \ hr$

64. The net equation on summing three equations:

$$Cl + O_3 \longrightarrow ClO + O_2$$
$$ClO + O \longrightarrow Cl + O_2$$
$$O_3 \longrightarrow O + O_2$$
$$\overline{\qquad\qquad\qquad\qquad\qquad}$$
$$2 \, O_3 \longrightarrow 3 \, O_2$$

The overall process consumes 2 molecules of ozone, and since Cl atoms are not consumed the process can be repeated many, many times. Since Cl is consumed (in the first step) and produced in a subsequent step, we denote **Cl as a catalyst. ClO is an intermediate**.

66. The Arrhenius equation may be written:

$$k = A \cdot e^{-\frac{E^*}{RT}}$$

If we assume that A will be constant for the two pathways, we can judge the effect of changing the activation energy by substitution of some values into the Arrhenius equation. Pick an arbitrary value for E*, say 50 kJ/mol, at a temperature of 298 K.

$$k_1 = A \cdot e^{-\frac{50 \text{ kJ/mol}}{8.31 \times 10^{-3} \frac{\text{kJ}}{\text{K} \cdot \text{mol}} \cdot 298 \text{ K}}}$$

and $k_1 = e^{-20.2}$ or a value of approximately 1.70×10^{-9}

Now solve for k using $E^* = 45$ kJ/mol

$$k_2 = A \cdot e^{-\frac{45 \text{ kJ/mol}}{8.31 \times 10^{-3} \frac{\text{kJ}}{\text{K} \cdot \text{mol}} \cdot 298 \text{ K}}}$$

and $k_2 = e^{-18.2}$ or a value of approximately 1.28×10^{-8}

The ratio of these two values is: $\dfrac{k_2}{k_1} = \dfrac{1.28 \times 10^{-8}}{1.70 \times 10^{-9}} = 7.5$

Note that this ratio of the rate constants is good for 298 K. The ratio is dependent upon temperature. For example, using the two activation energies chosen here (50 and 45 kJ/mol) at 1000 K, the ratio of the k's is approximately 1.8. If we assume that the reaction is first order, we could write Rate $= k$ [A]. With k_2 being 7.5 times greater than k_1, then $Rate_2$ would be about 7.5 times greater than $Rate_1$.

68. For the first order process we can calculate the rate constant, k, from the equation :

$$\ln \left(\frac{[A]}{[A]_0}\right) = - kt$$

Since we know that the concentration has **decreased by 20%** in 6 hours,

$$\ln \left(\frac{[0.8]}{[1.0]}\right) = - k \cdot (6 \text{ hours}) \text{ and } k = 3.7 \times 10^{-2} \text{ hr}^{-1}$$

To calculate the half-life, $t_{1/2}$, we use a modified form of the previous expression since [A]—after one half-life— is exactly 1/2 the original concentration of A:

$$t_{1/2} = \frac{0.693}{k} = \frac{0.693}{3.7 \times 10^{-2} \text{ hr}^{-1}} = 1.9 \times 10^1 \text{ hr}$$

70. Determine k at 338.0 K

The Arrhenius equation may be written:

$$\ln k_2 - \ln k_1 = -\frac{E_a}{R}\left[\frac{1}{T_2} - \frac{1}{T_1}\right]$$

so $\ln k_2 - \ln (0.0900 \text{ min}^{-1}) = -\dfrac{103 \frac{\text{kJ}}{\text{mol}}}{8.31 \times 10^{-3} \frac{\text{kJ}}{\text{K} \cdot \text{mol}}}\left[\dfrac{1}{338.0 \text{ K}} - \dfrac{1}{328.0 \text{ K}}\right]$

$$\ln k_2 = -\frac{103 \frac{\text{kJ}}{\text{mol}}}{8.31 \times 10^{-3} \frac{\text{kJ}}{\text{K} \cdot \text{mol}}}\left[\frac{1}{338.0 \text{ K}} - \frac{1}{328.0 \text{ K}}\right] + \ln(0.0900 \text{ min}^{-1})$$

$$= 1.118 + (-2.408)$$

$$= -1.290$$

$$k_2 = 0.275 \text{ min}^{-1}$$

72.   a. What mass of CH_3OCH_3 remains after 125 min? 145?

$$\ln \frac{[CH_3OCH_3]}{[CH_3OCH_3]_0} = -k\,t \quad \text{and } k = \frac{0.693}{25.0 \text{ min}}$$

$$\ln \frac{[CH_3OCH_3]}{[CH_3OCH_3]_0} = -\frac{0.693}{25.0 \text{ min}} \cdot 125 \text{ min}$$

$$\frac{[CH_3OCH_3]}{[CH_3OCH_3]_0} = 0.0313$$

So 3.13 % remains or 3.13 % x 8.00 g = 0.250 g

After 145 min:

$$\ln \frac{[CH_3OCH_3]}{[CH_3OCH_3]_0} = -\frac{0.693}{25.0 \text{ min}} \cdot 145 \text{ min} = -4.02$$

$$\frac{[CH_3OCH_3]}{[CH_3OCH_3]_0} = 0.0180$$

So 1.80 % remains or 1.80 % x 8.00 g = 0.144 g

b. Time to decrease 7.6 ng to 2.25 ng

$$\ln \frac{2.25 \text{ ng}}{7.60 \text{ ng}} = -\frac{0.693}{25.0 \text{ min}} \cdot t$$

$$43.9 \text{ min} = t$$

c. Fraction remaining after 150 minutes?

   While we could substitute $t = 150$ min into the equation, we note that 150 minutes is exactly 6 half-lives, and the fraction remaining after n half-lives is $\left(\frac{1}{2}\right)^n$.

   So $\left(\frac{1}{2}\right)^6$ or $\frac{1}{64}$ of the original amount ($\sim 1.5\ \%$)

74. a. Calculate ΔH for

1. $O_2(g) \longrightarrow O_2(\text{adsorbed})$: $(-37) - (0) = -37$ kJ
2. $O_2(\text{adsorbed}) \longrightarrow 2\ O(\text{adsorbed})$: $(-251) - (-37) = -214$ kJ
3. $O_2(g) \longrightarrow 2\ O(\text{adsorbed})$: $(-251) - (0) = -251$ kJ

   or the sum of $\Delta H_1 + \Delta H_2$

b. What is E_a for $O_2(\text{adsorbed}) \longrightarrow 2\ O(\text{adsorbed})$?

   The energy difference between $O_2(\text{adsorbed}$ (-37 kJ) and the transition state (-17 kJ) is 20 kJ.

Conceptual Questions

76. While all situations (grain elevators, coal mines, flour mills, and rock quarries) involve finely divided particles, the first three situations possess the possibility of violent reaction of those particles with O_2 (to form gaseous products), while particulate rocks don't react with O_2 to any appreciable extent.

78. Finely divided rhodium (and any other metal for that matter) has a larger surface area than a block of the metal of the same mass. Hydrogenation reactions depend on adsorption of H_2 on the catalyst—hence the greater the surface area, the greater the locations for such adsorption to occur.

Summary Questions

80.   a. Volume of basement = 12 m x 7.0 m x 3.0 m = 252 m^3 or 0.252 L
           Pressure of Rn = 1.0 x 10^{-6} mmHg

We can calculate the moles of Rn with the Ideal Gas Law (assuming the basement is at 25 °C   or 298 K).

$$n = \frac{PV}{RT} = \frac{(1.0 \times 10^{-6} \text{ torr})(0.252 \text{ L})}{(62.4 \frac{\text{L} \cdot \text{torr}}{\text{K} \cdot \text{mol}})(298 \text{ K})} = 1.4 \times 10^{-11} \text{ mol Rn}$$

and multiplying by Avogadro's number — 8.2 x 10^{12} atoms Rn

So the number of atoms/L is: $\dfrac{8.2 \times 10^{12}}{0.252 \text{ L}} = 3.3 \times 10^{13} \dfrac{\text{atoms}}{\text{L}}$

b. With $t_{1/2}$ of 3.82 days, the number of atoms/Liter after 30. days will be:

$$\ln \frac{[\text{Rn}]}{[\text{Rn}]_0} = -\frac{0.693}{3.82 \text{ days}} \cdot 30. \text{ days} = -5.44$$

The fraction of radon is: $\dfrac{[\text{Rn}]}{[\text{Rn}]_0} = 4.3 \times 10^{-3}$

So the number of $\dfrac{\text{atoms}}{\text{L}}$ will be $(4.3 \times 10^{-3})(3.2 \times 10^{13} \dfrac{\text{atoms}}{\text{L}})$

$$= 1.4 \times 10^{11} \frac{\text{atoms}}{\text{L}}$$

Chapter 16:
Principles of Reactivity : Chemical Equilibria

Writing and Manipulating Equilibrium Constant Expressions

7. Equilibrium constant expressions:

a. $K_c = \dfrac{[H_2O]^2[O_2]}{[H_2O_2]^2}$
    b. $K_c = \dfrac{[PCl_5]}{[PCl_3][Cl_2]}$

c. $K_c = \dfrac{[CO_2]}{[CO][O_2]^{1/2}}$
    d. $K_c = \dfrac{[CO]^2}{[CO_2]}$

e. $K_c = \dfrac{[CO_2]}{[CO]}$

9. Comparing the two equilibria:

$$SO_2\ (g) + \frac{1}{2}O_2\ (g) \rightleftharpoons SO_3\ (g) \qquad K_1$$

$$2\ SO_3\ (g) \rightleftharpoons 2\ SO_2\ (g) + O_2\ (g) \qquad K_2$$

the expression that relates K_1 and K_2 is (e) : $K_2 = \dfrac{1}{K_1^2}$

11. Adding

$Fe(s) + CO_2(g) \rightleftharpoons FeO(s) + CO(g)$	$K_c = 1.5$
$H_2O(g) + CO(g) \rightleftharpoons H_2(g) + CO_2(g)$	$K_c = 1.6$

gives the overall $\quad Fe(s) + H_2O(g) \rightleftharpoons H_2(g) + FeO(s) \qquad K_{net}$

and $K_{net} = K_c \cdot K_c = 1.5 \times 1.6 = 2.4$

13. The changes are two-fold:

1. The desired reaction has CO_2 as a product not a reactant. If $K_1 = 6.66 \times 10^{-12}$ then the equilibrium for $CO(g) + 1/2\ O_2(g) \rightleftharpoons CO_2(g)$ would be $\dfrac{1}{K_1}$ or 1.50×10^{11}.

2. The coefficients for the desired reaction are all twice as large as the reversed reaction. So for $2\ CO(g) + O_2(g) \rightleftharpoons 2\ CO_2(g)$ the new K would be $(\dfrac{1}{K_1})^2$ or 2.25×10^{22}.

The Reaction Quotient

15. Calculate Q

$$Q = \frac{[SO_3]^2}{[SO_2]^2[O_2]} = \frac{(6.9 \times 10^{-3} \text{ M})^2}{(5.0 \times 10^{-3} \text{ M})^2(1.9 \times 10^{-3} \text{ M})} = 1.0 \times 10^3$$

Since Q >>Kc, SO$_3$ will need to form more SO$_2$ and O$_2$ to have Q approach K
(Products $\rightarrow$ Reactants)

17. Calculate Q

$$Q = \frac{[NO]^2[Cl_2]}{[NOCl]^2} = \frac{(2.5 \times 10^{-3} \text{ M})^2(2.0 \times 10^{-3} \text{ M})}{(5.0 \times 10^{-3} \text{ M})^2} = 5.0 \times 10^{-4}$$

Since Q < K, the **reaction is not at equilibrium**, and Reactants $\rightarrow$ Products will
increase Q—so that Q approaches K.

Calculating Equilibrium Constants

19. For $K = \dfrac{[SO_3]^2}{[SO_2]^2[O_2]} = \dfrac{(4.13 \times 10^{-3} \text{ M})^2}{(3.77 \times 10^{-3} \text{ M})^2(4.30 \times 10^{-3} \text{ M})} = 279$

21. For the reaction: H$_2$ (g) + CO$_2$ (g) $\rightleftharpoons$ H$_2$O (g) + CO (g)

    a. Calculate K at 986 °C if at equilibrium :P[CO] = P[H$_2$O] = 0.11 atm

$$P[H_2] = P[CO_2] = 0.087 \text{ atm}$$

$$Kc = \frac{[P_{CO}][P_{H_2O}]}{[P_{CO_2}][P_{H_2}]} = \frac{(0.11 \text{ M})^2}{(0.87 \text{ M})^2} = 1.6$$

    b. Substituting the equilibrium concentrations of H$_2$ and CO$_2$ (0.025 M) into the
equilibrium, we obtain

$$Kc = \frac{[CO][H_2O]}{(0.025)^2} = 1.6 \text{ and since } [CO] = [H_2O]$$

$$[CO] = [H_2O] = 3.2 \times 10^{-2} \text{ M}$$

and the amounts of CO and H$_2$O are $3.2 \times 10^{-2} \dfrac{\text{mol}}{\text{L}} \cdot 2.0 \text{ L} = 6.4 \times 10^{-2}$ M

23. K_c for the equation: $H_2O\ (g)\ \rightleftharpoons\ H_2\ (g) + \frac{1}{2}O_2\ (g)$

If $[H_2O]_i\ =\ 2.0\ M$ and 10. % dissociates to H_2 and O_2.

	H_2O	H_2	O_2
Initial	2.0 M	0	0
Change	- 0.20	+ 0.20	+ 0.10
Equilibrium	1.8	+ 0.20	+ 0.10

and $K_c = \dfrac{[H_2][O_2]^{1/2}}{[H_2O]} = \dfrac{(0.20)(0.10)^{1/2}}{(1.8)} = 0.035$

25.

	SO_3	SO_2	O_2
Initial	0.375 M	0	0
Change	- 0.145	+ 0.145	+ 0.0725
Equilibrium	+0.230	+ 0.145	+ 0.0725

and $K_c = \dfrac{[SO_2]^2[O_2]}{[SO_3]} = \dfrac{(+0.145)^2(0.0725)}{(+0.230)^2} = 0.029$ (to 2 sf)

27. K_p at 25 °C for the system: $N_2H_6CO_2\ (s)\ \rightleftharpoons\ 2\ NH_3\ (g) + CO_2\ (g)$

Total Pressure = 0.116 atm = $P(NH_3) + P(CO_2)$

The reaction produces 2 mol NH_3 and 1 mol O_2 for each mol of ammonium carbamate that sublimes. The partial pressure of ammonia will be twice as great as that of carbon dioxide.

Pressure = 0.116 atm = $2 \cdot P(CO_2) + P(CO_2)$

0.116 atm = $3 \cdot P(CO_2)$

and $P(NH_3)$ = $2 \cdot P(CO_2)$ = 0.0773 atm.

Substitution into the equilibrium constant expression yields:

$K_P = P_{CO_2} \cdot P_{NH_3}^2 = (0.0387) \cdot (0.0773)^2 = 2.31 \times 10^{-4}$

29. For the system butane $\rightleftharpoons$ isobutane K_c = 2.5

    Equilibrium concentrations may be found:

	butane	isobutane
Initial	0.034	0
Change	- x	+ x
Equilibrium	0.034 - x	x

    $$K_c = \frac{x}{0.034 - x} = 2.5 \text{ and } x = 2.4 \times 10^{-2}$$

    and [isobutane] $= 2.4 \times 10^{-2}$ M, [butane] $= 0.034 - x = 1.0 \times 10^{-2}$ M

31. Given at equilibrium

    $$\frac{[CO][Br_2]}{[COBr_2]} = 0.190$$

    If $[COBr_2] = 6.0 \times 10^{-3}$ then $[CO][Br_2] = (0.190)(6.0 \times 10^{-3})$

    and $[CO] = [Br_2] = 0.034$ M

33.   a. $K_c = \frac{[NO_2]^2}{[N_2O_4]} = 5.88 \times 10^{-3}$

    Initial $[N_2O_4] = \frac{20.0 \text{ g } N_2O_4}{5.00 \text{ L}} \cdot \frac{1 \text{ mol } N_2O_4}{92.01 \text{ g } N_2O_4} = 0.0435$ M

    and $\frac{(2x)^2}{(0.0435 - x)} = 5.88 \times 10^{-3}$

    Note that each mole of N_2O_4 consumed (x) produces 2 mol of NO_2 (2x)

    Solving $4x^2 = 5.88 \times 10^{-3} (0.0435 - x)$

    $\qquad\quad 4x^2 = 2.56 \times 10^{-4} - 5.88 \times 10^{-3}$

    and $4x^2 + 5.88 \times 10^{-3}x - 2.56 \times 10^{-4} = 0$

    Solving via the quadratic equation yields $x = 7.30 \times 10^{-3}$

    So the concentration of NO_2 at equilibrium 1.46×10^{-2} and the number of moles of NO_2 at equilibrium is 0.0730 mol.

    b. The percentage of N_2O_4 that dissociated:

    At equilibrium $[N_2O_4] = 4.35 \times 10^{-2} - 7.30 \times 10^{-3}$

    The concentration of N_2O_4 that was consumed (i.e. dissociated) is 7.30×10^{-3}

    The % dissociated is $\frac{7.30 \times 10^{-3}}{4.35 \times 10^{-2}}$ x   100 = 16.7 %

35. The initial concentration of $COBr_2$ is

$$\frac{1.06 \text{ g } COBr_2}{8.0 \text{ L}} \cdot \frac{1 \text{ mol } COBr_2}{187.8 \text{ g } COBr_2} = 7.06 \times 10^{-4} \text{ M}$$

a. Concentration of reactants and products at equilibrium

$$\frac{[CO][Br_2]}{[COBr_2]} = 0.190$$

Since 1 mol of CO (and of Br_2) forms for each mol of $COBr_2$, if **a** mol/L of $COBr_2$ forms products then

$$\frac{(a)(a)}{7.06 \times 10^{-4} - a} = 0.190$$

and $a^2 = 0.190(7.06 \times 10^{-4} - a)$.

Using the quadratic: $a = 7.03 \times 10^{-4} = [CO] = [Br_2]$

The concentration of $COBr_2$ is:

$$[COBr_2] = [7.06 \times 10^{-4}] - [7.03 \times 10^{-4}]$$
$$= 3 \times 10^{-6}$$

b. The total pressure is

$$P = \frac{nRT}{V} \text{ and since we have the molarities } (\frac{n}{V}) \text{ for the three gases.}$$

$$P = (7.03 \times 10^{-4} + 7.03 \times 10^{-4} + 3 \times 10^{-6})(0.0821 \frac{L \cdot atm}{K \cdot mol}) \cdot 346 \text{ K} = 0.400 \text{ atm}$$

37. For the equilibrium the equilibrium constant is $\frac{[I]^2}{[I_2]} = 3.76 \times 10^{-3}$

With an initial concentration of 0.500 M for I_2, the equation indicates that 2 mol of I form for each mol of I_2 that reacts. The equilibrium concentrations can then be substituted into the equilibrium expression.

$$\frac{(2x)^2}{0.500 - x} = 3.76 \times 10^{-3}$$

or $4x^2 = 3.76 \times 10^{-3}(0.500 - x)$

$$4x^2 + 3.76 \times 10^{-3}x - 1.88 \times 10^{-3} = 0$$

$$x = 2.12 \times 10^{-2}$$

So at equilibrium:

$$[I_2] = 0.500 - 0.0212 = 0.479 \text{ M}$$
$$[I] = 2(2.12 \times 10^{-2}) = 0.0424 \text{ M}$$

Disturbing a Chemical Equilibrium : Le Chatelier's Principle

39. Write the equilibrium expression to include the heat term :
$$2\,NOBr(g) + 16.1\,kJ \;\rightleftharpoons\; 2\,NO(g) + Br_2(g)$$

a. Adding more $Br_2(g)$: The equilibrium will **shift left** as Q becomes bigger than K, and more NOBr will need to form

b. Removing $NOBr(g)$: The equilibrium will **shift left** as the system attempts to compensate for the decreased amount of $NOBr(g)$

c. Decreasing the temperature : As T drops, the exothermic reaction will be favored and the **equilibrium will shift left**.

d. Increasing the container volume : If T is held constant, increasing the container volume will result in a decrease in Pressure, and the system will shift toward the side of the equilibrium with the greater total number of gas molecules (**shift right**).

41. K_c for butane $\rightleftharpoons$ iso-butane is 2.5.

a. Equilibrium concentration if 0.50 mol/L of iso-butane is added:

	butane	iso-butane
Original concentration	1.0	2.5
Change immediately after addition	1.0	2.5 + 0.50
Change (going to equilibrium)	+ x	- x
Equilibrium concentration	1.0 + x	3.0 - x

$$K_c = \frac{[\text{iso-butane}]}{[\text{butane}]} = \frac{3.0 - x}{1.0 + x} = 2.5$$

$$3.0 - x = 2.5\,(1.0 + x)$$

$$\text{and } 0.14 = x$$

The equilibrium concentrations are:

$$[\text{butane}] = 1.0 + x = 1.1\ M \text{ and } [\text{iso-butane}] = 3.0 - x = 2.9\ M$$

b. Equilibrium concentrations if 0.50 mol /L of butane is added:

	butane	iso-butane
Original concentration	1.0	2.5
Change immediately after addition	1.0 + 0.50	2.5
Change (going to equilibrium)	- x	+ x
Equilibrium concentration	1.5 - x	2.5 + x

$$K_c = \frac{[\text{iso-butane}]}{[\text{butane}]} = \frac{2.5 + x}{1.5 - x} = 2.5$$

$$2.5 + x = 2.5\,(1.5 - x)$$

$$\text{and } x = 0.36$$

The equilibrium concentrations are:

[butane] = 1.5 - x = 1.1 M and [iso-butane] = 2.5 + x = 2.9 M

43. The proposed rate law would be: Rate = $k_2[H][CO]$

But H is an intermediate, produced by the dissociation of H_2 with Rate_1 = $k_1[H_2]$ and consumed by the combination of H atoms in a reaction with Rate'_y = $k'_1[H]^2$
and since this step is an equilibrium

$$\text{Rate}_1 = \text{Rate}'_1$$

so $k_1[H_2]$ = $k'_1[H]^2$ and $\dfrac{k_1[H_2]}{k'_1}$ = $[H]^2$

and we can write the ratios of k_1 / k'_1 as K_c

so $K_c[H_2]$ = $[H]^2$

and $(K_c[H_2])^{1/2}$ = $[H]$

Substituting into the proposed rate law:

Rate = $k_2(\,K_c[H_2])^{1/2}\,[CO]$ and since $K_c^{1/2} \cdot k_2$ is another constant (call it k_0)

Rate = $k_0[H_2]^{1/2}[CO]$

45. The rate law indicates a bimolecular dependence on NO and unimolecular dependence on H_2.

    Mechanism 1 is unimolecular in both NO and H_2.

    Mechanism 2 is unimolecular in H_2 and the intermediate N_2O_2. N_2O_2 is produced by a reaction that can be written Rate $= K[NO]^2$. (see Example 16.10 p 782) and would therefore have the rate equation

$$Rate = K[NO]^2 \cdot k[H_2]$$
$$or \quad K'[NO]^2[H_2]$$

General Questions

47. $K_c = \dfrac{[NO]^2}{[N_2][O_2]} = 1.7 \times 10^{-3}$ at 2300 K.

    a. What is K_p for the reaction?

$$K_p = K_c \cdot (RT)^{\Delta n} \quad \text{equation 16.2}$$

    Since $\Delta n = 0$

$$N_2 + O_2 \rightleftharpoons 2\,NO$$

    2 mol gas total $\rightleftharpoons$ 2 mol gas total

$$K_p = 1.7 \times 10^{-3}$$

    b. K_c for the reaction when all coefficients are halved?

$$K_c = \frac{[NO]^2}{[N_2][O_2]} \qquad\qquad K'_c = \frac{[NO]}{[N_2]^{1/2}\,[O_2]^{1/2}}$$

    so $K_c = (K'_c)^2$ or $K'_c = (K_c)^{1/2}$

    $K'_c = (1.7 \times 10^{-3})^{1/2}$ or 4.1×10^{-2}

    c. K_c for the reversed reaction is the reciprocal of the original K_c

$$\frac{1}{1.7 \times 10^{-3}} = 5.9 \times 10^2$$

49. Begin by converting K_c to K_p

    since $N_2 + O_2 \rightleftharpoons 2\,NO$, $\Delta n = 0$

        2 mol gas   2 mol gas

    (see question 47)

    and $K_p = K_c = 1.7 \times 10^{-3}$

If the system is at equilibrium Q $= 1.7 \times 10^{-3}$

$$Q = \frac{P_{NO}^2}{P_{N_2} \cdot P_{O_2}} = \frac{(4.2 \times 10^{-3} \text{ atm})^2}{(0.50 \text{ atm})(0.25 \text{ atm})} = 1.4 \times 10^{-4}$$

The system **is not at equilibrium**.

Since Q < K, the system would proceed right (reactants $\longrightarrow$ products) towards equilibrium.

51.

	NOCl	NO	Cl$_2$
Initial	2.00 M	0	0
change	-0.66 M	+0.66 M	+0.33 M
Equilibrium	1.34 M	+0.66 M	+0.33 M

$$\text{So Kc} = \frac{[0.66]^2[0.33]}{[1.34]^2} = 8.0 \times 10^{-2}$$

53. $P_{total} = P_{NO_2} + P_{N_2O_4} = 1.5$ atm ; $P_{N_2O_4} = 1.5 - P_{NO_2}$

$$K_p = \frac{P_{N_2O_4}}{P_{NO_2}^2} = 6.75$$

$$K_p = \frac{(1.5 - P_{NO_2})}{P_{NO_2}^2} = 6.75 \quad \text{or} \quad 1.5 - P_{NO_2} = 6.75\, P_{NO_2}^2$$

and rearranging $6.75\, P_{NO_2}^2 + P_{NO_2} - 1.5 = 0$

and solving for P_{NO_2} (with the quadratic equation) yields:

$$P_{NO_2} = 0.40 \text{ atm and } P_{N_2O_4} = 1.1 \text{ atm}$$

55. $K_c = \dfrac{[dimer]}{[monomer]^2} = 3.2 \times 10^4$

	monomer	dimer
Initial pressure	5.4×10^{-4}	0
Equilibrium	x	$\frac{1}{2}(5.4 \times 10^{-4} - x)$

$$\frac{\frac{1}{2}(5.4 \times 10^{-4} - x)}{x^2} = 3.2 \times 10^4$$

or $2.7 \times 10^{-4} - 0.5 x = 3.2 \times 10^4 x^2$

rearranging $3.2 \times 10^4 x^2 + 0.5 x - 2.7 \times 10^{-4} = 0$

and solving via the quadratic equation yields $x = 8.4 \times 10^{-5}$

a. The % of acetic acid converted to the dimer is
$$\frac{5.4 \times 10^{-4} - 8.4 \times 10^{-5}}{5.4 \times 10^{-4}} \times 100 = 84\%$$

b. As the temperature increases, the equilibrium would **shift to the left**, producing more monomeric acetic acid.

57. $K_c = \dfrac{[SO_3]^2}{[SO_2]^2[O_2]} = 279$; $\qquad K_p = \dfrac{P_{SO_3}^2}{P_{SO_2}^2 \cdot O_2} = ?$

Remember that $K_p = K_c \cdot (RT)^{\Delta n}$
and $\Delta n = 2 - 3$ or -1

$$K_p = 279 \cdot \frac{1}{(0.821 \frac{L \cdot atm}{K \cdot mol})(1000K)} = 3.40$$

59. For zinc carbonate dissolving in water, the equilibrium constant expression is:

$$K_c = [Zn^{2+}][CO_3^{2-}] = 1.5 \times 10^{-11} \qquad \text{[Solids are omitted]}$$
At equilibrium $[Zn^{2+}] = [CO_3^{2-}]$ so $[Zn^{2+}]^2 = [CO_3^{2-}]^2 = 1.5 \times 10^{-11}$
and $[Zn^{2+}] = [CO_3^{2-}] = 3.9 \times 10^{-6}$ M

61. For the system shown, the relationship between K_p and K_c is:

$$K_p = K_c \cdot (RT)^{\Delta n}$$
Convert the given K_p to the more convenient (in this case) K_c.
Since $\Delta n = (2 - 1)$, $K_c = K_p \cdot (RT)^{-1}$

$$K_c = \frac{1.2 \times 10^{-10}}{(RT)^{\Delta n}} = \frac{1.2 \times 10^{-108}}{(0.082057 \frac{L \cdot atm}{K \cdot mol} \cdot 1800\ K)} = 8.12 \times 10^{-13}$$

With an original concentration of 0.10 M, if we assume that x mol/L of O_2 dissociates, the amount of O present at equilibrium will be 2x.

$$K_c = \frac{[O]^2}{[O_2]} = \frac{(2x)^2}{(0.10 - x)} = 8.12 \times 10^{-13}$$

The magnitude of K allows us to simplify the denominator to ~ 0.10, and solving gives

$$x = 1.4 \times 10^{-7} \text{ and } [O] = 2.8 \times 10^{-7}\ M.$$

The number of O atoms present:

$$\frac{2.8 \times 10^{-7}\ mol}{1\ L} \cdot \frac{10.\ L}{1} \cdot \frac{6.022 \times 10^{23}\ O\ atoms}{1\ mol\ O\ atoms} = 1.7 \times 10^{18}\ O\ atoms$$

63. $N_2O_4(g) \rightleftharpoons 2\ NO_2(g)$

   a. Calculate the fraction dissociated when $P_{total} = 1.50$ atm

$$K_p = 0.113 = \frac{P_{NO}^2}{P_{N_2O_4}}$$

$P_{total} = 1.50\ atm = P_{NO} + P_{N_2O_4}$ so $P_{N_2O_4} = 1.50\ atm - P_{NO}$

$$K_p = 0.113 = \frac{P_{NO}^2}{[1.50 - P_{NO}]}$$

$$P_{NO}^2 + 0.113 \cdot P_{NO} - 0.170 = 0$$

$P_{NO} = 0.360$ atm   Therefore, $P_{N_2O_4} = 1.14$ atm at equilibrium

Initially, $P_{N_2O_4}$ must have been 1.14 atm + 1/2 $\cdot$ 0.366 atm = 1.32 atm

Fraction of N_2O_4 dissociated $= \frac{1.32\ atm - 1.14\ atm}{1.32\ atm} = 0.14$

   b. Calculate the fraction dissociated when $P_{total} = 1.00$ atm

$$K_p = 0.113 = \frac{P_{NO}^2}{[1.00 - P_{NO}]}$$ Solving exactly as above,

$P_{NO} = 0.284$ atm   Therefore, $P_{N_2O_4} = 0.716$ atm at equilibrium

Initially, $P_{N_2O_4}$ must have been 0.716 atm + 1/2 • 0.284 atm = 0.166

Note that when the total pressure dropped, the fraction dissociated **increased**.

65.   a. The initial concentration of SO_2Cl_2 is:

$$\frac{6.70 \text{ g } SO_2Cl_2}{1.00 \text{ L}} \cdot \frac{1 \text{ mol } SO_2Cl_2}{135.0 \text{ g } SO_2Cl_2} = 0.0496 \text{ M}$$

As the SO_2Cl_2 dissociates into SO_2 and Cl_2 (x mol/L) equilibrium occurs and we write:

$$K_c = \frac{x^2}{(0.0496 - x)} = 0.045 \text{ or } x^2 = 0.045 (0.0496 - x)$$

Solving this equation via the quadratic equation gives x = 2.98 x 10^{-2} M or 3.0 x 10^{-2} M.

At equilibrium, $[SO_2] = [Cl_2] = 3.0$ x 10^{-2} M   and $[SO_2Cl_2] = (4.96 - 3.0)$ x 10^{-2}

or 2.0 x 10^{-2} (2 sf).

The fraction of SO_2Cl_2 that has dissociated is

$$\frac{4.96 \text{ x } 10^{-2} - 1.96 \text{ x } 10^{-2} \text{ atm}}{4.96 \text{ x } 10^{-2}} = 0.60.$$

b. Solving for the concentration of Cl_2

$$\frac{P}{RT} = \frac{n}{V} = \frac{1.00 \text{ atm}}{(0.082057 \frac{L \cdot atm}{K \cdot mol})(648 \text{ K})} = 1.88 \text{ x } 10^{-2} \text{ M}$$

Substituting into the equilibrium expression:

$$K_c = \frac{(x)(x + 1.88 \text{ x } 10^{-2})}{(0.0496 - x)} = 0.045$$

$$x^2 + 0.0188 x = 0.00223 - 0.045 x$$

$$x^2 + 0.0638 x - 0.00223 = 0$$

$$x = 2.51 \text{ x } 10^{-2} \text{ or } 2.5 \text{ x } 10^{-2} \text{ (to 2 sf)}$$

The equilibrium concentrations are :

$[SO_2Cl_2] = 2.5$ x 10^{-2} M

$[SO_2] = 2.5$ x 10^{-2} M

$[Cl_2] = 4.4$ x 10^{-2} M

The fraction of SO_2Cl_2 dissociated is : $\dfrac{4.96 \text{ x } 10^{-2} - 2.51 \text{ x } 10^{-2}}{4.96 \text{ x } 10^{-2}} = 0.50$

215

    c. LeChatelier's principle predicts that the addition of chlorine would shift the position of equilibrium to the left, preserving more SO_2Cl_2--a prediction which is confirmed by these calculations.

67.   a. Step 2 is rate-determining.

        The rate law for this step is : Rate $= k[O_3][O]$

    b. The equilibrium in the first step can be written

$$K = \frac{[O_2][O]}{[O_3]}, \text{ and solving for [O]: } \frac{K[O_3]}{[O_2]} = [O]$$

Substituting unto the rate law (part a)

$$Rate = k[O_3]\frac{K[O_3]}{[O_2]} = K'\frac{[O_3]^2}{[O_2]}$$

69. $K = \frac{[COHb][O_2]}{[O_2Hb][CO]} = 200$

    If $\frac{[COHb]}{[O_2Hb]} = 1$ then $\frac{[O_2]}{[CO]} = 200$

    If $[O_2] = 0.2$ atm then $\frac{0.2 \text{ atm}}{[CO]} = 200$

    and $\frac{0.2 \text{ atm}}{200} = [CO] = 1 \times 10^{-3}$ atm.

Chapter 17:
Principles of Reactivity : The Chemistry of Acids and Bases

The Brønsted Concept of Acids and Bases

12. <u>Conjugate Base of:</u>   <u>Formula</u>   <u>Name</u>

  a. HCN CN^- cyanide ion
  b. HSO_4^- SO_4^{2-} sulfate ion
  c. HF F^- fluoride ion
  d. HNO_2 NO_2^- nitrite ion
  e. HCO_3^- CO_3^{2-} carbonate ion

14. Products of acid-base reactions:

  a. HNO_3 + H_2O $\rightarrow$ H_3O^+ + NO_3^-
    acid   base   conjugate acid conjugate base

  b. HSO_4^- + H_2O $\rightarrow$ H_3O^+ + SO_4^{2-}
    acid   base   conjugate acid conjugate base

  c. H_3O^+ + F^- $\rightarrow$ HF   + H_2O
    acid   base   conjugate acid conjugate base

16. The equation for potassium carbonate dissolving in water:
    K_2CO_3 (aq) $\rightarrow$ 2 K^+ (aq) + CO_3^{2-} (aq)

  Soluble salts--like K_2CO_3--dissociate in water. The carbonate ion formed in this process is a base, and reacts with the acid, water.
    CO_3^{2-} (aq) + H_2O (ℓ) $\rightarrow$ HCO_3^- (aq) + OH^- (aq)

  The production of the hydroxide ion, a strong base, in this second step is responsible for the <u>basic</u> nature of solutions of this carbonate salt.

18. HPO_4^{2-} (aq) + H_2O (ℓ) $\rightleftharpoons$ PO_4^{3-} (aq) + H_3O^+(aq) (Acid)
  HPO_4^{2-} (aq) + H_2O (ℓ) $\rightleftharpoons$ $H_2PO_4^-$ (aq) + OH^-(aq) (Base)

20.   a. $HCOOH \ (aq) + H_2O \ (\ell) \ \rightleftharpoons \ HCOO^- \ (aq) + H_3O^+ \ (aq)$
           acid   base   conjugate   conjugate
                                                      of HCOOH   of H_2O

       b. $H_2S \ (aq) + NH_3 \ (aq) \ \rightleftharpoons \ NH_4^+ \ (aq) + HS^- \ (aq)$
           acid   base   conjugate   conjugate
                                                      of NH_3 of H_2S

       c. $HSO_4^- \ (aq) + OH^- \ (aq) \ \rightleftharpoons \ SO_4^{2-} \ (aq) + H_2O \ (\ell)$
           acid   base   conjugate   conjugate
                                                      of HSO_4^- of OH^-

22.   a. strongest acid: HF   largest K_a
           weakest acid: NH_4^+ smallest K_a

       b. conjugate base of HF: F^-

       c. acid with weakest conjugate base: HF   conjugate of strongest acid → weakest base.

       d. acid with strongest conjugate base: NH_4^+ conjugate of weakest acid → strongest base

24.   a. strongest base: NH_3 (largest K_b)   weakest base: C_5H_5N (smallest K_b)

       b. conjugate acid of C_5H_5N: $C_5H_5NH^+$

       c. base with strongest conjugate acid: C_5H_5N conjugate of weakest base → strongest acid
           base with weakest conjugate acid: NH_3 conjugate of strongest base → weakest acid

26. The substance which has the smallest value for K_a will have the strongest conjugate base. An
     examination of Table 17.4 shows that--of these three substances--HClO has the smallest
     K_a, and ClO^- will therefore be the strongest conjugate base.

Writing Acid-Base Reactions

28. Since both NH_4Cl and NaH_2PO_4 are soluble in water, we should view this in two steps:
     [1] dissociation into component ions: $NH_4Cl \ \rightarrow \ NH_4^+ \ (aq) + Cl^- \ (aq)$
                                          $NaH_2PO_4 \rightarrow \ Na^+ \ (aq) + H_2PO_4^- \ (aq)$

[2] transfer of proton:

$$NH_4^+ (aq) + H_2PO_4^- (aq) \rightarrow NH_3 (aq) + H_3PO_4 (aq)$$

One might suggest that the proton would be transferred from the ammonium ion to the dihydrogen phosphate ion, as shown above. However, examination of Table 17.2 in your text shows H_3PO_4 to be more acidic than NH_4^+, making this process unlikely.

30. a. $CO_3^{2-} (aq) + H_2S (aq) \rightleftharpoons HCO_3^- (aq) + HS^- (aq)$

H_2S is a stronger acid than HCO_3^- and CO_3^{2-} a stronger base than HS^-, hence the equilibrium lies to the **right**.

   b. $HCN (aq) + SO_4^{2-} (aq) \rightleftharpoons CN^- (aq) + HSO_4^- (aq)$

The HSO_4^- ion is a stronger acid than HCN and CN^- a stronger base than SO_4^{2-}, so the equilibrium lies to the **left**.

   c. $CN^- (aq) + NH_3 (aq) \rightleftharpoons HCN (aq) + NH_2^- (aq)$

Since HCN is a stronger acid than NH_3 and NH_2^- is a stronger base than OH^-, the equilibrium lies to the **left**.

   d. $SO_4^{2-} (aq) + CH_3COOH (aq) \rightleftharpoons HSO_4^- (aq) + CH_3COO^- (aq)$

As in part (c) the stronger acid (HSO_4^-) and base (CH_3COO^-) are on the right side of the equilibrium, hence this equilibrium would lie to the **left**.

pH Calculations

32. Since pH = 3.40, $[H_3O^+] = 10^{-pH}$ or $10^{3.40}$ or 4.0×10^{-4} M

Since the $[H_3O^+]$ is greater than 1×10^{-7} (pH<7), the solution is acidic.

34. pH of a solution of 0.0013 M HNO_3:

Since HNO_3 is considered a strong acid, a solution of 0.0013 M HNO_3 has $[H_3O^+] = 0.0013$ or 1.3×10^{-3}.

$$pH = -\log[H_3O^+] = -\log[1.3 \times 10^{-3}] = 2.89$$

The hydroxide ion concentration is readily determined since $[H_3O^+] \cdot [OH^-] = 1.0 \times 10^{-14}$

$$[OH^-] = \frac{1.0 \times 10^{-14}}{1.3 \times 10^{-3}} = 7.7 \times 10^{-12} \text{ M}$$

36.   The pH of a solution of $Ca(OH)_2$ can be calculated by remembering that this base dissolves in water to provide two OH^- ions for each formula unit.

$$[Ca(OH)_2] = 0.0015 \text{ M} \Rightarrow [OH^-] = 0.0030 \text{ M}$$

$$pOH = -\log[OH^-] = -\log[0.0030] = 2.52 \text{ and}$$

$$pH = 14.00 - pOH = 11.48$$

38.   Interconversions:

	pH	$[H_3O^+]$ M	$[OH^-]$ M	Solution character
a.	**1.00**	1.0×10^{-1}	1.0×10^{-13}	acidic
b.	**10.50**	3.2×10^{-11}	3.2×10^{-4}	basic
c.	4.89	**1.3×10^{-5}**	7.7×10^{-10}	acidic
d.	10.36	4.3×10^{-11}	**2.3×10^{-4}**	basic

Using pH to Calculate Ionization Constants

40.   a. With a pH = 3.80 the solution has a $[H_3O^+] = 10^{-3.80}$ or 1.6×10^{-4} M

b. Writing the equation for the unknown acid, HA, in water we obtain:

$$HA \text{ (aq)} + H_2O \text{ (}\ell\text{)} \rightleftharpoons H_3O^+ \text{ (aq)} + A^- \text{ (aq)}$$

$[H_3O^+] = 1.6 \times 10^{-4}$ implying that $[A^-]$ is also 1.6×10^{-4}. Therefore the equilibrium concentration of acid, HA, is $(2.5 \times 10^{-3} - 1.6 \times 10^{-4})$ or $\approx 2.3 \times 10^{-3}$.

$$K_a = \frac{[H_3O^+][A]}{[HA]} = \frac{(1.6 \times 10^{-4})^2}{2.3 \times 10^{-3}} = 1.1 \times 10^{-5}$$

We would classify this acid as a <u>moderately weak acid.</u>

42.   a. If we write the formula for the acid as HA, then [HA] = 0.015 M initially.
Since pH = 2.67, $[H_3O^+] = 10^{-2.67}$ or 2.1×10^{-3} M.

b. Substituting into the equilibrium expression for a weak monoprotic acid we obtain:

$$K_a = \frac{[H_3O^+][A^-]}{[HA]} = \frac{(2.1 \times 10^{-3})^2}{(0.015 - 0.0021)} = 3.6 \times 10^{-4}$$

44. With a pH = 9.11, the solution has a pOH of 4.89 and $[OH^-] = 10^{-4.89} = 1.3 \times 10^{-5}$ M. At equilibrium, $[H_2NOH] = [H_2NOH] - [OH^-] = (0.025 - 1.3 \times 10^{-5})$ or approximately 0.025 M.

$$K_b = \frac{[H_3NOH^+][OH^-]}{[H_2NOH]} = \frac{(1.3 \times 10^{-5})^2}{0.025} = 6.6 \times 10^{-9}$$

Using Ionization Constants to Calculate pH

46.   The equilibrium concentrations of the species may be found as follows:

If we assume that some small amount (x) of the acid HA dissociates, the equilibrium concentration of the acid HA will be $[HA]_i$ - x. The amount of H_3O^+ will be equal to the amount of molecular acid which dissociates, x. This will also be equal to the concentration of the anion, A^-, present at equilibrium. Substitution into the K_a expression yields:

$$K_a = \frac{[A^-][H_3O^+]}{[HA]} = \frac{(x)^2}{(0.040 - x)} = 4.0 \times 10^{-9}$$

Since the K_a for the acid is small, assume that the denominator may be simplified to yield:

$$\frac{x^2}{0.040} = 4.0 \times 10^{-9}$$

and $x = 1.3 \times 10^{-5}$ (to two significant figures)

$[H_3O^+] = [A^-] = 1.3 \times 10^{-5}$ M and [HA] = 0.040 M.

48.   Using the same logic as in question 46 above, we can write:

$$HCN\ (aq)\ +\ H_2O\ (\ell)\ \rightleftharpoons\ H_3O^+\ (aq)\ +\ CN^-\ (aq)$$

Initial concentration:	0.025	0	0
Change (going to eq.)	-x	+x	+x
Eq. concentrations	0.025 - x	x	x

Substituting these values into the K_a expression for HCN:

$$K_a = \frac{[CN^-][H_3O^+]}{[HCN]} = \frac{(x)^2}{(0.025 - x)} = 4.0 \times 10^{-10}$$

Assuming that the denominator may be approximated as 0.025 M, we obtain:

$$\frac{x^2}{0.025} = 4.0 \times 10^{-10}$$

and $x = 3.2 \times 10^{-6}$. Since x represents $[H_3O^+]$ the pH $= -\log(3.2 \times 10^{-6})$ or 5.50.

50.   The $[C_2H_5COOH]_{initial}$ is: $\dfrac{0.588 \text{ g acid}}{0.250 \text{ L}} \cdot \dfrac{1 \text{ mol acid}}{74.08 \text{ g acid}} = 3.17 \times 10^{-2}$ M

Substituting into the K_a expression:

$$K_a = \frac{[C_2H_5COO^-][H_3O^+]}{[C_2H_5COOH]} = \frac{(x)^2}{(0.0317 - x)} = 1.4 \times 10^{-5}$$

We can write a K_a expression for propanoic acid as we did for HCN in 48 above, using the K_a for propanoic acid, 1.3×10^{-5}. If we simplify the denominator to be approximately equal to 0.0317, we can write:

$$x^2 = 1.3 \times 10^{-5} \cdot 0.0317 = 4.12 \times 10^{-7} \quad \text{and} \quad x = 6.4 \times 10^{-4}.$$

Since x represents $[H_3O^+]$ the pH = $- \log(6.4 \times 10^{-4})$ or 3.19.

52.   a. Chlorobenzoic acid has a larger K_a than benzoic acid, so it is the stronger acid.

b. A 0.010 M solution of benzoic acid would produce a lower concentration of hydronium ions than would a similar solution of chlorobenzoic, and would be therefore less acidic— i.e. have a higher pH.

54. For the equilibrium

$$MOH \rightleftharpoons M^+ + OH^- \qquad K_b = 5.0 \times 10^{-4}$$

	[MOH]	[M^+]	[OH^-]
initial concentration	0.15	0	0
change	-x	+x	+x
Equilibrium	0.15 - x	x	x

Substituting into the equilibrium expression: $\dfrac{x^2}{0.15 - x} = 5.0 \times 10^{-4}$

Solving via the quadratic equation yields: x = 8.41×10^{-3}

$[M^+]$ = $[OH^-]$ = 8.41×10^{-3} M (8.4×10^{-3} M to 2 sf)

and $[MOH]$ = $0.15 - 8.41 \times 10^{-3}$ = 0.14 M

56. For $CH_3NH_2 + H_2O \rightleftharpoons CH_3NH_3^+ + OH^-$ $K_b = 5.0 \times 10^{-4}$

Using the approach of problem 54, the equilibrium expression can be written:

$$\frac{x^2}{0.25 - x} = 5.0 \times 10^{-4}$$

Solving the expression **with approximations** gives x = 1.1×10^{-2}

$[OH^-] = 1.1 \times 10^{-2}$

pOH would then be 1.96 and pH 12.04 (to 2 sf)

58. pH of 1.0×10^{-3} M HF; $K_a = 7.2 \times 10^{-4}$

Using the method of question 54, we can write the expression

$$K_a = \frac{x^2}{1.0 \times 10^{-3} - x} = 7.2 \times 10^{-4}$$

The concentration of the HF and K_a preclude the use of our usual approximations
$(1.0 \times 10^{-3} - x \approx 1.0 \times 10^{-3})$

$$x^2 = 7.2 \times 10^{-4}(1.0 \times 10^{-3} - x)$$
$$x^2 = 7.2 \times 10^{-7} - 7.2 \times 10^{-4} x$$

Using the quadratic equation we solve for x:

and x = 5.6×10^{-4} = $[F^-]$ = $[H_3O^+]$ and pH = 3.25

Acid-Base Properties of Salts

60. Most of the salts shown are sodium salts. Since Na^+ does not hydrolyze, we can estimate the acidity (or basicity) of such solutions by looking at the extent of reaction of the anions with water (hydrolysis).

The Al^{3+} ion is acidic, as is the $H_2PO_4^-$ ion. From Table 17.4 we see that the K_a for the hydrated aluminum ion is greater than that for the $H_2PO_4^-$ ion, making the **Al^{3+} solution more acidic —lower pH** than that of $H_2PO_4^-$. All the other salts will produce basic solutions and since the S^{2-} ion has the largest K_b, we will anticipate that the **Na_2S solution will be most basic—i.e. have the highest pH.**

62. Hydrolysis of the NH_4^+ produces H_3O^+ according to the equilibrium:

$$NH_4^+ (aq) + H_2O (l) \rightleftharpoons H_3O^+ (aq) + NH_3 (aq)$$

With the ammonium ion acting as an acid, to donate a proton, we can write the K_a expression:

$$K_a = \frac{[NH_3][H_3O^+]}{[NH_4^+]} = \frac{K_w}{K_b} = \frac{1.0 \times 10^{-14}}{1.8 \times 10^{-5}} = 5.6 \times 10^{-10}$$

The concentrations of both terms in the numerator are equal, and the concentration of ammonium ion is 0.20 M. (Note the approximation for the **equilibrium** concentration of NH_4^+ to be equal to the **initial** concentration.)

Substituting and rearranging we get

$$[H_3O^+] = \sqrt{0.20 \cdot 5.6 \times 10^{-10}} = 1.1 \times 10^{-5}$$

and the pH = 4.98.

64. The hydrolysis of CN^- produces OH^- according to the equilibrium:

$$CN^- (aq) + H_2O (l) \rightleftharpoons HCN (aq) + OH^- (aq)$$

Calculating the concentrations of Na^+ and CN^- :

$$[Na^+]_i = [CN^-]_i = \frac{10.8 \text{ g NaCN}}{0.500 \text{ L}} \cdot \frac{1 \text{ mol NaCN}}{49.01 \text{ g NaCN}} = 4.41 \times 10^{-1} \text{ M}$$

$$\text{Then } K_b = \frac{K_w}{K_a} = \frac{1.0 \times 10^{-14}}{4.0 \times 10^{-10}} = 2.5 \times 10^{-5} = \frac{[HCN][OH^-]}{[CN^-]}$$

Substituting the $[CN^-]$ concentration into the K_b expression and noting that:

$$[OH^-]_e = [HCN]_e \quad \text{we may write}$$

$$[OH^-]_e = [(2.5 \times 10^{-5})(4.41 \times 10^{-1})]^{\frac{1}{2}} = 3.3 \times 10^{-3} \text{ M}$$

$$[H_3O^+] = \frac{1.0 \times 10^{-14}}{3.3 \times 10^{-3}} = 3.0 \times 10^{-12} \text{ M}$$

Conjugate Acid-Base Pairs

66. a. The Ka for the conjugate acid of a base can be calculated if we remember the relationship:

$$K_a \cdot K_b = K_w$$

$$K_a \text{ (anilinium)} = \frac{K_w}{K_{b(aniline)}} = \frac{1.0 \times 10^{-14}}{4.0 \times 10^{-10}} = 2.5 \times 10^{-5}$$

b. The pH of 0.080 M anilinium hydrochloride

$$Ka = \frac{[C_6H_5NH_2][H^+]}{[C_6H_5NH_3^+]} = 2.5 \times 10^{-5}$$

Using the normal approach:

$$\frac{x^2}{0.080 - x} = 2.5 \times 10^{-5}$$

Solving for x yields $x = 1.4 \times 10^{-3}$

The approximation yields 1.4×10^{-3} while the quadratic equation gives 1.40×10^{-3}

So $[H^+] = 1.4 \times 10^{-3}$ and a pH of 2.85.

68. The sodium salt would exist in water as sodium and saccharide ions. The weakly acidic nature of saccharin would indicate some reaction of the saccharide ion with water.

$$C_7H_4NO_3S^- + H_2O \rightleftharpoons HC_7H_4NO_3S + OH^-$$

$$\text{with a } K_b = \frac{1.0 \times 10^{-14}}{2.1 \times 10^{-12}} = 4.8 \times 10^{-3}$$

$$\frac{[HC_7H_4NO_3S][OH^-]}{[C_7H_4NO_3S^-]} = \frac{(x)(x)}{0.10 - x} = 4.8 \times 10^{-3}$$

The approximation $(0.10 - x \approx 0.10)$ gives a value for x of 2.2×10^{-2}.

Using the quadratic equation, we get $x = 1.96 \times 10^{-2}$ (2.0×10^{-2} to 2 sf) as $[OH^-]$ and pOH of 1.71 and a pH of 12.29

Polyprotic Acids and Bases

70.   a. pH of 0.45 M H_2SO_3:

The equilibria for the diprotic acid are:

$$K_{a1} = \frac{[HSO_3^-][H_3O^+]}{[H_2SO_3]} = 1.2 \times 10^{-2} \text{ and } K_{a2} = \frac{[SO_3^{2-}][H_3O^+]}{[HSO_3^-]} = 6.2 \times 10^{-8}$$

For the first step of dissociation:

	H_2SO_3	HSO_3^-	H_3O^+
Initial concentration	0.45		
Change	-x	+x	+x
Equilibrium concentration	0.45 - x	+x	+x

$$K_{a1} = \frac{(x)(x)}{(0.45-x)} = 1.2 \times 10^{-2}$$

We must solve this expression with the quadratic equation since $(0.45 < 100 \cdot K_{a1})$. The equilibrium concentrations for HSO_3^- and H_3O^+ ions are found to be 0.0677 M. The further dissociation is indicated by K_{a2}. Using the equilibrium concentrations obtained in the first step, we substitute into the K_{a2} expression.

	HSO_3^-	SO_3^{2-}	H_3O^+
Initial concentration	0.0677	0	0.0677
Change	-x	+x	+x
Equilibrium concentration	0.0677 - x	+x	0.0677 + x

$$K_{a2} = \frac{[SO_3^{2-}][H_3O^+]}{[HSO_3^-]} = \frac{(+x)(0.0677 + x)}{(0.0677 - x)} = 6.2 \times 10^{-8}$$

We note that x will be small in comparison to 0.0677, and we simplify the expression:

$$K_{a2} = \frac{(+x)(0.0677)}{(0.0677)} = 6.2 \times 10^{-8}$$

In summary the **concentrations of HSO_3^- and H_3O^+ ions have been virtually unaffected** by the second dissociation.

Then $[H_3O^+] = 0.0677$ M and pH = 1.17

b. The equilibrium concentration of SO_3^{2-} :

From the K_{a2} expression above: $[SO_3^{2-}] = 6.2 \times 10^{-8}$ M

72.   a. Concentrations of OH^-, $N_2H_5^+$, and $N_2H_6^{2+}$ in 0.010 M N_2H_4:

The K_{b1} equilibrium allows us to calculate $N_2H_5^+$ and OH^- formed by the reaction of N_2H_4 with H_2O. $K_{b1} = \dfrac{[N_2H_5^+][OH^-]}{[N_2H_4]} = 8.5 \times 10^{-7}$

	N_2H_4	$N_2H_5^+$	OH^-
Initial concentration	0.010		
Change	-x	+x	+x
Equilibrium concentration	0.010 - x	+x	+x

Substituting into the K_{b1} expression : $\dfrac{(x)(x)}{0.010 - x} = 8.5 \times 10^{-7}$

We can simplify the denominator ($0.010 > 100 \cdot K_{b1}$).

$$\frac{(x)(x)}{0.010} = 8.5 \times 10^{-7} \text{ and } x = 9.2 \times 10^{-5} = [N_2H_5^+] = [OH^-]$$

The second equilibrium (K_{b2}) indicates further reaction of the $N_2H_5^+$ ion with water. The step should consume some $N_2H_5^+$ and produce more OH^-. The magnitude of K_{b2} indicates that the equilibrium "lies to the left" and we anticipate that not much $N_2H_6^{2+}$ (or additional OH^-) will be formed by this interaction.

	$N_2H_5^+$	$N_2H_6^{2+}$	OH^-
Initial concentration	9.2×10^{-5}		9.2×10^{-5}
Change	$-x$	$+x$	$+x$
Equilibrium concentration	$9.2 \times 10^{-5} - x$	x	$9.2 \times 10^{-5} + x$

$$K_{b2} = \frac{[N_2H_6^{2+}][OH^-]}{[N_2H_5^+]} = 8.9 \times 10^{-16} = \frac{x \cdot (9.2 \times 10^{-5} + x)}{(9.2 \times 10^{-5} - x)}$$

Simplifying yields $\dfrac{x \cdot (9.2 \times 10^{-5})}{(9.2 \times 10^{-5})} = 8.9 \times 10^{-16}$; $x = 8.9 \times 10^{-16} = [N_2H_6^{2+}]$

In summary we see that the second stage produces a negligible amount of OH^- and consumes very little $N_2H_5^+$ ion. The equilibrium concentrations are:

$[N_2H_5^+]$: 9.2×10^{-5} M; $[N_2H_6^{2+}]$: 8.9×10^{-16} M; $[OH^-]$: 9.2×10^{-5} M

b.   The pH of the 0.010 M solution: $[OH^-] = 9.2 \times 10^{-5}$ M   pOH = 4.04

and pH = 14.0 - 4.04 = 9.96

Lewis Acids and Bases

74.   a. Mn^{2+} electron deficient   Lewis acid
      b. $:NH_2(CH_3)$ electron rich   Lewis base
      c. H_2NOH electron rich   Lewis base
      d. SO_2 donates an electron pair Lewis base
      e. $Zn(OH)_2$ accepts electron pairs   Lewis acid

76. **BH_3 is a Lewis acid** in that it seeks to complete the electron octet at B by forming a new bond with the lone pair at the nitrogen atom of NH_3.

78. ICl_3 accepts an electron pair from the chloride ion in forming the ICl_4^- ion. This behavior is that of a **Lewis acid**. The Lewis dot structure for ICl_3 is:

$$:\overset{\bullet\bullet}{\underset{\bullet\bullet}{Cl}} - \overset{\bullet\bullet}{I} - \overset{\bullet\bullet}{\underset{\bullet\bullet}{Cl}}:$$
$$|$$
$$:\overset{\bullet\bullet}{\underset{\bullet\bullet}{Cl}}:$$

ICl_3 will have a T-shape and ICl_4^- will be a square planar complex.

(6 electron pairs around the I atom, four of which are bonding pairs to Cl atoms.)

I in ICl_3 will be sp^3d hybridized and in ICl_4^-, sp^3d^2.

General Questions

80. $H^- (s) + H_2O (\ell) \rightarrow OH^- (aq) + H_2 (g)$

The resulting solution of sodium hydroxide would be basic. The hydride ion will act as a Lewis base and water will act as a Lewis acid.

82. $HC_9H_7O_4 (aq) + H_2O (aq) \rightleftharpoons C_9H_7O_4^- (aq) + H_3O^+ (aq)$

$$K_a = \frac{[C_9H_7O_4^-][H_3O^+]}{[HC_9H_7O_4]} = 3.27 \times 10^{-4}$$

The initial concentration of aspirin is:

$$2 \text{ tablets} \cdot \frac{0.325 \text{ g}}{1 \text{ tablet}} \cdot \frac{1 \text{ mol } HC_9H_7O_4}{180.2 \text{ g } HC_9H_7O_4} \cdot \frac{1}{0.200 \text{ L}} = 1.80 \times 10^{-2} \text{ M } HC_9H_7O_4$$

Substituting into the K_a expression:

$$K_a = \frac{[H_3O^+]^2}{1.80 \times 10^{-2}} = 3.27 \times 10^{-4} = \frac{x^2}{1.80 \times 10^{-2} - x}$$

Since the $100 \cdot K_a \approx$ [aspirin], the quadratic equation will provide a "good" value. Using the quadratic equation, $[H_3O^+] = 2.27 \times 10^{-3}$ M and pH = 2.644.

84. The reaction of H_2S and $NaCH_3CO_2$ involves the reaction of H_2S with $CH_3CO_2^-$ ion:

$$H_2S(aq) + CH_3CO_2^- (aq) \rightleftharpoons HS^- (aq) + CH_3CO_2H(aq)$$

  acid   base   conjugate   conjugate

                                      base   acid

From Table 17.3, we see that CH_3CO_2H is a stronger acid than H_2S and HS^- is a stronger base than $CH_3CO_2^-$. Hence the **reaction does not occur** to any appreciable extent.

86. $0.50 \text{ g Ca(OH)}_2 \cdot \dfrac{1 \text{ mol}}{74.1 \text{ g}} = 0.0067$ mol and when dissolved in water to make 1.0 L,

the concentration of $Ca(OH)_2$ is 0.0067 M.

However,

$[OH]^- = 2 \cdot 0.0067 \text{ M} = 0.014 \text{ M}$

(When $Ca(OH)_2$ dissolves, there are 2 OH^- per formula)

$pOH = 1.87$ and so $pH = 12.13$.

88. If the $pH = 3.44$ then $[H_3O^+] = 10^{-pH} = 10^{-3.44} = 3.6 \times 10^{-4}$ M

$$Acid(aq) + H_2O(\ell) \rightleftharpoons Conjugate\ Base\ (aq) + H_3O^+(aq)$$

	[Acid]	[Conjugate Base]	[H3O+]
initial	0.010	0	0
change	-0.00036	+0.00036	+0.00036
Equilibrium	0.010	0.00036	0.00036

$$K_a = \frac{[Conjugate\ Base][H_3O^+]}{[Acid]} = \frac{(0.00036)^2}{0.010 - 0.00036} = 1.4 \times 10^{-5}$$

90.   a. The K_b for butylamine is:

$$K_b = \frac{K_w}{K_a} = \frac{1.0 \times 10^{-14}}{2.3 \times 10^{-11}} = 4.3 \times 10^{-4}$$

b. With a K_a of 2.3×10^{-11}, the butylammonium ion would go immediately above the phosphate ion (PO_4^{3-}) and below $Ni(H_2O)_5OH^+$. The **PO_4^{3-} ion is a stronger base** than butylamine (as are all the other bases below it in the list.)

92. pH of aqueous solutions of

		reaction	pH
a.	$NaHSO_4$	hydrolysis of HSO_4^- produces H_3O^+	< 7
b.	NH_4Br	hydrolysis of HSO_4^- produces H_3O^+	< 7
c.	$KClO_4$	no hydrolysis occurs	= 7
d.	Na_2CO_3	hydrolysis of CO_3^{2-} produces OH^-	> 7
e.	$(NH_4)_2S$	hydrolysis of S^{2-} produces OH^-	> 7

reaction		pH
f. $NaNO_3$	no hydrolysis occurs	= 7
g. Na_2HPO_4	hydrolysis of HPO_4^{2-} produces OH^-	> 7
h. LiBr	no hydrolysis occurs	= 7
i. $FeCl_3$	hydrolysis of Fe^{3+} produces H_3O^+	< 7

94. $$NH_2OH(aq) + H_2O(\ell) \rightleftharpoons NH_3OH^+(aq) + OH^-(aq)$$

	$[NH_2OH]$	$[NH_3OH^+]$	$[OH^-]$
initial	0.051	0	0
change	- x	+ x	+ x
Equilibrium	0.051 - x	x	x

$$K_b = \frac{[NH_3OH^+][OH^-]}{[NH_2OH]} = \frac{x^2}{0.051 - x} = 6.6 \times 10^{-9}$$

Making the very reasonable assumption that x is very small relative to 0.051, we find
$x = 1.8 \times 10^{-5} = [NH_3OH^+] = [OH^-]$. Therefore, the hydroxide ion concentration
gives pOH = 4.74 and pH = 9.26.

96.   Of the solutions listed:

a. Acidic

0.1 M CH_3COOH	a weak acid
0.1 M NH_4Cl	a salt of a strong acid and weak base

b. Basic

0.1 M NH_3	a weak base
0.1 M Na_2CO_3	a salt of a strong base and a weak acid
0.1 M $NaCH_3COO$	a salt of a strong base and a weak acid

NaCl   neutral—NaCl is the salt of a strong acid and a strong base.
$NH_4CH_3CO_2$ neutral—both NH_4^+ and $CH_3CO_2^-$ ions hydrolyze—but to
                       the same extent, hence the solution remains neutral.

c. The most acidic:

CH_3COOH K_a for acetic acid is greater than the corresponding K_a of the

ammonium ion.

98. The solutions may be arranged in order of increasing pH by examining K_a's and K_b's in Table 17.4.

$$HCl < CH_3COOH < NaCl < NH_3 < NaCN < NaOH$$

100. For oxalic acid $K_{a1} = 5.9 \times 10^{-2}$ and $K_{a2} = 6.4 \times 10^{-5}$

Beginning with 0.25 M $H_2C_2O_4$, let's calculate the equilibrium concentrations *taking into account only the first step:*

Representing oxalic as H_2A

$$H_2A + H_2O \rightleftharpoons HA^- + H_3O^+ \qquad K_a = 5.9 \times 10^{-2}$$

At equilibrium $\dfrac{[HA^-][H_3O^+]}{[H_2A]} = 5.9 \times 10^{-2}$

Following the normal procedure, if a $\dfrac{mol}{L}$ of H_2A form products:

$$\dfrac{(a)(a)}{0.25 - a} = 5.9 \times 10^{-2}$$

Using the quadratic equation to solve; $a = 0.0955$ (0.096 to 2 sf)

So $[HA^-] = [H_3O^+] = 0.096$ and $[H_2A] = 0.15$

Using the method of successive approximations gives 0.0948 and a rounded value of 0.095 M.

Now examining the second step

$$HA^- + H_2O \rightleftharpoons A^{2-} + H_3O^+ \qquad K_a = 6.4 \times 10^{-5}$$

Using initial concentrations of HA^- and H_3O^+ from step 1, assume that b $\dfrac{mol}{L}$ of HA^- form products

$$\dfrac{(b)(b + 0.096)}{(0.096 - b)} = 6.4 \times 10^{-5}$$

Solving the equation exactly with the quadratic equation gives $b = 6.38 \times 10^{-5}$ (6.4×10^{-5} to 2 sf). What does this tell us? The second step produces a negligible amount of additional hydronium ion.

The pH $= -\log(0.096 \text{ M})$ or 1.02 , and

$$[HC_2O_4^-] = 0.096 \text{ M} \quad \text{and} \quad [C_2O_4^{2-}] = 6.4 \times 10^{-5} \quad (\text{the } K_{a2})$$

102. I_2 + I^- $\rightleftharpoons$ I_3^-
            Lewis   Lewis
            acid   base

In forming the triiodide ion, the iodide ion donates a pair of electrons to one of the iodine atoms in the I_2 molecule.

104.   a. As the K_a values increase, the concentration of Hydronium ion that the acid would produce increases—reducing the pH of the solution. So the **nitropyridinium hydrochloride (x = NO_2) would have the lowest pH** and the **methylpyridinium hydrochloride (x = CH_3) would provide the highest pH.**

b. Strongest and weakest Brønsted base:
Remembering the relationship that $K_a \cdot K_b = K_w$, it follows that the conjugate acid with the largest K_a (nitropyridinium hydrochloride) will have the conjugate base (**nitropyridine**) with the smallest K_b (**weakest base**) and similarly **methyl pyridine** would be the **strongest base.**

Conceptual Questions

106.   a. As hydrogen atoms are successively replaced by the very electronegative chlorine atoms, an increase in acidity is seen. This is quite understandable if you remember that an increasing "pull" on the electrons in the "carboxylate" end of the molecule--brought about by an **increasing number of chlorine atoms**--will weaken the O-H bond and **increase the acidity of the specie.**

b. The acid with the largest K_a (Cl_3CCO_2H) would be the strongest acid and would have the lowest pH. The acid with the smallest K_a (CH_3CO_2H) would have the highest pH.

108.   a. The reaction of perchloric acid with sulfuric acid:

$$HClO_4 + H_2SO_4 \rightleftharpoons ClO_4^- + H_3SO_4^+$$

b. Lewis dot structure for sulfuric acid:

c. Sulfuric acid has two oxygen atoms capable of "donating" an electron pair, e.g. to a proton, H^+, thereby acting as a base.

111.   a. Volume of aniline to prepare 150.0 g of sulfanilic acid:

$$150.0 \text{ g NH}_2\text{C}_6\text{H}_4\text{SO}_3\text{H} \cdot \frac{100 \text{ g}}{85 \text{ g}} \cdot \frac{1 \text{ mol NH}_2\text{C}_6\text{H}_4\text{SO}_3\text{H}}{173.2 \text{ g NH}_2\text{C}_6\text{H}_4\text{SO}_3\text{H}} \cdot$$

$$\frac{1 \text{ mol aniline}}{1 \text{ mol NH}_2\text{C}_6\text{H}_4\text{SO}_3\text{H}} \cdot \frac{93.13 \text{ g aniline}}{1 \text{ mol aniline}} \cdot \frac{1.00 \text{ mL aniline}}{1.02 \text{ g aniline}} = 93 \text{ mL}$$

b. All of the angles are 109 °, since each central atom of the indicated bonds has four electron-pair groups around it.

c. $NH_2C_6H_4SO_3^-(aq) + H_2O(\ell) \rightleftharpoons NH_2C_6H_4SO_3H(aq) + OH^-(aq)$

$$K_b = \frac{[NH_2C_6H_4SO_3H][OH^-]}{[NH_2C_6H_4SO_3^-]} = \frac{1.0 \times 10^{-14}}{5.9 \times 10^{-4}} = 1.7 \times 10^{-11}$$

The concentration of the salt is:

$$1.25 \text{ g} \cdot \frac{1 \text{ mol salt}}{195.2 \text{ g}} \cdot \frac{1}{0.100 \text{ L}} = 0.0640 \text{ M}$$

Substituting into the K_b expression:

$$\frac{x^2}{0.0640 - x} = 1.7 \times 10^{-11}$$

$$x = [OH^-] = 1.05 \times 10^{-6} \text{ M}$$

$$pOH = 5.98 \text{ and } pH = 8.02$$

233

Chapter 18:
Principles of Reactivity : Reactions Between Acids and Bases

Acid and Base Reactions

8. The value of the equilibrium constant for:
$$C_6H_5CO_2H + OH^- \longrightarrow H_2O + C_6H_5CO_2^-$$

Note that this equilibrium can be represented as the sum of two reactions

$$C_6H_5CO_2H + H_2O \longrightarrow C_6H_5CO_2^- + H_3O^+ \qquad K_a$$

$$H_3O^+ + OH^- \rightleftharpoons 2\ H_2O \qquad \frac{1}{K_w}$$

_____

$$C_6H_5CO_2H + OH^- \longrightarrow H_2O + C_6H_5CO_2^- \qquad K_{net}$$

and $K_{net} = K_a \cdot \dfrac{1}{K_w} = 6.3 \times 10^{-5} \cdot \dfrac{1}{1.0 \times 10^{-14}} = 6.3 \times 10^9$

This equilibrium **lies** predominantly **to the right**. We take advantage of this when we react weak acids with strong bases.

10. The net reaction is:

$$CH_3COOH\ (aq) + NaOH\ (aq) \longrightarrow CH_3COO^-\ (aq) + Na^+\ (aq) + H_2O\ (\ell)$$

The addition of 22.0 mL of 0.10 M NaOH (2.2 mmol NaOH) to 22.0 mL of 0.10 M CH_3COOH (2.2 mmol CH_3COOH) produces water and the soluble salt, sodium acetate (2.2 mmol $CH_3COO^-\ Na^+$). The acetate ion is the anion of a weak acid and reacts with water according to the equation:

$$CH_3COO^-\ (aq) + H_2O\ (\ell) \longrightarrow CH_3COOH\ (aq) + OH^-\ (aq)$$

The equilibrium constant expression may be written:

$$K_b = \frac{[CH_3COOH][OH^-]}{[CH_3COO^-]} = 5.6 \times 10^{-10}$$

The concentration of acetate ion is: $\dfrac{2.2\ mmol}{(22.0 + 22.0)\ ml} = 0.050\ M$

Equilibrium concentrations:

	$CH_3CO_2^-$	CH_3COOH	OH^-
Initial concentration	0.050		
Change (going to equilibrium)	-x	+x	+x
Equilibrium concentration	0.050 - x	+x	+x

$$K_b = \frac{[CH_3COOH][OH^-]}{[CH_3COO^-]} = \frac{x^2}{0.050 - x} = 5.6 \times 10^{-10}$$

Simplifying $(100 \cdot K_b << 0.050)$ we get $\quad \dfrac{x^2}{0.050} = 5.6 \times 10^{-10}$

$$x = 5.3 \times 10^{-6} = [OH^-]$$

The hydrogen ion concentration is related to the hydroxyl ion concentration by the equation:

$$K_w = [H_3O^+][OH^-] = 1.0 \times 10^{-14}$$

$$[H_3O^+] = \frac{1.0 \times 10^{-14}}{[OH^-]} = \frac{1.0 \times 10^{-14}}{5.3 \times 10^{-6}} = 1.9 \times 10^{-9}$$

$$pH = 8.72$$

The pH is greater than 7, as we expect for a salt of a strong base and weak acid.

12.

pH of solution	Reacting Species	Reaction controlling pH
a. >7	CH_3CO_2H/KOH	Hydrolysis of $CH_3CO_2^-$
b. <7	HCl/NH_3	Hydrolysis of NH_4^+
c. =7	HNO_3/NaOH	No hydrolysis

14. $\quad \dfrac{0.515 \text{ g } C_6H_5OH}{0.100 \text{ L}} \cdot \dfrac{1 \text{ mol } C_6H_5OH}{94.11 \text{ g } C_6H_5OH} = 5.47 \times 10^{-2} \text{ M } C_6H_5OH$

At the equivalence point 5.47×10^{-3} mol of NaOH will have been added. Phenol is a monoprotic acid. One mol of phenol reacts with one mol of sodium hydroxide. The volume of 0.123 NaOH needed to provide this amount of base is:

$$\text{moles} = M \times V$$

$$5.47 \times 10^{-3} \text{ mol NaOH} = 0.123 \text{ mol NaOH} \times V$$

or 44.5 mL of the NaOH solution.

The total volume would be (100. + 44.5) or 145 mL solution.

Sodium phenoxide is a soluble salt hence the initial concentration of both sodium and phenoxide ions will be equal to:

$$\frac{5.47 \times 10^{-3} \text{ mol}}{0.145 \text{ L}} = 3.79 \times 10^{-2} \text{ M}$$

The phenoxide ion however is the conjugate ion of a weak acid and undergoes hydrolysis.

$$C_6H_5O^- \text{ (aq)} + H_2O \text{ (ℓ)} \rightleftharpoons C_6H_5OH \text{ (aq)} + OH^- \text{ (aq)}$$

	$C_6H_5O^-$	C_6H_5OH	OH^-
Initial	3.79×10^{-2} M		
Change	-x	+x	+x
Equilibrium	3.79×10^{-2} M - x	x	x

$$K_b = \frac{[C_6H_5OH][OH^-]}{[C_6H_5O^-]} = \frac{1.0 \times 10^{-14}}{1.3 \times 10^{-10}} = \frac{(x)(x)}{3.79 \times 10^{-2} - x} = 7.7 \times 10^{-5}$$

Since $7.7 \times 10^{-5} \cdot 100 < 3.79 \times 10^{-2}$ we simplify: $\frac{(x)(x)}{3.79 \times 10^{-2}} = 7.7 \times 10^{-5}$

$$x = 1.7 \times 10^{-3}$$

At the equivalence point: $[OH^-] = 1.7 \times 10^{-3}$

$$\text{and } [H_3O^+] = \frac{1.0 \times 10^{-14}}{1.7 \times 10^{-3}} = 5.9 \times 10^{-12}$$

$$\text{and pH} = 11.23$$

16. At the equivalence point the moles of acid = moles of base.

$$(0.03678 \text{ L}) (0.0105 \text{ M HCl}) = 3.86 \times 10^{-4} \text{ mol HCl}$$

If this amount of base were contained in 25.0 mL of solution, **the concentration of NH$_3$ in the original solution was 0.0154 M.**

At the equivalence point NH$_4$Cl will hydrolyze according to the equation:

$$NH_4^+ \text{ (aq)} + H_2O \text{ (ℓ)} \rightleftharpoons NH_3 \text{ (aq)} + H_3O^+ \text{ (aq)}$$

$$K_a = \frac{[NH_3][H_3O^+]}{[NH_4^+]} = \frac{1.0 \times 10^{-14}}{1.8 \times 10^{-5}} = 5.6 \times 10^{-10}$$

The salt (3.86×10^{-4} mol) is contained in (25.0 + 36.78) 61.78 mL. Its concentration will

be $\frac{3.69 \times 10^{-4} \text{ mol}}{0.06178 \text{ L}}$ or 6.25×10^{-3} M.

Substituting into the K_a expression :

$$\frac{[H_3O^+]^2}{6.25 \times 10^{-3}} \ = \ 5.6 \times 10^{-10} \quad \text{and} : [H_3O^+] \ = \ \mathbf{1.9 \times 10^{-6}}$$

$$\text{and } \mathbf{pH \ = \ 5.72}.$$

Since $[H_3O^+][OH^-] \ = \ 1.0 \times 10^{-14}$ then $[OH^-] = \dfrac{1.0 \times 10^{-14}}{1.9 \times 10^{-6}} \ = \ \mathbf{5.3 \times 10^{-9}}$

and the $\mathbf{[NH_4^+] \ = \ 6.25 \times 10^{-3} \ M}$

18. Examine the equilibria in each case:

a. $NH_3 \ (aq) + H_2O \ (\ell) \ \rightleftharpoons \ NH_4^+ \ (aq) + OH^- \ (aq)$

As the NH_4Cl dissolves, ammonium ions are liberated, shifting the position of equilibrium to the left—reducing OH^-, and **decreasing the pH**.

b. $CH_3COOH + H_2O \ \rightleftharpoons \ CH_3COO^- + H_3O^+$

As sodium acetate dissolves, the additional acetate ion will shift the position of equilibrium to the left—reducing H_3O^+, and **increasing the pH**.

c. $KOH \ \rightarrow \ K^+ + OH^-$

KOH is a strong base, and as such is totally dissociated. Since the added KCl does not hydrolyze to any appreciable extent—**no change in pH occurs**.

20. As in study question 18, examine the equilibria involved:

a. Adding HCl will consume NH_3, **lowering the pH**.

b. Adding NaOH will consume acetic acid, **raising the pH**.

22. The pH of the buffer solution is:

$$K_b = \frac{[NH_4^+][OH^-]}{[NH_3]} \ = \ \frac{(0.20)[OH^-]}{(0.20)} \ = \ 1.8 \times 10^{-5}$$

and solving for hydroxyl ion yields: $[OH^-] = 1.8 \times 10^{-5}$

$$pOH = 4.74 \quad pH = 9.26$$

24. The original pH will be:

$$\frac{[NH_4^+][OH^-]}{[NH_3]} = 1.8 \times 10^{-5} \quad \text{and} \quad \frac{x^2}{0.12 - x} = 1.8 \times 10^{-5}$$

Assuming $(0.12 - x \approx 0.12)$, $x = 1.47 \times 10^{-3}$

$[OH^-] = 1.47 \times 10^{-3}$ and pOH $= 2.83$ with pH $= 11.17$

Adding 2.2 g of NH_4Cl (0.041 mol) to 250 mL will produce an immediate increase of 0.16 M NH_4^+ (0.041 mol/0.250 L).

Substituting into the equilibrium expression
$$\frac{(x + 0.16)(x)}{0.12 - x} = 1.8 \times 10^{-5}$$
Assuming $(x + 0.16 \approx 0.16)$ and $(0.12 - x \approx 0.12)$
$$\frac{0.16(x)}{0.12} = 1.8 \times 10^{-5} \quad x = 1.35 \times 10^{-5} = [OH^-]$$

Note the hundred-fold decrease in $[OH^-]$ over the initial ammonia solution, as predicted by LeChatelier's principle (See question 20a)

So pOH $= 4.87$ and pH $= 9.13$ (lower than original)

26. We'll assume that any solid ammonium chloride will not change the volume of the solution from 500. mL.

For the equilibrium of an ammonia solution.
$$\frac{[NH_4^+][OH^-]}{[NH_3]} = 1.8 \times 10^{-5}$$
For $[NH_3]$ we substitute 0.10 M.

If pH $= 9.0$, then pOH $= 5.0$ and $[OH^-] = 1.0 \times 10^{-5}$
$$\frac{[NH_4^+][1.0 \times 10^{-5}]}{[0.10]} = 1.8 \times 10^{-5} \quad \text{and} [NH_4^+] = 0.18 \text{ M}$$

and in 500. mL, we need $= 0.090$ mol
That mass of NH_4Cl is

$$0.090 \text{ mol } NH_4^+ \cdot \frac{1 \text{ mol } NH_4Cl}{1 \text{ mol } NH_4^+} \cdot \frac{53.5 \text{ g } NH_4Cl}{1 \text{ mol } NH_4Cl} \text{ or } 4.8 \text{ g } NH_4Cl$$

28.   pH of buffer with 12.2 g C_6H_5COOH and 7.20 g C_6H_5COONa in 250. mL of solution:

1. Calculate concentrations for the acid and its salt:

$$12.2 \text{ g } C_6H_5COOH \cdot \frac{1 \text{ mol } C_6H_5COOH}{122.1 \text{ g } C_6H_5COOH} \cdot \frac{1}{250 \text{ L}} = 0.400 \text{ M } C_6H_5COOH$$

$$7.20 \text{ g } C_6H_5COONa \cdot \frac{1 \text{ mol } C_6H_5COONa}{144.1 \text{ g}C_6H_5COONa} \cdot \frac{1}{250 \text{ L}} = 0.200 \text{ M } C_6H_5COONa$$

2. Substitute equilibrium concentrations into the equilibrium constant expression:

$$C_6H_5COOH \text{ (aq)} + H_2O(\ell) \rightleftharpoons C_6H_5COO^- \text{ (aq)} + H_3O^+ \text{ (aq)}$$

	$\underline{C_6H_5COOH}$	$\underline{C_6H_5COO^-}$	$\underline{H_3O^+}$
Initial concentration	0.400	0.200	
Change (going to equilibrium)	-x	+x	+x
Equilibrium concentration	0.400 - x	0.200 + x	+x

$$K_a = \frac{[C_6H_5COO^-][H_3O^+]}{[C_6H_5COOH]} = \frac{(0.200 + x)(x)}{0.400 - x} = 6.3 \times 10^{-5}$$

Simplifying the equation: $\dfrac{(0.200)(x)}{(0.400)} = 6.3 \times 10^{-5}$

$$x = 1.26 \times 10^{-4} = [H_3O^+] \text{ and } pH = 3.90$$

Diluting this solution to 0.500 L **will not change the pH**. Notice that the concentrations of both acid and salt are changed, leaving the ratio of salt : acid and the pH unchanged.

30. The best combination to provide a buffer solution of pH 9 is b, the NH_3/NH_4^+ system. Note that K_a (for NH_4^+) is approximately 10^{-10} . Buffer systems are good when the desired pH is ± 1 unit from pK_a (10 in this case).
The HCl and NaCl don't form a buffer . The acetic acid/sodium acetate system would form an acidic buffer ($pK_a \approx 5$) in the pH range 4 - 6.

32.   a. Initial pH

Need to know concentration of the conjugate pairs:
Since we have the ratio of the conjugate pairs in the equilibrium expression, we can calculate **moles** of the conjugate pairs.

$CH_3CO_2H = 0.250 \text{ L} \cdot 0.150 \text{ M} = 0.0375 \text{ mol}$

$NaCH_3CO_2 = 4.95 \text{ g} \cdot \dfrac{1 \text{ mol}}{82.07 \text{ g}} = 0.0603 \text{ mol}$

Substituting into the K_a expression:

$$\frac{[CH_3CO_2^-][H_3O^+]}{[CH_3CO_2H]} = 1.8 \times 10^{-5} = \frac{[0.0603][H_3O^+]}{[0.0375]}$$

and solving for $[H_3O^+] = 1.1 \times 10^{-5}$ M ; pH = 4.95

b. pH after 80. mg NaOH is added to 100. mL of the buffer. The amount of the conjugate pairs in 100/250 of the buffer is $(100/250)(0.0375 \text{ mol}) = 0.015$ mol CH_3CO_2H

and $(100/250)(0.0603 \text{ mol}) = 0.0241$ mol $CH_3CO_2^-$

80. mg NaOH $\cdot \dfrac{1 \text{ mmol NaOH}}{40.0 \text{ mg NaOH}} = 2.0$ mmol NaOH or 0.0020 mol NaOH

This base would **consume** an equivalent amount of $\mathbf{CH_3CO_2H}$ and **produce** an equivalent amount of $\mathbf{CH_3CO_2^-}$. After that process: 0.013 mol CH_3CO_2H and 0.0261 mol $CH_3CO_2^-$ are present. Substituting into the K_a expression as in part a

$$\frac{(0.0261)[H_3O^+]}{(0.013)} = 1.8 \times 10^{-5}$$

and $[H_3O^+] = 9.0 \times 10^{-6}$ and pH = 5.05

34.   a. The pH of the buffer solution is:

$$K_b = \frac{[NH_4^+][OH^-]}{[NH_3]} = \frac{(0.250)[OH^-]}{(0.500)} = 1.8 \times 10^{-5}$$

and solving for hydroxyl ion yields: $[OH^-] = 3.6 \times 10^{-5}$
        pOH = 4.44   pH = 9.56

b. pH after addition of 0.0100 mol HCl:

The basic component of the buffer (NH_3) will react with the HCl, producing more ammonium ion. The composition of the solution is then:

	NH_3	NH_4Cl
Moles present (before HCl is added)	0.250	0.125
Change (reaction)	- 0.0100	+ 0.0100
Following reaction	0.240	+ 0.135

The amounts of NH_3 and NH_4Cl following the reaction with HCl are only slightly different from those amounts prior to reaction. Converting these numbers into molar concentrations (volume is 500. mL) and substituting the concentrations into the K_b expression yields:

$$K_b = \frac{[NH_4^+][OH^-]}{[NH_3]} = \frac{(0.270)[OH^-]}{(0.480)} = 1.8 \times 10^{-5}$$

$[OH^-] = 3.2 \times 10^{-5}$; pOH = 4.50, and the new pH = 9.50.

Using the Henderson-Hasselbalch Equation

36. The pK_a for acetic acid is = $-\log(K_a)$ or 4.74

$$pH = pK_a + \log \frac{[conjugate\ base]}{[acid]}$$

$$= 4.74 + \log \frac{0.075}{0.050} = 4.74 + 0.18 \ \ or \ \ 4.92$$

38.   a. $pH = pK_a + \log \frac{[conjugate\ base]}{[acid]}$

$$= 3.74 + \log \frac{0.035}{0.050} = 3.74 - 0.15 \ \ or \ \ 3.59$$

   b. Increasing pH gives 4.09

Substituting into the Henderson-Hasselbalch equation

$$4.09 = 3.74 + \log \frac{[conjugate\ base]}{[acid]}$$

$$+0.35 = \log \frac{[conjugate\ base]}{[acid]}$$

$$-0.35 = \log \frac{[acid]}{[conjugate\ base]}$$

$$0.45 = \frac{[acid]}{[conjugate\ base]}$$

Titration Curves and Indicators

40.   The titration of 0.10 M NaOH with 0.10 M HCl (a strong base vs a strong acid)

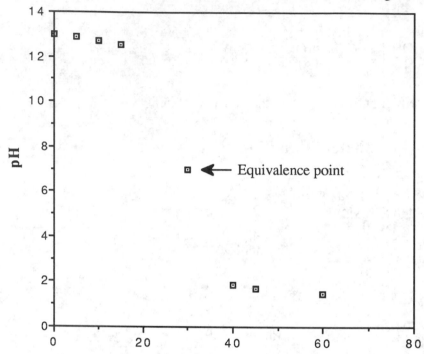

The initial pH of a 0.10 M NaOH would be pOH = - log[0.1]
pOH = 1 and pH = 13.

When 15.0 mL of 0.10 M HCl have been added, one-half of the NaOH initially present will be consumed, leaving 0.5 (0.030 L • 0.10 mol/L) or 1.50×10^{-3} mol NaOH in 45.0 mL—therefore a concentration of 0.0333 M NaOH. pOH = 1.48 and pH = 12.52.

At the equivalence point (30.0 mL of the 0.10 M acid are added) there is only NaCl present. Since this salt does not hydrolyze, the pH at that point is exactly 7.0. The total volume present at this point is 60 mL.

Once a total of 60.0 mL of acid are added, there is an excess of 3.0×10^{-3} mol of HCl. Contained in a total volume of 90.0 mL of solution, the [HCl] = 0.0333 and the pH =1.5.

42.   a. pH of 25.0 mL of 0.11 M NH_3:

For the weak base, NH_3, the equilibrium in water is represented as:

$$NH_3 \text{ (aq)} + H_2O \text{ (}\ell\text{)} \rightleftharpoons NH_4^+ \text{ (aq)} + OH^- \text{ (aq)}$$

The slight dissociation of NH_3 would form equimolar amounts of NH_4^+ and OH^- ions.

$$K_b = \frac{[NH_4^+][OH^-]}{[NH_3]} = \frac{(x)(x)}{0.11 - x} = 1.8 \times 10^{-5}$$

Simplifying, we get : $\frac{x^2}{0.11} = 1.8 \times 10^{-5}$ $x = 1.4 \times 10^{-3} = [OH^-]$

pOH = 2.85 and pH = 11.15

Addition of HCl will consume NH_3 and produce NH_3^+ (the conjugate) according to the net equation: $NH_3 \text{ (aq)} + H^+ \text{ (}\ell\text{)} \rightleftharpoons NH_4^+ \text{ (aq)}$

The strong acid will drive this equilibrium to the right so we will assume this reaction to be complete. Let us first calculate the moles of NH_3 initially present:

$$(0.0250 \text{ L}) (0.11 \frac{\text{mol } NH_3}{\text{L}}) = 0.0028 \text{ mol } NH_3$$

Reaction with the HCl will produce the conjugate acid, NH_4^+. The task is two-fold. First calculate the amounts of the conjugate pair present. Second substitute the concentrations into the K_b expression. [One time-saving hint: **The ratio of concentrations** and the **ratio of the amounts (moles)** will have the same numerical value. One can substitute the amounts of the conjugate pair into the K_b expression.]

$$K_b = \frac{[NH_4^+][OH^-]}{[NH_3]} = 1.8 \times 10^{-5}$$

Present initially: 2.75 millimol NH_3

mL of 0.10 M HCl added	millimol HCl added	millimol NH_3 after reaction	millimol NH_4^+ after reaction	$[OH^-]$ after reaction	pH
5.00	0.50	2.25	0.50	8.1×10^{-5}	9.91
11.0	1.1	1.65	1.1	2.7×10^{-5}	9.43
12.5	1.25	1.50	1.25	2.2×10^{-5}	9.33
15.0	1.5	1.25	1.5	1.5×10^{-5}	9.18
20.0	2.0	0.75	2.0	6.8×10^{-6}	8.83
22.0	2.2	0.55	2.2	4.5×10^{-6}	8.65

b. When 28.0 mL of the 0.10 M HCl has been added (total solution volume = 53.0 mL), the reaction is at the equivalence point. All the NH_3 will be consumed, leaving the salt, NH_4Cl. The NH_4Cl (2.75 millimol) has a concentration of 5.2×10^{-2} M. This salt, being formed from a weak base and strong acid, undergoes hydrolysis.

$$NH_4^+(aq) + H_2O \, (\ell) \; \rightleftharpoons \; NH_3 \, (aq) + H_3O^+ \, (aq)$$

	NH_4^+	NH_3	H_3O^+
Initial concentration	5.2×10^{-2}		
Change (going to equilibrium)	-x	+x	+x
Equilibrium concentration	5.2×10^{-2} - x	+x	+x

$$K_a = \frac{[NH_3][H_3O^+]}{[NH_4^+]} = 5.6 \times 10^{-10} = \frac{x^2}{5.2 \times 10^{-2} - x}$$

$$\approx \frac{x^2}{5.2 \times 10^{-2}} = 5.6 \times 10^{-10} \quad \text{and } x = 5.4 \times 10^{-6} = [H_3O^+]$$

and pH = 5.27 (equivalence point)

c. The midpoint of the titration occurs when 13.75 (13.8) mL of the acid have been added. At that point the amount of base and salt present are equal. An examination of the K_b expression will show that under these conditions the $[OH^-] = K_b$. Hence a pOH of 4.74 and a pH = 9.26.

d. From the Figure 17.7 we see that one indicator to use is Methyl Red. This indicator would be yellow prior to the equivalence point and red past that point. Bromcresol green would also be suitable, being blue prior to the equivalence point and yellow-green after.

e. All points except the one for pH after the addition of 30.0 mL have been listed above. Addition of acid in excess of 27.50 mL will result in a solution which is essentially a strong acid.

After the addition of 30.0 mL, substances present are:

  millimol HCl added: 3.00
  millimol NH_3 present: 2.75
  excess HCl present :   0.25 millimol

This HCl is present in a total volume of 55.0 mL of solution, hence the calculation for a strong acid proceeds as follows:

$$[H_3O^+] = \frac{0.25 \text{ millimol HCl}}{55.0 \text{ milliliters}} = 4.5 \times 10^{-3} \text{ M} \quad \text{and pH} = 2.34.$$

The graph for these data is:

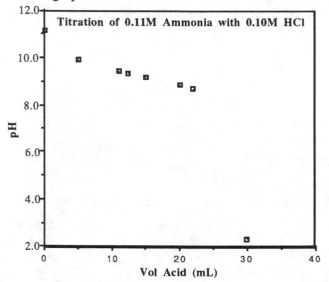

A Summary

mL acid	pH
0.00	11.15
5.00	9.91
11.0	9.43
12.5	9.33
15.0	9.18
20.0	8.83
22.0	8.65
30.0	2.34

44. Suitable indicators for titrations:

   a. HCl with pyridine: A solution of pyridinium chloride would have a pH of approximately 3. A suitable indicator would be **thymol blue**.

   b. NaOH with formic acid: The salt formed at the equivalence point is sodium formate. Hydrolysis of the formate ion would give rise to a basic solution (pH $\approx$ 8.5). **Phenolphthalein** would be a suitable indicator.

   c. Hydrazine and HCl: The salt, hydrazinium hydrochloride, hydrolyzes. The hydrazinium ion would produce an acidic solution (pH $\approx$ 4.5). A suitable indicator would be **bromcresol green or methyl orange**.

General Questions

46.   a. $CH_3CH_2COOH(aq) + OH^-(aq) \longrightarrow CH_3CH_2COO^-(aq) + H_2O(\ell)$

b. Since the equilibrium constant is large for this reaction (see question 8), the reaction proceeds by having all the base consumed, consuming 0.40 mol of acid (leaving 0.60 mol of acid) and producing 0.40 mol of the conjugate base (propanoate ion).

c. The pH of the solution:

$$\frac{[CH_3CH_2COO^-][H_3O^+]}{[CH_3CH_2COOH]} = 1.3 \times 10^{-5}$$

$$[H_3O^+] = 1.3 \times 10^{-5} \cdot \frac{0.60}{0.40}$$

$$[H_3O^+] = 2.0 \times 10^{-5} \quad \text{and pH} = 4.71$$

d. This mass of NaOH corresponds to 0.01 mol of base. If we add this amount to the solution, we'll have 0.59 mol of acid and 0.41 mol of salt present. Substituting these data into the expression above we get

$$[H_3O^+] = 1.3 \times 10^{-5} \frac{(0.59)}{(0.41)} = 1.9 \times 10^{-5}$$

and a pH of 4.7 **The pH would remain the same**.

48. For the equilibrium : $NH_3(aq) + H_2O(\ell) \rightleftharpoons NH_4^+(aq) + OH^-(aq)$

The NH_3 would have the equilibrium:

$$K_b = \frac{[NH_4^+][OH^-]}{[NH_3]} = \frac{(x)(x)}{(0.20 - x)} = 1.8 \times 10^{-5}$$

The 0.040 M NaOH would change $[OH^-]$ to $x + 0.040$

$$\frac{(x)(x + 0.040)}{(0.20 - x)} = 1.8 \times 10^{-5}$$

Assuming that x is negligible with respect to 0.20 and 0.040, we can write

$$\frac{(x)(0.040)}{(0.20)} = 1.8 \times 10^{-5} \quad \text{and} \quad x = 9.0 \times 10^{-5}$$

So $x = [NH_4^+] = 9.0 \times 10^{-5}$; $[NH_3] = 0.20$

and $[Na^+] = [OH^-] = 0.040$

50. pH of 0.160 M CH3COOH:

The equilibrium may be written: $CH_3COOH + H_2O \rightarrow CH_3COO^- + H_3O^+$

$$K_a = \frac{[CH_3COO^-][H_3O^+]}{[CH_3COOH]} = 1.8 \times 10^{-5}$$

$$= \frac{[H_3O^+]^2}{(0.160)} = 1.8 \times 10^{-5} \quad \text{and} \quad [H_3O^+] = 1.70 \times 10^{-3} \quad \text{and pH} = 2.77$$

pH after 56.8 g of NaCH3COO is added:

The concentration of sodium acetate is :

$$\frac{56.8 \text{ g NaCH}_3\text{COO}}{1.50 \text{ L}} \cdot \frac{1 \text{ mol NaCH}_3\text{COO}}{82.03 \text{ g NaCH}_3\text{COO}} = 0.462 \text{ M NaCH}_3\text{COO}$$

Assuming that there is no volume change when the sodium acetate is added, the concentrations can be substituted into the K_a expression:

$$K_a = \frac{(x + 0.462)(H_3O^+)}{0.160 - x} = 1.8 \times 10^{-5}$$

Simplifying: $\frac{0.462[H_3O^+]}{0.160} = 1.8 \times 10^{-5}$

$$[H_3O^+] = 6.2 \times 10^{-6} \qquad \text{and pH} = 5.20$$

52. At the equivalence point, the number of moles of acid and base are equal. 25.0 mL of 0.120 M HCO_2H contain 0.00300 moles. The volume of 0.105 M NaOH which contains

that amount of HCO_2H is $\frac{(0.025)(0.120)}{0.105} = 0.0286$ L.

The total volume of solution is (0.025 L + 0.0286) 0.0536 L. The concentration of the salt formed when the NaOH reacts with HCO_2H is then $\frac{0.00300 \text{ mol}}{0.0536 \text{ L}}$ or 0.0560 M.

Substituting into the K_b expression for the formate ion gives:

$$K_b = \frac{[HCO_2H][OH^-]}{[HCO_2^-]} = \frac{[OH^-]^2}{0.0560} = \frac{1.0 \times 10^{-14}}{1.8 \times 10^{-4}}$$

$$[OH^-] = 1.7 \times 10^{-6} \text{ and pH} = 8.25$$

54.   a. The pH of the buffer can be calculated.

The Ka for acetic acid is 1.8×10^{-5}

$$pH = pK_a + \log \frac{[\text{conjugate base}]}{[\text{acid}]}$$

$$pH = 4.74 + \log\frac{0.88}{0.50} = 4.99$$

b. pH after 0.10 mol of NaOH is added to 1.0 L of buffer 0.10 mol of NaOH will consume 0.10 mol of acid and produce 0.10 mol of the conjugate base, with a resulting pH

$$pH = 4.74 + \log\frac{0.98}{0.40} = 5.13$$

56. Buffer capacity:

The initial pH is 4.74 (See 54a for the Henderson-Hasselbalch equation.) Use the Henderson-Hasselbalch equation to calculate the **ratio if the pH increased by 1 unit** :

$$5.74 = 4.74 + \log\frac{(acetate)}{(acetic\ acid)}$$
$$1.00 = \log\frac{(acetate)}{(acetic\ acid)}$$
$$10^1 = \frac{acetate}{acetic\ acid}$$

Beginning with a ratio of 1 : 1 we now must change the ratio to 10 : 1. If we increase the amount of acetate (and decrease the amount of acetic acid) by some a mol/L we can write

$$10 = \frac{0.10 + a}{0.10 - a}$$
$$10(0.10 - a) = 0.10 + a$$
$$1.0 - 10a = 0.10 + a$$
$$0.90 = 11a$$
$$\frac{0.90}{11} = a = 0.082\ mol/L$$

The buffer capacity is therefore 0.082 mol. (or 0.08 to 1 sf)

58. a. The pH of 0.0256 M $C_6H_5NH_2$ is:

$$K_b = \frac{[C_6H_5NH_3^+][OH^-]}{[C_6H_5NH_2]} = 4.2 \times 10^{-10}$$

and since $[C_6H_5NH_3^+] = [OH^-]$ then $\frac{[OH^-]^2}{0.0256} = 4.2 \times 10^{-10}$

and $[OH^-] = 3.3 \times 10^{-6}$ pOH $= 5.48$ and pH $= 8.52$

b. At the equivalence point, moles acid = moles base--and the amount of base present is:
    (0.025 L)(0.0256 M $C_6H_5NH_2$) $= 6.4 \times 10^{-4}$ mol base

To furnish that amount of HCl from a solution that is 0.0195 M HCl requires 0.0328 L.

At the equivalence point there are $\dfrac{6.4 \times 10^{-4} \text{ mol salt}}{0.0578 \text{ L}}$ = 0.0111 M salt

The pH of this salt solution is controlled by the equilibrium:

$$C_6H_5NH_3^+ \text{ (aq)} + H_2O \text{ (}\ell\text{)} \rightleftharpoons C_6H_5NH_2 \text{ (aq)} + H_3O^+ \text{ (aq)}$$

$$K_a = \frac{[C_6H_5NH_2][H_3O^+]}{[C_6H_5NH_3^+]} = \frac{1.0 \times 10^{-14}}{4.2 \times 10^{-10}} = \frac{[H_3O^+]^2}{0.0111} = 2.38 \times 10^{-5}$$

$$[H_3O^+] = 5.1 \times 10^{-4} \text{ and pH} = 3.29$$

c. At the midpoint of the titration, half the base will be consumed--producing an equal number of moles of salt.

moles of base consumed = $1/2(6.4 \times 10^{-4} \text{ mol})$ = 3.2×10^{-4} mol

moles of salt produced = 3.2×10^{-4} mol

Substituting into the K_b expression as in part (a) gives:

$$\frac{[3.2 \times 10^{-4}][OH^-]}{(3.2 \times 10^{-4})} = 4.2 \times 10^{-10} \qquad pOH = 9.38$$

$$\text{and } [H_3O^+] = 2.38 \times 10^{-5} \qquad pH = 4.62$$

Note that in the K_b expression, you can substitute the **number of moles of each specie** for the **concentration of each specie** since the moles and their concentration will be in the same proportion.

d. With an equivalence point of 3.29, this titration could be followed with **bromphenol blue**, which would be yellow at a pH below the equivalence point and blue above the equivalence point.

e. The addition of acid results in (1) consumption of aniline and (2) production of the salt, anilinium chloride. The solution containing both the weak base and its conjugate salt is a buffer. The pH can be calculated:

$$K_b = \frac{[C_6H_5NH_3^+][OH^-]}{[C_6H_5NH_2]} = 4.2 \times 10^{-10}$$

rearranging : $[OH^-] = 4.2 \times 10^{-10} \cdot \dfrac{[C_6H_5NH_2]}{[C_6H_5NH_3^+]}$

Note that as in part (c) the mathematical expression has the **ratio** of base to acid--allowing you to substitute the number of moles of base and salt instead of their actual concentrations.

	vol acid	moles acid added	moles aniline	moles salt produced
1.	5.00	9.75×10^{-5}	5.43×10^{-4}	9.75×10^{-5}
2.	10.0	1.95×10^{-4}	4.45×10^{-4}	1.95×10^{-4}
3.	15.0	2.93×10^{-4}	3.47×10^{-4}	2.93×10^{-4}
4.	20.0	3.90×10^{-4}	2.50×10^{-4}	3.90×10^{-4}
5.	24.0	4.68×10^{-4}	1.72×10^{-4}	4.68×10^{-4}
6.	30.0	5.85×10^{-4}	5.50×10^{-5}	5.85×10^{-4}

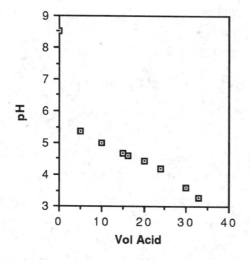

Point	[OH$^-$]	pOH	pH
1	2.34×10^{-9}	8.63	5.37
2	9.58×10^{-10}	9.02	4.98
3	4.97×10^{-10}	9.30	4.70
4	2.69×10^{-10}	9.57	4.43
5	1.54×10^{-10}	9.81	4.19
6	3.95×10^{-11}	10.40	3.60

60. The ratio of the two buffer components can be calculated from the equilibrium expression:

$$K_a = \frac{[H_3O^+][HPO_4^{2-}]}{[H_2PO_4^-]} = 6.2 \times 10^{-8}$$

With pH = 7.40, $[H_3O^+] = 4.0 \times 10^{-8}$. Substituting this value into the K_a expression and rearranging gives:

$$\frac{[HPO_4^{2-}]}{[H_2PO_4^-]} = \frac{6.2 \times 10^{-8}}{4.0 \times 10^{-8}} = 1.6$$

The requested ratio, however, is the reciprocal of $\dfrac{[HPO_4{}^{2-}]}{[H_2PO_4{}^-]}$. Taking the reciprocal gives a ratio of 0.65.

62.   According to the equation on p 751 of the text :

$$[H_3O^+] = \left(\frac{K_w \cdot K_a}{K_b}\right)^{0.5}$$

$$[H_3O^+] = \left(\frac{(1.0 \times 10^{-14})(6.3 \times 10^{-5})}{1.8 \times 10^{-5}}\right)^{0.5} = 1.9 \times 10^{-7} \text{ M}$$

and pH = 6.73

64.   a. What is the pH of the [NH$_3$OH]Cl solution before the titration begins?

$$NH_3OH^+(aq) + H_2O(\ell) \rightleftharpoons NH_2OH(aq) + H_3O^+(aq)$$

$$K_a = \frac{K_w}{K_b} = \frac{1.0 \times 10^{-14}}{6.6 \times 10^{-9}} = 1.5 \times 10^{-6}$$

$$K_a = \frac{[NH_2OH][H_3O^+]}{[NH_3OH^+]} = \frac{[H_3O^+]^2}{0.155 - [H_3O^+]} = 1.5 \times 10^{-6}$$

$$[H_3O^+] = 4.8 \times 10^{-4} \text{ M} \quad pH = 3.31$$

b. What is the pH at the equivalence point?

0.025 L • 0.155 mol/L = 0.0039 mol NH$_3$OH$^+$

0.0039 mol NH$_3$OH$^+$ requires 0.0039 mol NaOH and produces 0.0039 mol NH$_2$OH.

The volume of NaOH that contains 0.0039 mol NaOH is:

$$0.0039 \text{ mol NaOH} \cdot \frac{1.00 \text{ L}}{0.108 \text{ mol}} = 0.036 \text{ L}$$

Total volume at the equivalence point = 61 mL   (25 + 36)

$$[NH_2OH] = \frac{0.0039 \text{ mol}}{0.061 \text{ L}} = 0.064 \text{ M}$$

$$K_b = \frac{[NH_3OH^+][OH^-]}{[NH_2OH]} = \frac{[OH^-]^2}{0.064 - [OH^-]} = 6.6 \times 10^{-9}$$

$$[OH^-] = 2.1 \times 10^{-5} \text{ M} \qquad pOH = 4.69 \text{ and } pH = 9.31$$

c. What is the pH at the mid-point of the titration?

At the midpoint of the titration, [NH$_2$OH] = [NH$_3$OH$^+$].

Therefore [H$_3$O$^+$] = K$_a$ for NH$_3$OH$^+$.

[H$_3$O$^+$] = K$_a$ = 1.5 x 10^{-6} M and so pH = 5.82

66. At the second equivalence point in the titration of oxalic acid, there is only $C_2O_4^{2-}$ ion (and water). This equilibrium can be written

$$C_2O_4^{2-} + H_2O \rightleftharpoons HC_2O_4^- + OH^-$$

This K_b can be calculated by $\dfrac{K_w}{K_{a2}}$ or $\dfrac{1.0 \times 10^{-14}}{6.4 \times 10^{-5}} = 1.6 \times 10^{-10}$

The K_b expression is then:

$$\frac{[HC_2O_4^-][OH^-]}{[C_2O_4^{2-}]} = 1.6 \times 10^{-10}$$

At this point we will have added ~200 mL of 0.100 M NaOH (Figure 18.7) reducing the $[C_2O_4^{2-}]$ to approximately 1/3 the original concentration (or ~0.0333 M). Substituting into the K_b expression

$$\frac{x^2}{0.0333 - x} = 1.6 \times 10^{-10} \text{ and}$$

$$x = 2.31 \times 10^{-6} = [OH^-] \text{ and pOH} = +5.64 \text{ and pH} = 8.36$$

Conceptual Questions

68. a. HB is the stronger acid.

At the equivalence point, the conjugate base of the weak acid is present.

$$B + H_2O \rightleftharpoons HB + OH^-$$

The stronger the acid, the farther this equilibrium will lie to the left, reducing OH^- —providing a lower pH.

b. Since HA is the weaker of the two acids, **A⁻ is the stronger base**.

70. a. Examine the equilibrium of acetic acid in water

$$CH_3COOH + H_2O \rightarrow CH_3COO^- + H_3O^+$$

LeChatelier's principle tells us that if we remove H_3O^+ (raise the pH) the equilibrium will shift to the right—with the effect of producing more acetate ion (and consuming acetic acid).

b. Predominant species at pH 4 is acetic acid (~ 85 %)

At pH 6, acetate ion predominates (~ 95 %)

c. Remembering the Henderson-Hasselbalch equation

$$pH = pK_a + \log \frac{[\text{conjugate base}]}{[\text{acid}]}$$

At the point in which the concentrations of acid and conjugate base are equal, the log term vanishes. So the pH $= pK_a$. The pK_a for acetic acid is 4.75, hence the pH of the solution is anticipated to be 4.75.

73.   a. Approximate bond angles for:

(i) C=C-C in the ring : $120°$ for sp^2 hybridized C

(ii) C-C=O: $120°$ for sp^2 hybridized C

(iii) C-O-H: $109°$ for sp^3 hybridized C

(iv) C-C-H: $120°$ for sp^2 hybridized C

b. C atoms in the ring are sp^2 hybridized as is the C atom in -COOH group.

c. pH of the solution containing 1.00 g of salicylic acid in 460 mL of water
The concentration of the salicylic acid is:

$$\frac{1.00 \text{ g salicylic acid}}{0.46 \text{ L}} \cdot \frac{1 \text{ mol salicylic acid}}{138.1 \text{ g salicylic acid}} = 1.6 \times 10^{-2} \text{ mol}$$

$$K_a = \frac{[A^-][H_3O^+]}{[HA]} = \frac{[x][x]}{1.6 \times 10^{-2} - x} = 1.1 \times 10^{-3}$$

Since $100 \cdot K_a \approx [HA]$, the quadratic equation must be used to find an exact solution.

$$x^2 = 1.73 \times 10^{-5} - 1.1 \times 10^{-3}(x)$$
$$x^2 + 1.1 \times 10^{-3}(x) - 1.73 \times 10^{-5} = 0$$
$$x = [H_3O^+] = 3.65 \times 10^{-3} \text{ and the pH} = 2.44$$

d. The percentage of the acid in the form of salicylate ion :

From the equilibrium expression:
$$\frac{[H_3O^+][\text{salicylate}]}{[\text{acid}]} = 1.1 \times 10^{-3}$$
At a pH = 2.0, $[H_3O^+] = 1 \times 10^{-2}$

Substituting into the equilibrium expression:

253

$$\frac{[1 \times 10^{-2}][\text{salicylate}]}{[\text{acid}]} = 1.1 \times 10^{-3} \text{ or}$$

$$\frac{[\text{salicylate}]}{[\text{acid}]} = \frac{1.1 \times 10^{-3}}{1 \times 10^{-2}} = 0.11$$

or 11 % of the acid as the salicylate ion.

e. The pH at the midpoint of the titration :

At the midpoint of the titration, the concentration of the acid and its conjugate ion are equal. Therefore $[H_3O^+] = K_a = 1.1 \times 10^{-3}$ and pH = 2.96

The pH at the equivalence point can be calculated after the concentration of the sodium salicylate is calculated.

$$0.0250 \text{ mL} \cdot 0.014 \text{ M salicylic acid} = 3.5 \times 10^{-4} \text{ moles}$$

The volume of 0.010 M NaOH needed to neutralize this amount is 35.0 mL.

The concentration of the salt is: $\dfrac{3.5 \times 10^{-4}}{(0.0250 + 0.0350)} = 5.8 \times 10^{-3}$ M

Substituting into the K_b expression for the salicylate ion:

$$K_b = \frac{[\text{salicylic acid}][OH^-]}{[\text{salicylate}]} = \frac{[OH^-]^2}{5.83 \times 10^{-3}} = \frac{1.0 \times 10^{-14} \leftarrow K_w}{1.1 \times 10^{-3} \leftarrow K_a}$$

and $[OH^-] = 2.3 \times 10^{-7}$; pOH = 6.64 and pH = 7.36

Chapter 19:
Principles of Reactivity: Precipitation Reactions

Solubility Guidelines

8. Two insoluble salts of

a.	Cl^-	AgCl and Hg_2Cl_2
b.	Zn^{2+}	ZnS and $Zn_3(PO_4)_2$
c.	Fe^{2+}	FeS and $Fe(OH)_2$

10. Using the table of solubility guidelines, predict water solubility for the following:

a.	$(NH_4)_2S$	Ammonium salts are **soluble**.
b.	$ZnCO_3$	Carbonates are generally **insoluble**.
c.	FeS	Sulfides are generally **insoluble**.
d.	$BaSO_4$	Sr^{2+}, Ba^{2+}, and Pb^{2+} form **insoluble** sulfates.

12. Solution producing precipitate
    a. $Ag^+ (aq) + I^- (aq) \rightarrow AgI (s)$
    b. $Pb^{2+} (aq) + 2 Cl^- (aq) \rightarrow PbCl_2 (s)$
    c. No precipitate is expected.

Writing Solubility Product Constant Expressions

14.

Salt dissolving	K_{sp} expression	K_{sp} values
a. $AgCN(s) \rightleftharpoons Ag^+(aq) + CN^-(aq)$	$K_{sp} = [Ag^+][CN^-]$	1.2×10^{-16}
b. $PbCO_3(s) \rightleftharpoons Pb^{2+}(aq) + CO_3^{2-}(aq)$	$K_{sp} = [Pb^{2+}][CO_3^{2-}]$	1.5×10^{-13}
c. $AuI_3(s) \rightleftharpoons Au^{3+}(aq) + 3 I^-(aq)$	$K_{sp} = [Au^{3+}][I^-]^3$	1.0×10^{-46}

Calculating K_{sp}

16. $K_{sp} = [Tl^+][Br^-] = (1.8 \times 10^{-3})(1.8 \times 10^{-3}) = 3.2 \times 10^{-6}$

18. Molar solubility of $Ca(OH)_2$:

$$\frac{0.93 \text{ g } Ca(OH)_2}{L} \cdot \frac{1 \text{ mol } Ca(OH)_2}{74.1 \text{ g } Ca(OH)_2} = 0.013 \text{ M}$$

When $Ca(OH)_2$ dissolves there is <u>one</u> Ca^{2+} and <u>two</u> OH^-, so

$$[Ca^{2+}][OH^-]^2 \;=\; (0.013)(0.026)^2 \;=\; 7.9 \times 10^{-6}$$

20. Molar solubility of Ag_2CO_3 :

$$\frac{34.9 \times 10^{-3} \text{ g}}{1.0 \text{ L}} \cdot \frac{1 \text{ mol } Ag_2CO_3}{275.7 \text{ g } Ag_2CO_3} = 1.27 \times 10^{-4} \text{ M}$$

When Ag_2CO_3 dissolves : $[Ag^+] \;=\; 2.53 \times 10^{-4}$ M ; $[CO_3^{2-}] \;=\; 1.27 \times 10^{-4}$ M

$$K_{sp} \;=\; [Ag^+][CO_3^{2-}] \;=\; (2.53 \times 10^{-4})^2(1.27 \times 10^{-4}) \;=\; 8.1 \times 10^{-12}$$

22.   With pH = 12.40, pOH = 1.60 and $[OH^-] = 2.51 \times 10^{-2}$ M

The stoichiometry of $Ca(OH)_2$ indicates that $[Ca^{2+}]$ is one-half $[OH^-]$, so

$[Ca^{2+}] = 1.26 \times 10^{-2}$ M; $K_{sp} = [Ca^{2+}][OH^-]^2 = (1.26 \times 10^{-2})(2.51 \times 10^{-2})^2$

$$K_{sp} \;=\; 7.9 \times 10^{-6}$$

Estimating Salt Solubility from K_{sp}

24.   a. The K_{sp} for AgCN is 1.2×10^{-16}. The equation for AgCN dissolving is:

$$AgCN \text{ (s)} \;\rightleftharpoons\; Ag^+ \text{ (aq)} + CN^- \text{ (aq)}$$

From the equation we see that $[Ag^+] = [CN^-]$.

The K_{sp} expression for AgCN is: $K_{sp} = [Ag^+][CN^-] = 1.2 \times 10^{-16}$

$$K_{sp} = [Ag^+]^2 = [CN^-]^2 = 1.2 \times 10^{-16}$$

$$[Ag^+] = [CN^-] = 1.1 \times 10^{-8} \text{ M}$$

These concentrations tell us that 1.1×10^{-8} mol/L of AgCN dissolve.

b. Solubility of AgCN in g/L:

$$1.1 \times 10^{-8} \frac{\text{mol AgCN}}{\text{L}} \cdot \frac{134 \text{ g AgCN}}{1 \text{ mol AgCN}} = 1.5 \times 10^{-6} \frac{\text{g AgCN}}{\text{L}}$$

26. $K_{sp} \;=\; [Ra^{2+}][SO_4^{2-}] \;=\; 4.2 \times 10^{-11}$ so $[Ra^{2+}] \;=\; [SO_4^{2-}] \;=\; 6.5 \times 10^{-6}$ M

$RaSO_4$ will dissolve to the extent of 6.5×10^{-6} mol/L

Express this as grams in 100. mL (or 0.1 L)

$$\frac{6.5 \times 10^{-6} \text{ mol } RaSO_4}{1 \text{ L}} \cdot \frac{0.1 \text{ L}}{1} \cdot \frac{322 \text{ g } RaSO_4}{1 \text{ mol } RaSO_4} = 2.09 \times 10^{-4} \text{ g}$$

Expressed as milligrams : 0.209 mg or 0.21 (to 2 sf)

28. The K_{sp} for MgF_2 = 6.4×10^{-9}

   K_{sp} = $[Mg^{2+}][F^-]^2$ = 6.4×10^{-9}

   if a mol/L of MgF_2 dissolve, $[Mg^{2+}]$ = a and $[F^-]$ = 2a

   $(a)(2a)^2$ = $4a^3$ = 6.4×10^{-9} and a = 1.17×10^{-3}

   a. The molar solubility is then 1.2×10^{-3} (to 2 sf)

   b. Solubility is g/L

   $$1.2 \times 10^{-3} \frac{\text{mol } MgF_2}{L} \cdot \frac{62.3 \text{ g } MgF_2}{1 \text{ mol } MgF_2} = 7.3 \times 10^{-2} \text{ g } MgF_2 /L$$

30. In each case the more soluble compound would have the greater Ksp. *This is true if— and only if— you compare compounds of similar stoichiometry!*

   a. AgSCN

   b. $SrSO_4$

   c. MgF_2

   d. Caution here! The vastly different Ksp's would lead you to assume that AgI is more soluble. We better check! Let a = molar solubility

   $[Ag^+][I^-]$ = a^2 = 1.5×10^{-16} a = 1.2×10^{-8}

   $[Hg^{2+}][I^-]^2$ = $4a^3$ = 4.0×10^{-29} a = 2.2×10^{-10}

   While our first assumption (AgI more soluble) is correct, the difference in solubility is not as great as the Ksp's might have initially indicated.

32.   Compounds in order of increasing solubility in H_2O:

Compound	K_{sp}
BaF_2	1.7×10^{-6}
$BaCO_3$	8.1×10^{-9}
Ag_2CO_3	8.1×10^{-12}
Na_2CO_3	not listed: Na_2CO_3 is very soluble in water and is certainly the most soluble salt of the 4.

To determine the relative solubilities find the molar solubilities. The equation representing BaF_2 dissolving in H_2O is: $BaF_2(s) \rightleftharpoons Ba^{2+}$ (aq) $+ 2 F^-$ (aq)

For each mol of BaF_2 dissolving, one mol of Ba^{2+} and two mol of F^- are produced. If x mol/L of BaF_2 dissolve, the equilibrium concentrations of Ba^{2+} and F^- are x and 2x respectively.

$$K_{sp} = [Ba^{2+}][F^-]^2 = (x)(2x)^2 = 1.7 \times 10^{-6}$$
$$4x^3 = 1.7 \times 10^{-6}$$
$$x = 7.5 \times 10^{-3}$$

The molar solubility of BaF_2 is then 7.5×10^{-3} M.

Similarly for $BaCO_3$:

$$K_{sp} = [Ba^{2+}][CO_3^{2-}] = (x)(x) = 8.1 \times 10^{-9} \text{ and } x = 9.0 \times 10^{-5}$$

The molar solubility of $BaCO_3$ is then 9.0×10^{-5} M.

and for Ag_2CO_3:

$$K_{sp} = [Ag^+]^2[CO_3^{2-}] = (2x)^2(x) = 8.1 \times 10^{-12}$$
$$4x^3 = 8.1 \times 10^{-12}$$
$$\text{and } x = 1.3 \times 10^{-4}$$

The molar solubility of Ag_2CO_3 is 1.3×10^{-4} M.

In order of increasing solubility: $BaCO_3$ < Ag_2CO_3 < BaF_2

34. The concentration corresponding to 5.0×10^{-3} g $NiCO_3$ in 1.0 L of water:

$$\frac{5.0 \times 10^{-3} \text{ g } NiCO_3}{1} \cdot \frac{1 \text{ mol } NiCO_3}{119 \text{ g } NiCO_3} = 4.2 \times 10^{-5} \text{ M}$$

Q for this solution would be $[Ni^{2+}][CO_3^{2-}] = (4.2 \times 10^{-5})^2 = 1.8 \times 10^{-9}$

Since Q < K_{sp} for $NiCO_3$ (6.6×10^{-9}), all the solid dissolves.

Precipitations

36.   Precipitation of a substance will occur if the reaction quotient exceeds the K_{sp} for the substance.

For $NiCO_3$: $K_{sp} = [Ni^{2+}][CO_3^{2-}] = 6.6 \times 10^{-9}$

a. If $[Ni^{2+}] = 2.4 \times 10^{-3}$ M and $[CO_3^{2-}] = 1.0 \times 10^{-6}$ M

The reaction quotient is $[Ni^{2+}][CO_3^{2-}] = (2.4 \times 10^{-3})(1.0 \times 10^{-6}) = 2.4 \times 10^{-9}$

and precipitation does not occur.

b. If $[Ni^{2+}] = 2.4 \times 10^{-3}$ M and $[CO_3^{2-}] = 1.0 \times 10^{-4}$ M

The reaction quotient is $[Ni^{2+}][CO_3^{2-}] = (2.4 \times 10^{-3})(1.0 \times 10^{-4}) = 2.4 \times 10^{-7}$

and precipitation occurs.

38.   If $Zn(OH)_2$ is to precipitate, the reaction quotient (Q) must exceed the K_{sp} for the salt.

4.0 mg of NaOH in 10. mL corresponds to a concentration of:

$$[OH^-] = \frac{4.0 \times 10^{-3} \text{ g NaOH}}{0.0100 \text{ L}} \cdot \frac{1 \text{ mol NaOH}}{40.0 \text{ g NaOH}} = 0.01 \text{ M}$$

The value of Q is :

$$[Zn^{2+}][OH^-]^2 = (1.6 \times 10^{-4})(1.0 \times 10^{-2})^2 = 1.6 \times 10^{-8}$$

The value of Q is greater than the K_{sp} for the salt (4.5×10^{-17}) , so $\mathbf{Zn(OH)_2}$

precipitates.

40.   1350 mg Mg^{2+}/L corresponds to a concentration of :

$$[Mg^{2+}] = \frac{1.350 \text{ g Mg}^{2+}}{1 \text{ L}} \cdot \frac{1 \text{ mol Mg}^{2+}}{24.305 \text{ g Mg}^{2+}} = 5.55 \times 10^{-2} \text{ M}$$

For $Mg(OH)_2$ to begin precipitation, Q must exceed K_{sp} for $Mg(OH)_2$ (1.5×10^{-11}).

$$Q = (5.55 \times 10^{-2})(OH^-)^2 = 1.5 \times 10^{-11} \text{ and } [OH^-] = 1.6 \times 10^{-5} \text{ M}$$

42. Calculate the concentration of barium ion and sulfate ion after mixing--but before reaction.

$$0.0012 \text{ M Ba}^{2+} \cdot \frac{50. \text{ mL}}{(50. + 25 \text{ mL})} = 8.0 \times 10^{-4} \text{ M Ba}^{2+}$$

$$1.0 \times 10^{-6} \text{ M SO}_4^{2-} \cdot \frac{25 \text{ mL}}{(50. + 25 \text{ mL})} = 3.3 \times 10^{-7} \text{ M SO}_4^{2-}$$

$$Q = [Ba^{2+}][SO_4^{2-}] = (8.0 \times 10^{-4})(3.3 \times 10^{-7}) = 2.7 \times 10^{-10}$$

and since $Q > K_{sp}$ (1.1×10^{-10}) $\mathbf{BaSO_4}$ **precipitates**.

44. Calculate the concentration of lead ion and chloride ion after mixing--but before reaction.

$$0.0010 \text{ M Pb}^{2+} \cdot \frac{10. \text{ mL}}{(10. + 5.0 \text{ mL})} = 6.7 \times 10^{-4} \text{ M Pb}^{2+}$$

$$0.015 \text{ M Cl}^- \cdot \frac{5.0 \text{ mL}}{(10. + 5.0 \text{ mL})} = 5.0 \times 10^{-3} \text{ M Cl}^-$$

$$Q = [Pb^{2+}][Cl^-]^2 = (6.7 \times 10^{-4})(5.0 \times 10^{-3})^2 = 1.7 \times 10^{-8}$$
and since $Q < K_{sp}$ (1.7×10^{-5}) **PbCl$_2$ does not precipitate.**

46. These metal sulfates are 1:1 salts. The K_{sp} expression has the general form:
$$K_{sp} = [M^{2+}][SO_4^{2-}]$$
To determine the $[SO_4^{2-}]$ necessary to begin precipitation, we can divide the equation by the metal ion concentration to obtain:
$$\frac{K_{sp}}{[M^{2+}]} = [SO_4^{2-}]$$
The concentration of the three metal ions under consideration are each 0.10 M. Substitution of the appropriate K_{sp} for the sulfates and 0.10 M for the metal ion concentration yields the sulfate ion concentrations in the table below. As the soluble sulfate is added to the metal ion solution, the sulfate ion concentration increases from zero molarity. The lowest sulfate ion concentration is reached first, with higher concentrations reached later. The order of precipitation is listed in the last column of the table below.

Compound	K_{sp}	Maximum $[SO_4^{2-}]$	Order of Precipitation
BaSO$_4$	1.1×10^{-10}	1.1×10^{-9}	1
SrSO$_4$	2.8×10^{-7}	2.8×10^{-6}	3
PbSO$_4$	1.8×10^{-8}	1.8×10^{-7}	2

48. The K_{sp} expressions for the hydroxides of Fe^{3+}, Pb^{2+}, and Al^{3+} are:

$$[Fe^{3+}][OH^-]^3 = 6.3 \times 10^{-38} \qquad [Pb^{2+}][OH^-]^2 = 2.8 \times 10^{-16}$$

$$[Al^{3+}][OH^-]^3 = 1.9 \times 10^{-33}$$

The solution in question contains the cations each in 0.1 M concentration. Substituting this value for the metal ion concentrations, and solving for the $[OH^-]$ yields:

Fe(OH)$_3$: $[OH^-] = (6.3 \times 10^{-37})^{1/3} = 8.6 \times 10^{-13}$ M

Pb(OH)$_2$: $[OH^-] = (2.8 \times 10^{-15})^{1/2} = 5.3 \times 10^{-8}$ M

Al(OH)$_3$: $[OH^-] = (1.9 \times 10^{-32})^{1/3} = 2.7 \times 10^{-11}$ M

The salts would precipitate in the order: Fe(OH)$_3$, Al(OH)$_3$, Pb(OH)$_2$.

Common Ion Effect

50. The equilibrium for AgSCN dissolving is: $AgSCN\ (s) \rightleftharpoons Ag^+\ (aq) + SCN^-\ (aq)$.

    As x mol/L of AgSCN dissolve in pure water, x mol/L of Ag^+ and x mol/L of SCN^- are produced.

    $$K_{sp} = [Ag^+][SCN^-] = x^2 = 1.0 \times 10^{-12} \qquad \text{and } x = 1.0 \times 10^{-6}\ M$$

    So 1.0×10^{-6} mol AgSCN/L dissolve in pure water.

    The equilibrium for AgSCN dissolving in NaSCN (0.010 M) is like that above. Equimolar amounts of Ag^+ and SCN^- ions are produced as the solid dissolves. However the [SCN^-] is augmented by the soluble NaSCN.

    $$K_{sp} = [Ag^+][SCN^-] = (x)(x + 0.010) = 1.0 \times 10^{-12}$$

    We can simplify the expression by assuming that $x + 0.010 \approx 0.010$.

    $$(x)(0.010) = 1.0 \times 10^{-12} \quad \text{and } x = 1.0 \times 10^{-10}\ M$$

    The solubility of AgSCN in 0.010 M NaSCN is 1.0×10^{-10} M -- reduced by four orders of magnitude from its solubility in pure water. LeChatelier strikes again!

52. The K_{sp} expression for Ag_3PO_4 is: $K_{sp} = [Ag^+]^3[PO_4^{3-}] = 1.3 \times 10^{-20}$

    a. The stoichiometry of the compound is such that when the solid dissolves,
    $$[Ag^+] = 3 \cdot [PO_4^{3-}]$$
    Substituting into the K_{sp} expression gives $(3 \cdot [PO_4^{3-}])^3[PO_4^{3-}] = 1.3 \times 10^{-20}$

    or $27 \cdot [PO_4^{3-}]^4 = 1.3 \times 10^{-20}$ and $[PO_4^{3-}] = 4.7 \times 10^{-6}$

    The molar amount of solid that dissolves will be equal to the concentration of phosphate ion. Expressing this concentration in mg/mL:

    $$\frac{4.7 \times 10^{-6}\ mol}{1\ L} \cdot \frac{419\ g\ Ag_3PO_4}{1\ mol} \cdot \frac{1000\ mg}{1\ g} \cdot \frac{1\ L}{1000\ mL} = 2.0 \times 10^{-3}\ mg/mL$$

    b. The solubility of the salt in 0.020 M $AgNO_3$:

    While the ratio of silver and phosphate ions from the solid dissolving remains the same, the equilibrium concentration of Ag^+ will be increased by the 0.020 M Ag^+ ion. If we represent that molar solubility of Ag_3PO_4 as x, the concentrations at equilibrium will be:

$$[PO_4^{3-}] = x \quad \text{and} \quad [Ag^+] = 3x + 0.020$$

Substituting the values into the K_{sp} expression:

$$[Ag^+]^3[PO_4^{3-}] = 1.3 \times 10^{-20} = (3x + 0.020)^3 x = 1.3 \times 10^{-20}$$

The amount of silver phosphate which dissolves is small, hence let's simplify the expression to: $(0.020)^3 x = 1.3 \times 10^{-20}$ and solving for x:

$$x = 1.6 \times 10^{-15} \text{ M, and expressing this concentration in mg/mL:}$$

$$\frac{1.6 \times 10^{-15} \text{ mol}}{1 \text{ L}} \cdot \frac{419 \text{ g Ag}_3\text{PO}_4}{1 \text{ mol}} \cdot \frac{1000 \text{ mg}}{1 \text{ g}} \cdot \frac{1 \text{ L}}{1000 \text{ mL}} = 6.8 \times 10^{-13} \text{ mg/mL}$$

Separations

54.   a. To determine the maximum concentration of oxalate ion before the Mg^{2+} salt begins to precipitate, substitute the concentration of Mg^{2+} into the K_{sp} expression:

$$K_{sp} = [Mg^{2+}][C_2O_4^{2-}] = 8.6 \times 10^{-5} \text{ and for a solution in which}$$

$[Mg^{2+}] = 0.020$ the maximum $[C_2O_4^{2-}] = \dfrac{8.6 \times 10^{-5}}{0.020} = 4.3 \times 10^{-3}$ M.

b. When the magnesium salt just begins to precipitate, the $[C_2O_4^{2-}] = 4.3 \times 10^{-3}$.

At that point the $[Ca^{2+}]$ would be:

$$K_{sp} = [Ca^{2+}][C_2O_4^{2-}] = 2.3 \times 10^{-9}$$

and the $[Ca^{2+}]$ would be $\dfrac{2.3 \times 10^{-9}}{4.3 \times 10^{-3}} = 5.3 \times 10^{-7}$ M.

56.   The more soluble salt will have the greater K_{sp}. The respective K_{sp}'s for the compounds are: $PbI_2 = 8.7 \times 10^{-9}$; $PbCO_3 = 1.5 \times 10^{-13}$

a. The K_{sp} expressions for these are:

$$K_{sp} = [Pb^{2+}][I^-]^2 = 8.7 \times 10^{-9} \quad \text{and } K_{sp} = [Pb^{2+}][CO_3^{2-}] = 1.5 \times 10^{-13}$$

Substituting the appropriate concentrations and solving for $[Pb^{2+}]$:

$$[Pb^{2+}] = \frac{8.7 \times 10^{-9}}{(0.10)^2} \qquad\qquad [Pb^{2+}] = \frac{1.5 \times 10^{-13}}{0.10}$$

$$[Pb^{2+}] = 8.7 \times 10^{-7} \text{ M} \qquad\qquad [Pb^{2+}] = 1.5 \times 10^{-12} \text{ M}$$

So **PbCO3 will begin to precipitate first**.

b. When PbI_2 begins to precipitate, $[Pb^{2+}] = 8.7 \times 10^{-7}$ and for a solution that is saturated in $PbCO_3$ we can write: $[Pb^{2+}][CO_3^{2-}] = 1.5 \times 10^{-13}$

Solving for $[CO_3^{2-}] = \dfrac{1.5 \times 10^{-13}}{8.7 \times 10^{-7}} = 1.7 \times 10^{-7} \text{ M}$

58.   Separate the following pairs of ions:

a. **Ba^{2+} and Na$^+$** : Since most sodium salts are soluble, it is simple to find a barium salt which is not soluble--e.g. the sulfate. Addition of dilute sulfuric acid should provide a source of SO_4^{2-} ions in sufficient quantity to precipitate the barium ions.

b. **Bi^{3+} and Cd^{2+}**: The hydroxide ion will serve as an effective reagent for selective precipitation of the two ions with the less soluble $Bi(OH)_3$ ($K_{sp} = 3.2 \times 10^{-40}$) precipitating well before the cadmium salt ($K_{sp} = 1.2 \times 10^{-14}$).

Simultaneous Equilibria and Complex Ions

60. The equation for AgBr dissolving is:

$\qquad$ AgBr (s) $\rightleftharpoons$ Ag$^+$ (aq) + Br$^-$ (aq). $\qquad$ $K_{sp} = [Ag^+][Br^-] = 3.3 \times 10^{-13}$

The formation of the silver ammine complex may be written:

$\qquad$ Ag$^+$ (aq) + 2 NH$_3$ (aq) $\rightleftharpoons$ [Ag(NH$_3$)$_2$]$^+$ (aq)

$$K_f = \frac{[Ag(NH_3)_2]^+}{[Ag^+][NH_3]^2} = 1.6 \times 10^7$$

Adding the two equations, Ag$^+$ (aq) on the "left" side of the second equation cancels with the same substance on the "right" side of the first equation giving an overall equation of:

$\qquad$ AgBr (s) + 2 NH$_3$ (aq) $\rightleftharpoons$ [Ag(NH$_3$)$_2$]$^+$ (aq) + Br$^-$ (aq)

$$K_{overall} = K_{sp} \cdot K_f = [Ag^+][Br^-] \cdot \frac{\{[Ag(NH_3)_2]^+\}}{[Ag^+][NH_3]^2} = \frac{\{[Ag(NH_3)_2]^+\}[Br^-]}{[NH_3]^2}$$

$$= (3.3 \times 10^{-13})(1.6 \times 10^7) = 5.3 \times 10^{-6}$$

62. The equation $\quad AgCl(s) + I^-(aq) \rightleftharpoons AgI(aq) + Cl^-$

can be obtained by adding two equations:

1. $AgCl(s) \rightleftharpoons Ag^+(aq) + Cl^-(aq)$ $K_{sp1} = 1.8 \times 10^{-10}$
2. $Ag^+(aq) + I^-(aq) \rightleftharpoons AgI(s)$ $\dfrac{1}{K_{sp2}} = 6.7 \times 10^{15}$

$$K_{net} = K_{sp1} \cdot \frac{1}{K_{sp2}} = 1.2 \times 10^6$$

The equilibrium lies to the right. This indicates that **AgI will form** if I^- is added to a saturated solution of AgCl.

64. Will 5.0 mL of 2.5 M NH_3 dissolve 1.0×10^{-4} mol of AgBr ?

The reaction that must occur for AgBr to dissolve is:

$$AgBr\,(s) + 2\,NH_3\,(aq) \rightleftharpoons [Ag(NH_3)_2]^+\,(aq) + Br^-\,(aq)$$

The reaction is the sum of (1) AgBr dissolving and (2) $[Ag(NH_3)_2]^+$ forming from Ag^+ and NH_3. Accordingly, $\quad K_{overall} = K_{sp} \cdot K_f$

$$K_{overall} = (8.3 \times 10^{-13})(1.6 \times 10^7) = 5.3 \times 10^{-6}$$

$$K_{overall} = \frac{\{[Ag(NH_3)_2]^+\}[Br^-]}{[NH_3]^2} = 5.3 \times 10^{-6}$$

Calculate the [NH3] necessary to dissolve the AgBr.

1. If the AgBr dissolves, the equilibrium amount of Br^- will be 1.0×10^{-4} M, and the complex will be 1.0×10^{-4} M. Substituting into the $K_{overall}$ expression:

$$K_{overall} = \frac{\{[Ag(NH_3)_2]^+\}[Br^-]}{[NH_3]^2} = 5.3 \times 10^{-6}$$

$$\frac{(1.0 \times 10^{-4})(1.0 \times 10^{-4})}{[NH_3]^2} = 5.3 \times 10^{-6}$$

$$[NH_3] = 0.043 \text{ M}$$

264

2. The total ammonia necessary is the ammonia to form the complex $(2 \cdot 0.020) =$ 4.0×10^{-2} M NH_3 and the ammonia to increase the NH_3 concentration to 0.043 M.

$$[NH_3] = 4.0 \times 10^{-2} + 0.043 = 0.083 \text{ M}.$$

The concentration of NH_3 diluted to 1.0 L is 0.012 M. We see that this concentration of NH_3 **would not be sufficient to dissolve the AgBr.**

Solubility and pH

66. Soluble in strong acid: $Ba(OH)_2$, $BaCO_3$

    For $Ba(OH)_2$ the addition of HCl for example would result in the acid/base reaction to form water, dissolving $Ba(OH)_2$.

    For $BaCO_3$, HCl would react to form CO_2 and H_2O, causing the carbonate salt to dissolve.

68. BiI_3 is not expected to be soluble in a strong acid.

    $Bi_2(CO_3)_3$ is expected to be soluble (See question 66 for an analogous situation).

    $BiPO_4$ is expected to be soluble, since the PO_4^{3-} is the conjugate base of a weak acid.

General Questions

70. Determining the more soluble of two substances with the same general composition (dipositive cation and dipositive anion) is done by noting that the substance with the greater K_{sp} is the more soluble. Aragonite ($K_{sp} = 6.0 \times 10^{-9}$) is slightly more soluble than calcite ($K_{sp} = 3.8 \times 10^{-9}$).

72. Determine if $Q > K_{sp}$ (for AgCl):

    $$K_{sp} = [Ag^+][Cl^-] = 1.8 \times 10^{-10} \quad Q = (1.0 \times 10^{-5})(2.0 \times 10^{-4}) = 2.0 \times 10^{-9}$$

    Since Q exceeds K_{sp}, AgCl will precipitate. (You are found out!)

74. Order of precipitation of Mg^{2+}, Ca^{2+}, and Ba^{2+} sulfates:

Compound	Ksp
$MgSO_4$	not listed
$CaSO_4$	2.4×10^{-5}
$BaSO_4$	1.1×10^{-10}

Since these salts are 1 : 1 in nature (a dipositive cation and a dinegative anion), a quick examination tells you that $BaSO_4$ is the least soluble and $MgSO_4$ is the most soluble of these three salts.

So, as the sulfate ion concentration increases with added sulfuric acid: $BaSO_4$ precipitates first, followed by $CaSO_4$, and eventually $MgSO_4$.

76. Calculate the concentrations of Ca^{2+} and OH^- after mixing (but before reaction)

$$0.010 \text{ M } Ca^{2+} \cdot \frac{15.0 \text{ mL}}{(15.0 + 25.0 \text{ mL})} = 3.75 \times 10^{-3} \text{ M } Ca^{2+}$$

$$0.0010 \text{ M } OH^- \cdot \frac{25.0 \text{ mL}}{(15.0 + 25.0 \text{ mL})} = 6.25 \times 10^{-4} \text{ M } OH^-$$

$$Q = [Ca^{2+}][OH^-]^2 = (3.75 \times 10^{-3})(6.25 \times 10^{-4})^2 = 1.46 \times 10^{-9}$$

Since K_{sp} (7.9×10^{-6}) > Q so **no $Ca(OH)_2$ precipitates**.

78. The equation for $Zn(OH)_2$ dissolving is : $Zn(OH)_2 \text{ (s)} \rightleftharpoons Zn^{2+} \text{ (aq)} + 2 OH^- \text{ (aq)}$

If a saturated solution of the base has a pH = 8.65 then pOH = 5.35 and $[OH^-] = 4.5 \times 10^{-6}$ M. The stoichiometry of the solid says that for two hydroxyl ions, one zinc ion is produced. So the concentration of $Zn^{2+} = 1/2 \cdot (4.5 \times 10^{-6}) = 2.2 \times 10^{-6}$.

The K_{sp} is then: $K_{sp} = [Zn^{2+}][OH^-]^2 = (2.2 \times 10^{-6})(4.5 \times 10^{-6})^2 = 4.5 \times 10^{-17}$.

80. The equation for $Mg(OH)_2$ dissolving is : $Mg(OH)_2(s) \rightleftharpoons Mg^{2+}(aq) + 2 OH^-(aq)$

If a saturated solution of the base has a pH = 10.49 then pOH = 3.51 and $[OH^-] = 3.1 \times 10^{-4}$ M. The stoichiometry of the solid says that for two hydroxyl ions, one magnesium ion is produced. So the concentration of $Mg^{2+} = 1/2(3.1 \times 10^{-4}) = 1.5 \times 10^{-4}$.

The Ksp is then: $[Mg^{2+}][OH^-]^2 = (1.5 \times 10^{-4})(3.1 \times 10^{-4})^2 = 1.5 \times 10^{-11}$

82. Calculate the initial concentrations:

$$\frac{2.00 \text{ g AgNO}_3}{0.0500 \text{ L}} \cdot \frac{1 \text{ mol AgNO}_3}{169.9 \text{ g AgNO}_3} = 0.235 \text{ M AgNO}_3$$

$$\frac{3.00 \text{ g K}_2CrO_4}{0.0500 \text{ L}} \cdot \frac{1 \text{ mol K}_2CrO_4}{194.2 \text{ g K}_2CrO_4} = 0.309 \text{ M K}_2CrO_4$$

Since NO_3^- and K^+ ions don't form precipitates, we can calculate their final concentrations easily.

$$[NO_3^-] = 0.235 \, \frac{\text{mol } AgNO_3}{L} \cdot \frac{1 \text{ mol } NO_3^-}{1 \text{ mol } AgNO_3} = \mathbf{0.235 \ M}$$

$$[K^+] = 0.235 \, \frac{\text{mol } K_2CrO_4}{L} \cdot \frac{2 \text{ mol } K^+}{1 \text{ mol } K_2CrO_4} = \mathbf{0.618 \ M}$$

The **initial concentration of Ag^+** is 0.235 M and of CrO_4^{2-} is 0.309 M. Since the K_{sp} for Ag_2CrO_4 is 9.0×10^{-12}, Ag_2CrO_4 will certainly precipitate. (Q > K)

$$Q = [Ag^+]^2[CrO_4^{2-}]$$
$$= (0.235)^2(0.309) = 1.7 \times 10^{-2}$$

The precipitation will remove 2 Ag^+ and 1 CrO_4^{2-} for each formula unit.

What amount of each will remain?

If "all" the Ag^+ precipitates, we would consume: 1.18×10^{-2} mol Ag^+

$$(2.00 \text{ g} \rightarrow 1.18 \times 10^{-2} \text{ mol}).$$

This would consume $1/2 \times 1.18 \times 10^{-2}$ mol CrO_4^{2-} (5.88×10^{-3} mol),

leaving (1.54×10^{-2} - 5.88×10^{-3}) 9.53×10^{-3} mol CrO_4^{2-} with a concentration of

$$\frac{9.53 \times 10^{-3} \text{ mol } CrO_4^{2-}}{0.050 \text{ L}}$$ or 0.191 M. The silver present in solution would then

be controlled by the solubility of Ag_2CrO_4

$$[Ag^+]^2[CrO_4^{2-}] = 9.0 \times 10^{-12}$$
$$[Ag^+]^2 = \frac{9.0 \times 10^{-12}}{0.191}$$
$$[Ag^+] = 6.9 \times 10^{-6} \text{ M}$$

84. a. The equilibrium expression for the equation may be written:

$$K = \frac{\{[Ag(S_2O_3)_2]^{3-}\}[Br^-]}{[S_2O_3^{2-}]^2}$$

NOTE: We have used { } to indicate concentration when complex ions are present.

The K_{sp} expression for AgBr is written: $K_{sp} = [Ag^+][Br^-] = 3.3 \times 10^{-13}$

The K_f for $[Ag(S_2O_3)_2]^{3-}$ may be written: $K_f = \frac{\{[Ag(S_2O_3)_2]^{3-}\}}{[Ag^+][S_2O_3^{2-}]^2} = 2.0 \times 10^{13}$

$K_{overall} = K_{sp} \cdot K_f = \frac{\{[Ag(S_2O_3)_2]^{3-}\}[Br^-]}{[S_2O_3^{2-}]^2} = (3.3 \times 10^{-13})(2.0 \times 10^{13}) = 6.6$

b. 1.0 g AgBr in 1.0 L corresponds to $\dfrac{1.0 \text{ g AgBr}}{1 \text{ L}} \cdot \dfrac{1 \text{ mol AgBr}}{188 \text{ g AgBr}} = 5.3 \times 10^{-3}$ M

Note that the concentration, 5.3×10^{-3} M, corresponds to the concentration of both silver complex and bromide ion. Note also that for each mole of silver complex, two moles of thiosulfate ion are consumed. The amount of $S_2O_3^{2-}$ consumed is then $(2 \cdot 5.3 \times 10^{-3})$ or 1.1×10^{-2} M.

Substituting into the $K_{overall}$ expression gives:

$$\frac{\{[Ag(S_2O_3)_2]^{3-}\}[Br^-]}{[S_2O_3^{2-}]^2} = \frac{(5.3 \times 10^{-3})(5.3 \times 10^{-3})}{[S_2O_3^{2-}]^2} = 6.6$$

and $[S_2O_3^{2-}]^2 = 4.3 \times 10^{-6}$ and $[S_2O_3^{2-}] = 2.1 \times 10^{-3}$

Note that this concentration is the equilibrium concentration for $S_2O_3^{2-}$.

(the result of $[S_2O_3^{2-}]$ initial - $[S_2O_3^{2-}]$ change)

$$2.1 \times 10^{-3} \text{ M} = [S_2O_3^{2-}] \text{ initial} - 1.1 \times 10^{-2} \text{ M}$$
$$1.3 \times 10^{-2} \text{ M} = [S_2O_3^{2-}] \text{ initial.}$$

This corresponds to 1.3×10^{-2} moles of $Na_2S_2O_3$.

The mass is: 1.3×10^{-2} mol $Na_2S_2O_3 \cdot \dfrac{158 \text{ g } Na_2S_2O_3}{1 \text{ mol } Na_2S_2O_3} = 2.0$ g $Na_2S_2O_3$

86. The equilibrium for ZnS dissolving in acid is a bit complicated.

There is the K_{sp}: $ZnS \rightleftharpoons Zn^{2+} + S^{2-}$, and the resulting hydrolytic equilibrium of S^{2-}

$$S^{2-} + H_2O \rightleftharpoons HS^- + OH^-$$

which provides an overall equation:

$$ZnS + 2 H_3O^+ \rightleftharpoons Zn^{2+} + H_2S + 2 H_2O \qquad K = 2 \times 10^{-4}$$

(p 903 of your text)

So $\dfrac{[Zn^{2+}][H_2S]}{[H_3O^+]^2} = 2 \times 10^{-4}$ and with pH = 1.50 $[H_3O^+] = 3.2 \times 10^{-2}$

$$\frac{[Zn^{2+}](0.1)}{(3.2 \times 10^{-2})^2} = 2 \times 10^{-4} \quad \text{and} \quad [Zn^{2+}] = 2.0 \times 10^{-6} \text{ M}$$

Conceptual Questions

88. The solubility of $Ni(OH)_2$ is affected by any equilibrium that affects $[Ni^{2+}]$ or $[OH^-]$. If we write the equilibrium equation $Ni(OH)_2$ dissolving,

$$Ni(OH)_2 \rightleftharpoons Ni^{2+} + 2\,OH^-$$

LeChatelier's principle tells us that **if we reduce[OH⁻]**, by adding acid (pH goes down), **more Ni(OH)$_2$ dissolves**, and conversely ; **if we increase [OH⁻]**, e.g. by adding a base like NaOH, the equilibrium will shift to the left, and **less Ni(OH)$_2$ dissolves**.

90. Define $AgCN(s) \rightleftharpoons Ag^+(aq) + CN^-(aq)$ as Equilibrium 1
    and $AgCN(s) + CN^-(aq) \rightleftharpoons [Ag(CN)_2^-]$ as Equilibrium 2

    Equilibrium 2 is a net equilibrium of two equilibria

Equilibrium 1 $AgCN(s) \rightleftharpoons Ag^+(aq) + CN^-(aq)$	$K = 1.2 \times 10^{-16}$
and $Ag^+(aq) + 2\,CN^-(aq) \rightleftharpoons [Ag(CN)_2]^-$	$K = 5.6 \times 10^{18}$
Equilibrium 2 $AgCN(s) + CN^-(aq) \rightleftharpoons [Ag(CN)_2^-]$	$K = 6.7 \times 10^2$

    The magnitude of this K indicates that as CN^- added, $Ag(CN)_2^-$ would be formed. The equilibrium **concentration of CN⁻ would definitely be greater**. **Silver ion** doesn't appear in the overall equilibrium (Eq 2), and its **concentration should not change**.

92.   a. Solubility of $BaCO_3$ affected by pressure of CO_2:   LeChatelier's principle tells us that adding CO_2 (**increasing pressure**) to the equilibrium mixture **will shift the equilibrium to the right**, resulting in **increased solubility of BaCO$_3$**.

    b. Solubility of $BaCO_3$ affected by pH decrease: Decreasing pH (**increasing [H$_3$O⁺]**) will consume additional HCO_3^- (and produce CO_2 and H_2O). The reduction in the amount of HCO_3^- will cause an **equilibrium shift to the right** —to attempt to "re-build" the amount of HCO_3^- present—**increasing the amount of BaCO$_3$ that dissolves**.

94.   a. $AlCl_3(aq) + H_3PO_4(aq) \rightarrow AlPO_4(s) + 3\ HCl(aq)$

   b. $152\ g\ AlCl_3 \cdot \dfrac{1\ mol\ AlCl_3}{133.3\ gAlCl_3} = 1.14\ mol\ AlCl_3$

   $0.750\ M\ H_3PO_4 \cdot 3.0\ L = 2.3\ mol\ H_3PO_4$

Examining the moles-available and moles-required ratios, we note that $AlCl_3$ is the limiting reagent, so the amount of $AlPO_4$ obtainable is:

$$1.14\ mol\ AlCl_3 \cdot \frac{1\ mol\ AlPO_4}{1\ mol\ AlCl_3} \cdot \frac{122.0\ AlPO_4}{1\ mol\ AlPO_4} = 139\ g\ AlPO_4$$

   c. $\dfrac{25.0\ g\ AlPO_4}{1\ L} \cdot \dfrac{1\ mol\ AlPO_4}{122.0\ g\ AlPO_4} = 0.205\ M\ AlPO_4$

Determine if this is a saturated solution:
$$K_{sp} = [Al^{3+}][PO_4^{3-}] = 1.3 \times 10^{-20}$$
$$[Al^{3+}] = [PO_4^{3-}] = 1.1 \times 10^{-10}\ M$$

Since 25.0 g of $AlPO_4$ corresponds to a greater concentration, the solution is saturated-- only 1.1×10^{-10} mol/L $AlPO_4$ will dissolve.

d. Since the phosphate ion is capable of reacting with H_3O^+ to form other ions (HPO_4^{2-}, $H_2PO_4^-$), addition of HCl will shift the phosphate equilibrium toward formation of these protonated anions--reducing $[PO_4^{3-}]$.

The reduction of $[PO_4^{3-}]$ will **increase the amount of** $AlPO_4$ which dissolves.

e. Concentrations after mixing:
$$[Al^{3+}] = 2.5 \times 10^{-3}\ M\ Al^{3+} \cdot \frac{1.50\ L}{(1.50\ L + 2.50\ L)} = 9.4 \times 10^{-4}\ M$$
$$[PO_4^{3-}] = 3.5 \times 10^{-2}\ M\ PO_4^{3-} \cdot \frac{2.50\ L}{(1.50\ L + 2.50\ L)} = 2.2 \times 10^{-2}\ M$$
$$Q = [Al^{3+}][PO_4^{3-}] = (9.4 \times 10^{-4})(2.2 \times 10^{-2}) = 2.1 \times 10^{-5}$$

Since $Q > K_{sp}$ (1.3×10^{-20}), **AlPO₄ precipitates**.

The Al^{3+} is the limiting reagent and the maximum amount of $AlPO_4$ that can form is :

$$\frac{9.4 \times 10^{-4}\ mol\ Al^{3+}}{L} \cdot \frac{4.0\ L}{1} \cdot \frac{1\ mol\ AlPO_4}{1\ mol\ Al^{3+}} \cdot \frac{122\ g\ AlPO_4}{1\ mol\ AlPO_4} = 0.46\ g\ AlPO_4$$

Chapter 20:
Principles of Reactivity: Entropy and Free Energy

Entropy

6. The sample with the higher entropy:
   a. CO_2 vapor at 0 °C -- The entropy of a substance is greatest in its gaseous form.
   b. Dissolved sugar -- The entropy of a dissolved substance is greater than the entropy of the pure solid.
   c. Mixture -- The entropy of each pure substance contained in separate beakers will be lower than the entropy of the solution formed by mixing the two liquids.

8. Compound with the higher entropy:
   a. $AlCl_3$ - Entropy increases with molecular complexity.
   b. CH_3CH_2I - Entropy increases with molecular complexity.
   c. NH_4Cl (aq) - Entropy of solutions is greater than that of solids.

10. Entropy changes:
   a. C (diamond) $\longrightarrow$ C (graphite)
   $$\Delta S° = 5.140 \frac{J}{K \cdot mol} (1 \text{ mol}) - 2.311 \frac{J}{K \cdot mol}(1 \text{ mol})$$
   $$= +3.363 \frac{J}{K}$$

   The increase in entropy reflects the greater order of diamond over graphite.

   b. Na (g) $\longrightarrow$ Na (s)
   $$\Delta S° = 51.21 \frac{J}{K \cdot mol} (1 \text{ mol}) - 153.112 \frac{J}{K \cdot mol} (1 \text{ mol})$$
   $$= -102.50 \frac{J}{K}$$

   The lower entropy of the solid state is evidenced by the negative sign.

   c. Br_2 (ℓ) $\longrightarrow$ Br_2 (g)
   $$\Delta S° = 245.463 \frac{J}{K \cdot mol} (1 \text{ mol}) - 152.2 \frac{J}{K \cdot mol}(1 \text{ mol})$$
   $$= 93.3 \frac{J}{K}$$

   The increase in entropy is expected with the transition to the disordered state of a gas.

12.   a. $\Delta S°$ for the transition of $(C_2H_5)_2O$ (ℓ) $\longrightarrow$ $(C_2H_5)_2O$ (g)

$$\Delta S° = \frac{\Delta Hvap}{T} = \frac{26.0 \times 10^3 \text{ J}}{308 \text{ K}} = 84.4 \frac{J}{K}$$

b. $\Delta S°$ for the transition of $(C_2H_5)_2O$ (g) $\longrightarrow$ $(C_2H_5)_2O$ (ℓ) is -84.4 J/K .

Note the reduction in entropy in the phase change from gas to liquid.

Reactions and Entropy Change

14.   For the reaction: 3 C (graphite) + 4 H_2 (g) $\longrightarrow$ C_3H_8 (g)

$\Delta S° = 1 \cdot S° \, C_3H_8 - [3 \cdot S° \, C \text{ (graphite)} + 4 \cdot S° \, H_2 \text{ (g)}]$

$= (1 \text{ mol})(269.9 \frac{J}{K \cdot mol}) -$

$[(3 \text{ mol})(5.740 \frac{J}{K \cdot mol}) + (4 \text{ mol})(130.684 \frac{J}{K \cdot mol})]$

$= -270.1 \frac{J}{K}$

16.   Calculate the standard molar entropy change for each substance from its elements:

a. H_2 (g) + $\frac{1}{2}$ O_2 (g) $\longrightarrow$ H_2O (ℓ)

$\Delta S° = (1 \text{ mol})(69.91 \frac{J}{K \cdot mol})$

$- [(1 \text{ mol})(130.684 \frac{J}{K \cdot mol}) + (\frac{1}{2} \text{ mol})(205.138 \frac{J}{K \cdot mol})] = -163.34 \text{ J/K}$

b. Mg(s) + O_2(g) + H_2(g) $\longrightarrow$ $Mg(OH)_2$(s)

$\Delta S° = (1 \text{ mol})(63.18 \frac{J}{K \cdot mol}) - [(1 \text{ mol})(32.68 \frac{J}{K \cdot mol}) +$

$(1 \text{ mol})(205.138 \frac{J}{K \cdot mol}) + (1 \text{ mol})(130.684 \frac{J}{K \cdot mol})] = -305.32 \frac{J}{K}$

c. Pb (s) + Cl_2 (g) $\longrightarrow$ $PbCl_2$ (s)

$\Delta S° = 1 \cdot S° \, PbCl_2 \text{ (s)} - [1 \cdot S° \, Pb \text{ (s)} + 1 \cdot S° \, Cl_2 \text{ (g)}]$

$\Delta S° = (1 \text{ mol})(136.0 \text{ J/K} \cdot \text{mol}) -$

$[(1 \text{ mol})(64.8 \frac{J}{K \cdot mol}) + (1 \text{ mol})(223.1 \frac{J}{K \cdot mol})] = -151.9 \text{ J/K}$

18.   a. $2 Al(s) + 3 Cl_2(g) \longrightarrow 2 AlCl_3(s)$

$\Delta S° = 2 \cdot S° \, AlCl_3(s) - [2 \cdot S° \, Al(s) + 3 \cdot S° \, Cl_2(g)]$

$\Delta S° = (2 \text{ mol})(110.67 \frac{J}{K \cdot mol}) - [(2 \text{ mol})(28.3 \frac{J}{K \cdot mol}) +$

$(1 \text{ mol})(223.066 \frac{J}{K \cdot mol})] = -504.5 \text{ J/K}$

or $-252.2 \frac{J}{K}$ per mol of $AlCl_3$

b. $C_2H_5OH(\ell) + 3 O_2(g) \longrightarrow 2 CO_2(g) + 3 H_2O(g)$

$\Delta S° = [2 \cdot S° \, CO_2(g) + 3 \cdot S° \, H_2O(g)] - [1 \cdot S° \, C_2H_5OH(l) + 3 \cdot S° \, O_2(g)]$

$\Delta S° = [(2 \text{ mol})(213.74 \frac{J}{K \cdot mol}) + (3 \text{ mol})(188.825 \frac{J}{K \cdot mol})] -$

$[(1 \text{ mol})(160.7 \frac{J}{K \cdot mol}) + (3 \text{ mol})(205.14 \frac{J}{K \cdot mol})] = +217.8 \text{ J/K}$

20. Sign of ΔH for $H_2O \longrightarrow H_2(g) + 1/2 \, O_2(g) = +$

This change will require the addition of energy.

Sign of $\Delta S = +$

The total number of gas molecules will increase—as will entropy. This reaction is *reactant-favored*, or nonspontaneous (at room temperature). Common observations tell us that water (in liquid or gaseous state) does not spontaneously revert to elemental oxygen and hydrogen. With both ΔH and ΔS having positive signs, temperature will play a role.

22.   a. $\Delta H = -, \Delta S = -$ Product-favored at lower T
       b. $\Delta H = +, \Delta S = -$ Reactant-favored

24.
$C_2H_6(g)$	$+$	$7/2 \, O_2(g)$	$\longrightarrow$	$2 \, CO_2(g)$	$+$	$3 \, H_2O(g)$	
-84.68		0		-393.509		-241.818	$\Delta H°_f$ (kJ/mol)
+229.60		+205.138		+213.74		+188.825	$S°$ (J/K·mol)

$\Delta H°_{rxn} = [2 \cdot \Delta H°_f \, CO_2(g) + 3 \cdot \Delta H°_f \, H_2O(g)] - [1 \cdot \Delta H°_f \, C_2H_6(g) +$

$7/2 \cdot \Delta H°_f \, O_2(g)]$

$= [(2 \text{ mol})(-393.509 \frac{kJ}{mol}) + (3 \text{ mol})(-241.818 \frac{kJ}{mol})] -$

$[(1 \text{ mol})(-84.68 \text{ kJ/mol}) + 0]$

$= -1427.79 \text{ kJ}$

$$\Delta S^\circ \text{ rxn} = [2 \cdot S^\circ \text{ CO}_2(g) + 3 \cdot S^\circ \text{ H}_2\text{O (g)}] - [1 \cdot S^\circ \text{ C}_2\text{H}_6 \text{ (g)} + 7/2 \cdot S^\circ \text{ O}_2 \text{ (g)}]$$

$$= [2 \text{ mol})(213.74 \frac{J}{K\cdot mol}) + (3 \text{ mol})(188.825 \frac{J}{K\cdot mol})] -$$

$$[(1 \text{ mol})(229.60 \frac{J}{K\cdot mol}) + 7/2 \text{ mol})(205.1381 \frac{J}{K\cdot mol})]$$

$$= 46.37 \text{ J/K}$$

$$\Delta S^\circ \text{surroundings} \qquad = \frac{\Delta H \text{ rxn}}{T} \qquad \text{(Assuming we're at 298 K)}$$

$$= \frac{1427.79 \text{ kJ}}{298 \text{ K}} \text{ (1000 J/kJ)} = +4790 \text{ J/K}$$

$$\text{so } \Delta S^\circ \text{system} + \Delta S^\circ \text{surroundings} = 46.37 \text{ J/K} + 4790 \text{ J/K}$$

$$= 4840 \text{ J/K (to 3 sf)}$$

Since ΔH° = - and ΔS° = +, the process is product-favored.

This calculation is consistent with our expectations. We know that hydrocarbons burn completely (in the presence of sufficient oxygen) to produce carbon dioxide and water.

Free Energy

26. Calculate $\Delta G^\circ \text{rxn}$ for

a. $\text{Sn(s)} + 2 \text{ Cl}_2(g) \longrightarrow \text{SnCl}_4(\ell)$

51.55	223.066	258.6	S° (J/K • mol)
0	0	-511.3	ΔH°f (kJ/mol)

$$\Delta H^\circ \text{rxn} = \Delta H^\circ f \text{ SnCl}_4(\ell) - [\Delta H^\circ f \text{ Sn(s)} + 2 \Delta H^\circ f \text{ Cl}_2(g)]$$

$$= [(1 \text{ mol})(-511.3 \frac{kJ}{mol})] - [(1 \text{ mol})(0) + (2 \text{ mol})(0)]$$

$$= -511.3 \text{ kJ}$$

$$\Delta S^\circ \text{rxn} = [(1 \text{mol})(258.6 \frac{J}{K\cdot mol})] - [(1 \text{ mol})(51.55 \frac{J}{K\cdot mol})$$

$$+ (2 \text{ mol})(223.066 \frac{J}{K\cdot mol})]$$

$$= -239.1 \text{ J/K}$$

$$\Delta G^\circ \text{rxn} = \Delta H^\circ \text{rxn} - T\Delta S^\circ \text{rxn}$$

$$= -511.3 \text{ kJ} - (298 \text{ K})(-240.1 \text{ J/K})(\frac{1.000 kJ}{1000 K})$$

$$= -440.1 \text{ kJ}$$

b. NH_3 (g) + HCl (g) $\rightarrow$ NH_4Cl (s)

    192.45   186.91   94.6   S° (J/K•mol)

    - 46.11   - 92.31   - 314.43 $\Delta H°_f$ (kJ/mol)

$\Delta S°_{rxn}$ = 1 • S° NH_4Cl (s) - [1 • S° NH_3 (g) + 1 • S° HCl (g)]

    = (1 mol)(94.6 J/K • mol) - [(1 mol)(192.45 J/K • mol) +

                             (1 mol)(186.9l J/K • mol)]

    = - 284.8 J/K

$\Delta H°_{rxn}$ = 1 • $\Delta H°_f$ NH_4Cl (s) - [l • $\Delta H°_f$ NH_3 (g) + l • $\Delta H°_f$ HCl (g)]

    = (1 mol)(- 314.43 kJ/mol) -

               [(1 mol)(- 46.11 kJ/mol) + (1 mol)(- 92.3l kJ/mol)]

    = - 176.01 kJ

$\Delta G°_{rxn}$ = $\Delta H°_f$ - T $\Delta S°_{rxn}$

    = - 176.01 kJ - (298 K)(- 284.76 J/K)$(\frac{1.000\ kJ}{1000\ J})$

    = - 176.01 kJ + 84.86 kJ

    = - 91.15 kJ

Part a corresponds to formation of one mole of a substance from its elements, each in their standard state--$\Delta G°_f$. The value obtained for $\Delta G°$ agrees with those in Appendix K. The values for $\Delta G°$ for **both equations** are negative, indicating that they **are product-favored**. Both reactions are enthalpy driven ($\Delta H < 0$).

28. Calculate the molar free energies of formation for:

a. CS_2 (g)   The reaction is: C (graphite) + 2 S (s,rhombic) $\longrightarrow$ CS_2 (g)

$\Delta H°_f$ = (1 mol)(117.36 $\frac{kJ}{mol}$) - [0 + 0] = 117.36 kJ

$\Delta S°$ = (1 mol)(237.84 $\frac{J}{K•mol}$)

    - [(1 mol)(5.740 $\frac{J}{K•mol}$) + (2 mol)(31.80 $\frac{J}{K•mol}$)] = 168.50 J/K

$\Delta G°_f$ = $\Delta H°_f$ - T$\Delta S°$

    = (1 mol)(117.36 $\frac{kJ}{mol}$) - (298 K)(168.50 $\frac{J}{K}$)($\frac{1.000\ kJ}{1000.\ J}$)

    = 67.15 kJ/mol   Appendix value: 67.12 kJ/mol

Reaction is reactant-favored.

b. N_2H_4 (ℓ)   The reaction is: $N_2(g) + 2\ H_2(g) \longrightarrow N_2H_4$ (ℓ)

$$\Delta H^\circ_f = (1\ mol)(50.63\ \frac{kJ}{mol}) - [0 + 0] = 50.63\ kJ$$

$$\Delta S^\circ = (1\ mol)(121.21\ \frac{J}{K \cdot mol}) -$$
$$[(1\ mol)(191.61\ \frac{J}{K \cdot mol}) + (2\ mol)(130.684\ \frac{J}{K \cdot mol})] = -331.77\ J/K$$

$$\Delta G^\circ_f = \Delta H^\circ_f - T\Delta S^\circ$$
$$= (1\ mol)(50.63\ kJ/mol) - (298\ K)(-331.77\ J/K)(\frac{1.000\ kJ}{1000.\ J})$$
$$= 149.50\ kJ/mol \qquad \text{Appendix value: } 149.34\ kJ/mol$$

Reaction is reactant-favored.

c. $COCl_2$ (g)   The reaction is: C (graphite) $+ \frac{1}{2}\ O_2$ (g) $+ Cl_2$ (g) $\longrightarrow COCl_2$ (g)

$$\Delta H^\circ_f = (1\ mol)(-218.8\ \frac{kJ}{mol}) - [0 + 0 + 0] = -218.8\ kJ$$

$$\Delta S^\circ = (1\ mol)(283.53\ \frac{J}{K \cdot mol}) - [(1\ mol)(5.740\ \frac{J}{K \cdot mol})$$
$$+ (\frac{1}{2}\ mol)(205.138\ \frac{J}{K \cdot mol}) + (1\ mol)(223.066\ \frac{J}{K \cdot mol})]$$
$$= -47.85\ J/K$$

$$\Delta G^\circ_f = \Delta H^\circ_f - T\Delta S^\circ$$
$$= (1\ mol)(-218.8\ kJ/mol) - (298\ K)(-47.85\ J/K)(\frac{1.000\ kJ}{1000.\ J})$$
$$= -204.5\ kJ/mol \qquad \text{Appendix value: } -204.6\ kJ/mol$$

The formation of $COCl_2$ is predicted to be product-favored.

30.   Formation of Fe_2O_3 (s): 2 Fe (s) + 3/2 O_2 (g) $\longrightarrow Fe_2O_3$ (s)

ΔG°_{rxn} for formation of 1.00 mole of Fe_2O_3 (s):

$$\Delta H^\circ_{rxn} = [1 \cdot \Delta H^\circ_f\ Fe_2O_3\ (s)] - [2 \cdot \Delta H^\circ_f\ Fe\ (s) + \frac{3}{2} \cdot \Delta H^\circ_f\ O_2\ (g)]$$
$$= (1\ mol)(-824.2\ \frac{kJ}{mol})] - 0 = -824.2\ kJ$$

$$\Delta S°_{rxn} = [1 \cdot S° \; Fe_2O_3 \; (s) \;] - [2 \cdot S° \; Fe \; (s) + \tfrac{3}{2} \cdot S° \; O_2 \; (g)]$$

$$= [(1 \; mol)(81.40 \; \tfrac{J}{K \cdot mol})] -$$

$$[(2 \; mol)(21.18 \; \tfrac{J}{K \cdot mol}) + (\tfrac{3}{2} \; mol)(205.14 \; \tfrac{J}{K \cdot mol})$$

$$= - \; 275.87 \; J/K$$

$$\Delta G°_{rxn} = \Delta H°_f - T \; \Delta S°_{rxn}$$

$$= (1 \; mol)(-824.2 \; \tfrac{kJ}{mol}) - (298 \; K)(- \; 275.87 \; \tfrac{J}{K})(\tfrac{1.000 \; kJ}{1000. \; J})$$

$$= - \; 742.0 \; kJ/mol \; (\text{in good agreement with } \Delta G°_{rxn} \text{ for } Fe_2O_3 \text{ in Appendix K})$$

Free energy change for formation of 454 g of Fe_2O_3 (s):

$$454 \; g \; of \; Fe_2O_3 \cdot \frac{1 \; mol \; Fe_2O_3}{160 \; g \; \; Fe_2O_3} \cdot \frac{-742.0 \; kJ}{1 \; mol \; Fe_2O_3} = \; -2.11 \; x \; 10^3 \; kJ$$

32. Calculate $\Delta G°_{rxn}$ for the following reactions:

a. $Ca \; (s) \; + \; Cl_2 \; (g) \; \longrightarrow \; CaCl_2 \; (s)$

$$\Delta G°_{rxn} = (1 \; mol)(- \; 748.1 \; \tfrac{kJ}{mol}) - [0 \; + \; 0]$$

$$= - \; 748.1 \; kJ$$

b. $2 \; HgO \; (s) \; \rightarrow \; 2 \; Hg \; (l) \; + \; O_2(g)$

$$\Delta G°_{rxn} = [0 \; + \; 0] \; - (2 \; mol)(-58.539 \; \tfrac{kJ}{mol})$$

$$= \; 117.08 \; kJ$$

c. $NH_3 \; (g) \; + 2 \; O_2 \; (g) \; \rightarrow \; HNO_3 \; (\ell) \; + \; H_2O \; (\ell)$

$$\Delta G°_{rxn} = [(1 \; mol)(- \; 80.71 \; \tfrac{kJ}{mol}) + (1 \; mol)(- \; 237.129 \; \tfrac{kJ}{mol})]$$

$$- [(1 \; mol)(- \; 16.45 \; \tfrac{kJ}{mol}) + 0 \;]$$

$$= - \; 301.39 \; kJ$$

Reactions in parts **a** and **c** are predicted to be product-favored.

34. Value for $\Delta G°_f$ for $BaCO_3(s)$:

$$\Delta G°_{rxn} = [\Delta G°_f BaO(s) + \Delta G°_f CO_2(g)] - [\Delta G°_f BaCO_3(s)]$$

$$+218.1 \text{ kJ} = [(1 \text{ mol})(-525.1 \frac{kJ}{mol}) + (1 \text{ mol})(-394.359 \frac{kJ}{mol})] - \Delta G°_f BaCO_3(s)$$

$$+218.1 \text{ kJ} = -919.5 \text{ kJ} - \Delta G°_f BaCO_3(s)$$

$$-1137.6 \text{ kJ/mol} = + \Delta G°_f BaCO_3(s)$$

36. Determine $\Delta H°$, $\Delta S°$, and $\Delta G°$ for:

	C_8H_{16} (g)	$+ H_2$ (g)	$\rightarrow$ C_8H_{18} (g)
$\Delta H°_f$ (kJ/mol)	- 82.93	0	- 208.45
S° (J/K • mol)	462.8	130.7	463.6

$$\Delta H°_{rxn} = [(1 \text{ mol})(- 208.45 \frac{kJ}{mol})] - [(1 \text{ mol})(- 82.93 \frac{kJ}{mol}) + 0]$$

$$= - 125.52 \text{ kJ}$$

$$\Delta S°_{rxn} = [(1 \text{ mol})(463.6 \frac{J}{K•mol})] -$$

$$[(1 \text{ mol})(462.8 \frac{J}{K•mol}) + (1 \text{ mol})(130.7 \frac{J}{K•mol})]$$

$$= - 129.9 \text{ J/K}$$

$$\Delta G°_{rxn} = \Delta H°_f - T \Delta S°_{rxn}$$

$$= - 125.52 \text{ kJ} - (298 \text{ K})(- 129.9 \text{ J/K})(\frac{1.000 \text{ kJ}}{1000 \text{ J}})$$

$$= - 86.81 \text{ kJ}$$

The negative value for $\Delta G°_{rxn}$ indicates that the reaction is **product-favored** under standard conditions.

Thermodynamics and Equilibrium Constants

38. Calculate K_p for the reaction:

$$\frac{1}{2} N_2(g) + \frac{1}{2} O_2(g) \longrightarrow NO (g) \qquad \Delta G°_f = + 86.57 \text{ kJ/mol NO}$$

$$\Delta G°_{rxn} = - RT \ln K_p$$

$$86.57 \times 10^3 \text{ J/mol} = - (8.314 \frac{J}{K•mol})(298 \text{ K}) \ln K_p$$

$$- 34.9 = \ln K_p$$

$$7 \times 10^{-16} = K_p$$

Note that the + value of $\Delta G°_f$ results in a value of K_p which is small--**reactants are favored**. A negative value would result in a large K_p -- a process in which the products were favored.

40. a. $\Delta G°_{rxn}$ = $\Delta G°_f$ C_2H_6 (g) - [$\Delta G°_f$ C_2H_4 (g) + $\Delta G°_f$ H_2(g)]

= (1 mol)(-32.82 $\frac{kJ}{mol}$) - [(1 mol)(68.15 $\frac{kJ}{mol}$) + (1 mol)(0)]

= -100.97 kJ

The reaction is predicted to be product-favored.

b. $\Delta G°$ = $-RTlnK_p$

-100.97 x 10^3 $\frac{J}{mol}$ = -(8.314 $\frac{J}{K \cdot mol}$)(298 K) lnK_p

40.754 = lnK_p

5.00 x 10^{17} = K_p

Note that a **negative ΔG** results in a **large K_p**.

42. a. $\Delta G°_{rxn}$ and K_p for steps :

step 1: $\Delta G°_{rxn}$ = [$\Delta G°_f$ Ti(s) + $\Delta G°_f$ CO_2(g)] - [$\Delta G°_f$ TiO_2(s) + $\Delta G°_f$ C(s)]

= [(1 mol)(0) + (1mol)(-394.359 $\frac{kJ}{mol}$)] -

[(1 mol)(-884.5 $\frac{kJ}{mol}$) + (1 mol)(0)]

= +490.1 kJ

step 2: $\Delta G°_{rxn}$ = [$\Delta G°_f$ $TiCl_4$(ℓ)] - [$\Delta G°_f$ Ti(s) + 2 $\Delta G°_f$ Cl_2(g)]

= [(1 mol)(-737.2 $\frac{kJ}{mol}$)] - [(1 mol)(0) + (2 mol)(0)]

= -737.2 kJ

step 1: $\Delta G°$ = $-RTlnK_p$

+490.1 x 10^3 J = -(8.314 $\frac{J}{K \cdot mol}$)(298 K) ln K_p

-197.8 = ln K_{p1} or $e^{-197.8}$

1.3 x 10^{-86} = K_{p1}

step 2: $\Delta G° = -RT\ln K_p$

$$-737.2 \times 10^3 \text{ J} = -(8.314 \frac{J}{K \cdot mol})(298 \text{ K}) \ln K_{p2}$$

$$297.5 = \ln K_{p2}$$
$$1.6 \times 10^{129} = K_{p2}$$

b. Overall : $\Delta G°_{rxn} = [\Delta G°_f \text{ TiCl}_4(\ell) + \Delta G°_f \text{ CO}_2(g)] - [\Delta G°_f \text{ TiO}_2(s)$

$$+ \Delta G°_f \text{ C}(s) + 2 \Delta G°_f \text{ Cl}_2(g)]$$

$$= [(1 \text{ mol})(-737.2 \frac{kJ}{mol}) + (1 \text{ mol})(-394.359 \frac{kJ}{mol})] -$$

$$[(1 \text{ mol})(-884.5 \frac{kJ}{mol}) + (1 \text{ mol})(0) + (2 \text{ mol})(0)]$$

$$= -247.1 \text{ kJ}$$

Note that $\Delta G°$ overall $= \Delta G°_1 + \Delta G°_2$

Overall $\Delta G° = -RT\ln K_p$

$$-247.1 \times 10^3 \text{ J} = -(8.314 \frac{J}{K \cdot mol})(298 \text{ K}) \ln K_p$$

$$99.7 = \ln K_p \text{ or } e^{-99.7}$$
$$2.06 \times 10^{43} = K_p \text{ or } K_{p1} \cdot K_{p2}$$

c. The overall negative free energy indicates the overall process is product-favored, even though the entropy is decreasing, so **the process is enthalpy driven**.

General Questions

44. As with many oxidations, we expect the reaction to be exothermic. Since the gaseous product is more complex (more atoms) than the reactant, we also anticipate the entropy to increase. The reaction should be product-favored.

S(s)	+	O$_2$(g)	$\longrightarrow$	SO$_2$(g)	
31.80		205.138		248.22	$S° (\frac{J}{K \cdot mol})$
0		0		-296.830	$\Delta H° (\frac{kJ}{mol})$

$$\Delta H°_{rxn} = [(1 \text{ mol})(-296.830 \frac{kJ}{mol})] - 0 = -296.83 \text{ kJ}$$

$$\Delta S°_{rxn} = [(1\ mol)(248.22\ \frac{J}{K \cdot mol})]\ -$$

$$[(1\ mol)(31.80\ \frac{J}{K \cdot mol}) + (1\ mol)(205.138\ \frac{J}{K \cdot mol})]$$

$$= 11.28\ J/K$$

46.

Process	$\Delta H°$	$\Delta S°$	$\Delta G°$
a. Electrolysis of H_2O (ℓ) to form gaseous H_2, O_2	+	+	+

The process does not proceed spontaneously ($\Delta G° > 0$). Two moles of water (ℓ) produce three moles of gas ($\Delta S° > 0$). Enthalpy of formation of a compound is greater than the enthalpy of formation for elements (in their standard states) ($\Delta H° > 0$).

Process	$\Delta H°$	$\Delta S°$	$\Delta G°$
b. Explosion of dynamite	-	+	-

The explosion of dynamite releases much heat ($\Delta H° < 0$). The explosion produces many moles of gas from a liquid ($\Delta S° > 0$). Unfortunately for many miners, the reaction proceeds spontaneously (and often very rapidly--but that's for "kinetics" to explain).

Process	$\Delta H°$	$\Delta S°$	$\Delta G°$
c. Combustion of gasoline	-	+	-

The combustion of gasoline in an automobile releases heat ($\Delta H° < 0$). There is an increase in the number of moles of gaseous products over moles of reactants ($\Delta S° > 0$). With a negative $\Delta H°$ and a positive $\Delta S°$, $\Delta G°$ should also be negative.

48. $BCl_3(g)\ +\ 3/2\ H_2(g) \longrightarrow B(s)\ +\ 3\ HCl(g)$

$$\Delta H°_{rxn} = [\Delta H°_f\ B(s)\ +\ 3\ \Delta H°_f\ HCl(g)]\ -\ [\Delta H°_f\ BCl_3(g)\ +\ 3/2\ \Delta H°_f\ H_2(g)]$$

$$= [(1\ mol)(0)\ +\ (3\ mol)(-92.307\ \frac{kJ}{mol})]\ -$$

$$[(1\ mol)(-403.8\ \frac{kJ}{mol})\ +\ (3/2\ mol)(0)]$$

$$= 126.9\ kJ$$

$$\Delta S° = [1 \cdot S° \text{ B (s)} + 3 \cdot S° \text{ HCl(g)}] - [1 \cdot S° \text{ BCl}_3 \text{ (g)} + \frac{3}{2} \cdot S° \text{ H}_2 \text{ (g)}]$$

$$= [(1 \text{ mol})(5.86 \frac{J}{K \cdot mol}) + (3 \text{ mol})(186.908 \frac{J}{K \cdot mol})] -$$

$$[(1 \text{ mol})(290 \frac{J}{K \cdot mol}) + (3/2 \text{ mol})(130.684 \frac{J}{K \cdot mol})]$$

$$= 81 \text{ J/K}$$

$$\Delta G°_{rxn} = 126.9 \text{ kJ} - (298 \text{ K})(0.081 \text{ kJ/K}) = 103 \text{ kJ}$$

This process is **not** product-favored. The process is quite endothermic, and the slight increase in entropy is not sufficient to make the reaction spontaneous.

50. If K for the process is 0.422 at 700 °C:

$$\Delta G°_{rxn} = -RT \ln K$$

$$\Delta G°_{rxn} = -(8.314 \frac{J}{K \cdot mol})(973 \text{ K}) \ln(0.422)$$

$$= 6.98 \times 10^3 \text{ J or } 6.98 \text{ kJ}$$

52.   a. For the reaction: $C_6H_6 \text{ (g)} + 3 H_2 \text{ (g)} \longrightarrow C_6H_{12} \text{ (g)}$

$$\Delta G°_{rxn} = \Delta H°_f - T \Delta S°_{rxn}$$

$$= -206.1 \text{ kJ} - (298 \text{ K})(-363.12 \text{ J/K} \cdot \text{mol})(\frac{1.000 \text{ kJ}}{1000 \text{ J}})$$

$$= -97.9 \text{ kJ}$$

The reaction is predicted to be product-favored. The reaction is enthalpy driven.

54.   a. Reaction: $C \text{ (s)} + H_2O \text{ (g)} \longrightarrow CO \text{ (g)} + H_2 \text{ (g)}$

$\Delta H°_f$ (kJ/mol)   0   - 241.82   - 110.52   0

S° (J/K • mol) 5.74   188.83   197.67   130.68

$$\Delta H°_{rxn} = [(1 \text{ mol})(-110.52 \frac{kJ}{mol}) + 0] - [0 + (1 \text{ mol})(-241.82 \frac{kJ}{mol})]$$

$$= 131.30 \text{ kJ}$$

$$\Delta S°_{rxn} = [(1 \text{ mol})(197.67 \frac{J}{K \cdot mol}) + (1 \text{ mol})(130.68 \frac{J}{K \cdot mol})] -$$

$$[(1 \text{ mol})(5.74 \frac{J}{K \cdot mol}) + (1 \text{ mol})(188.83 \frac{J}{K \cdot mol})]$$

$$= 133.78 \text{ J/K}$$

$$\Delta G°_{rxn} = \Delta H°_f - T \Delta S°_{rxn}$$
$$= 131.30 \text{ kJ} - (298 \text{ K})(133.78 \text{ J/K})(\frac{1.000 \text{ kJ}}{1000 \text{ J}})$$
$$= 91.4 \text{ kJ}$$

b. K_p for the reaction: $\Delta G° = - RT \ln K_p$
$$91.4 \times 10^3 \text{ J} = - (8.314 \frac{J}{K \cdot mol})(298 \text{ K}) \ln K_p$$
$$- 36.9 = \ln K_p$$
$$9.5 \times 10^{-17} = K_p$$

c. The reaction is not product-favored at 25 °C ($\Delta G° > 0$).
Temperature at which reaction becomes product-favored:
$$\Delta G°_{rxn} = \Delta H° - T \Delta S°_{rxn}$$
with the $\Delta H°$ and $\Delta S°$ calculated for this equation (from part a), find the temperature at which $\Delta G° = 0$.
$$0 = 131.30 \text{ kJ} - T \cdot (133.78 \times 10^{-3} \text{ kJ/K})$$
$$- 131.30 \text{ kJ} = - T \cdot (133.78 \times 10^{-3} \text{ kJ/K})$$
$$981.3 \text{ K} = T$$
At any T greater than this (708.3 °C) $\Delta G°$ will be negative.

56. For the reaction: $CH_3OH (\ell) \longrightarrow CH_4 (g) + \frac{1}{2} O_2 (g)$

	CH_3OH	CH_4	O_2
$\Delta H°_f$ (kJ/mol)	-238.7	-74.8	0
S° (J/K · mol)	126.8	186.264	205.138

a. Entropy change for the reaction:
$$\Delta S°_{rxn} = [(1 \text{ mol})(186.264 \frac{J}{K \cdot mol}) + (\frac{1}{2} \text{ mol})(205.138 \frac{J}{K \cdot mol})] - [(1 \text{ mol})(126.8 \frac{J}{K \cdot mol})]$$
$$= 162.0 \text{ J/K}$$
The **increase** in entropy is **anticipated** with the production of gases from a liquid.

b. Is the reaction product-favored?
Calculate $\Delta H°_{rxn}$:
$$\Delta H°_{rxn} = [(1 \text{ mol})(- 74.8 \frac{kJ}{mol}) + 0] - [(1 \text{ mol})(- 238.7 \frac{kJ}{mol})]$$
$$= 163.9 \text{ kJ}$$

Calculate $\Delta G°_{rxn}$:

$$\Delta G°_{rxn} = \Delta H° - T\Delta S°$$

$$= 163.9 \text{ kJ} - (298 \text{ K})(162.0 \text{ J/K})(\frac{1.000 \text{ kJ}}{1000 \text{ J}})$$

$$= 115.6 \text{ kJ}$$

The reaction is **not product-favored** at 25 °C. This is not surprising especially when one considers the many industrial and commercial uses of methanol.

c. Temperature at which the reaction becomes product-favored:

Calculate the temperature at which $\Delta G = 0$:

$$0 = 163.9 \text{ kJ} - T(0.1620 \text{ kJ/K})$$

$$- 163.9 \text{ kJ} = - T(0.1620 \text{ kJ/K})$$

$$1012 \text{ K} = T \quad \text{or} \quad 739 \text{ °C}$$

Any temperature greater than 739 °C would be sufficient to make the reaction product-favored.

58. $2 \text{ Ag}_2\text{O(s)} \longrightarrow 4 \text{ Ag(s)} + \text{O}_2\text{(g)}$

121.3	42.55	205.138	$S° (\frac{J}{K \cdot mol})$
-31.05	0	0	$\Delta H° (\frac{kJ}{mol})$

$$\Delta H°_{rxn} = [(4 \text{ mol})(0) + (1 \text{ mol})(0)] - [(2 \text{ mol})(-31.05 \frac{kJ}{mol})]$$

$$= +62.10 \text{ kJ}$$

$$\Delta S° = [(4 \text{ mol})(42.55 \frac{J}{K \cdot mol}) + (1 \text{ mol})(205.138 \frac{J}{K \cdot mol})] - [(2 \text{ mol})(121.3 \frac{J}{K \cdot mol})$$

$$= +132.7 \text{ J/K}$$

$$\Delta G°_{rxn} = (+62.10 \text{ kJ}) - (298 \text{ K})(+132.7 \text{ J/K})(\frac{1.000 kJ}{1000 \text{ J}})$$

$$= 22.5 \text{ kJ}$$

The decomposition is not product-favored at 25 °C.

Temperature for process to become product-favored:

$$\Delta G° = \Delta H° - T\Delta S°$$

At what T will $\Delta G°$ be zero? If $\Delta G° = 0$ then $\Delta H° = T\Delta S°$

$$62.1 \text{ kJ} = T(132.7 \text{ J/K})$$

$$\frac{62.1 \times 10^3 \text{ J}}{132.7 \text{ J/K}} = T = 468 \text{ K or } 195 \text{ °C}$$

Conceptual Questions

60. $\Delta S°$ for

	$HCl(g)$	+	$H_2O(\ell)$	$\longrightarrow$	$HCl(aq)$
$S° \left(\frac{J}{K \cdot mol}\right)$	186.908		69.91		56.5

$$\Delta S° = (1 \text{ mol})(56.5 \frac{J}{K \cdot mol}) - [(1 \text{ mol})(186.908 \frac{J}{K \cdot mol}) + (1 \text{ mol})(69.91 \frac{J}{K \cdot mol})]$$

$$= -200.3 \text{ J/K}$$

We anticipate that the **entropy should decrease** as the gas dissolves in water.

62. $CaCO_3(s) \rightleftharpoons CaO(s) + CO_2(g)$

92.9	39.75	213.74	$S° \left(\frac{J}{K \cdot mol}\right)$
-1206.92	-635.09	-393.509	$\Delta H° \left(\frac{kJ}{mol}\right)$

Calculations show $\Delta H°_{rxn} = +178.32$ kJ

and $\Delta S°_{rxn} = +160.59$ J/K

Since both $\Delta H°$ and $\Delta S°$ have + signs, the **reaction will be product-favored at high T.**

Summary Questions

64.   a. What is K_p at 25 °C?

$$\Delta G°_{rxn} = 2 \Delta G° NO(g) - [\Delta G° N_2(g) + \Delta G° O_2(g)]$$

$$= (2 \text{ mol})(86.55 \frac{kJ}{mol}) = 173.10 \text{ kJ}$$

$$\Delta G° = -RT\ln K_p$$

$$173.10 \times 10^3 \text{ J} = -(8.314 \frac{J}{K \cdot mol})(298 \text{ K}) \ln K_p$$

$$-69.87 = \ln K_p$$

$$4.54 \times 10^{-31} = K_p \qquad \text{The reaction is \textbf{not product-favored} at 25 °C.}$$

b. Calculate ΔG_{rxn} at 700. °C

$$N_2(g) \; + \; O_2(g) \longrightarrow \; 2\,NO(g)$$

191.61	205.138	210.76	$S°\;(\frac{J}{K\cdot mol})$
0	0	90.25	$\Delta H°\;(\frac{kJ}{mol})$

$\Delta H°rxn \; = \; (2\,mol)(90.25\,\frac{kJ}{mol}) \; = \; 180.5\,kJ$

$\Delta S°rxn \; = \; [(2\,mol)(210.76\,\frac{J}{K\cdot mol})] \; - \; [(1\,mol)(191.61\,\frac{J}{K\cdot mol})$

$$+ \; (1\,mol)(205.138\,\frac{J}{K\cdot mol})] \; = \; 24.77\,J/K$$

$\Delta G° \; = \; \Delta H° \; - \; T\Delta S$

$\quad = \; 180.5\,kJ \; - \; (973\,K)(24.77\,J/K)(\frac{1.000\,kJ}{1000J})$

$\quad = \; 156.4\,kJ$

K_p at 700.

$\Delta G° \; = \; \Delta H° \; - \; T\Delta S$

$156.4 \times 10^3 J \; = \; -(8.314\,\frac{J}{K\cdot mol})(973\,K)\ln K_p$

$\quad -19.33 \; = \; \ln K_p$

$4.0 \times 10^{-9} \; = \; K_p$ \quad The reaction **is not product-favored** at 700 °C.

c. Given K_p at 700 °C $\; = \; 4.0 \times 10^{-9}$

What are equilibrium pressures?

$$\frac{P_{NO}^2}{P_{N_2} \cdot P_{O_2}} \; = \; 4.0 \times 10^{-9}$$

If the initial pressures of both N_2 and O_2 are 1.00 atm, then the equilibrium pressures can be calculated:

If x atm of N_2 and O_2 form NO at equilibrium;

$$P_{N_2} \; = \; 1.00 \; - \; x \; \text{ and } \; P_{O_2} \; = \; 1.00 \; - \; x$$

and $P_{NO} \; = \; 2x$

$$\frac{(2x)^2}{(1.00 - x)^2} \; = \; 4.0 \times 10^{-9}$$

Taking the square root of both sides

$$\frac{2x}{1.00 - x} \; = \; 6.33 \times 10^{-5}$$

and $2x = 6.3 \times 10^{-5} = P_{NO}$ and $P_{N_2} = P_{O_2} = 1.00$ atm

66.   a. Question 58 uses the decomposition of Ag_2O to form elemental Ag and O_2. The ΔH°_{rxn}, ΔS°_{rxn}, and ΔG°_{rxn} for the reaction in this question will be identical in magnitude to those calculated in 58, but the signs will be reversed.

$\Delta H^{\circ}_{rxn} = -62.10$ kJ

$\Delta S^{\circ}_{rxn} = -132.7$ J/K

$\Delta G^{\circ}_{rxn} = -22.5$ kJ

b. To calculate P_{O_2} we'll need K_p.

$\Delta G^{\circ}_{rxn} = -RT\ln K_p$

-22.5×10^3 J $= -(8.314 \frac{J}{K \cdot mol})(298 \text{ K}) \ln K_p$

$+9.08 = K_p$

$K_p = \dfrac{1}{P_{O_2}}$ (Remember solids are omitted from equilibrium expressions.)

$\dfrac{1}{P_{O_2}} = 8.79 \times 10^3$ and $P_{O_2} = 1.1 \times 10^{-4}$ atm

c. T at which $P_{O_2} = 1.00$ atm

If $P_{O_2} = 1.00$ atm, then $K_p = \dfrac{1}{1.00 \text{ atm}} = 1.00$

Let's calculate the ΔG° at which $K_p = 1.00$

$\Delta G^{\circ}_{rxn} = -RT\ln K_p$

The difficulty is that if T is not 25 °C, we need to calculate ΔG°, $(\Delta H - T\Delta S)$, so

$\Delta H^{\circ} - T\Delta S^{\circ} = -RT \ln(1.00)$

or $\Delta H^{\circ} = -RT \ln(1.00) + T\Delta S^{\circ}$

or $\dfrac{\Delta H^{\circ}}{T} = -R \ln(1.00) + \Delta S^{\circ}$

Using ΔH° and ΔS° data (part a)

$\dfrac{-62.10 \text{ kJ}}{T} = -(8.314 \frac{J}{K \cdot mol})(0) + -132.7$ J/K

or $\dfrac{-62.10 \text{ kJ}}{T} = -132.7$ J/K

and solving for T : 467.8 K or 194.7 °C.

287

Chapter 21:
Electron Transfer Reactions

Balancing Equations for Redox Reactions

12. Balance the following:

	reactant is	overall process is
a. $Cr\,(s) \rightarrow Cr^{3+}\,(aq) + 3\,e^-$	reducing agent	oxidation
b. $AsH_3\,(g) \rightarrow As\,(s) + 3\,H^+\,(aq) + 3\,e^-$	reducing agent	oxidation
c. $VO_3^-\,(aq) + 6\,H^+\,(aq) + 3\,e^- \rightarrow$	oxidizing agent	reduction
$\quad V^{2+}\,(aq) + 3\,H_2O\,(\ell)$		

Note: e^- are used to balance charge; H^+ balances only H atoms; H_2O balances both H and O atoms.

14. Balance the following (in acidic solution)

	reactant is	overall process is
a. $Cr_2O_7^{\,2-}\,(aq) + 14\,H^+\,(aq) + 6\,e^-$	oxidizing agent	reduction
$\quad \rightarrow 2\,Cr^{\,3+}\,(aq) + 7\,H_2O\,(\ell)$		
b. $CH_3CHO\,(aq) + H_2O\,(\ell) \rightarrow$	reducing agent	oxidation
$\quad CH_3COOH\,(aq) + 2\,H^+\,(aq) + 2\,e^-$		
c. $Bi^{3+}\,(aq) + 3\,H_2O\,(\ell) \rightarrow$		
$\quad HBiO_3\,(aq) + 5\,H^+\,(aq) + 2\,e^-$	reducing agent	oxidation

16. Balance the following (in basic solution)

	reactant is	overall process is
a. $Sn\,(s) + 4\,OH^-\,(aq) \rightarrow$	reducing agent	oxidation
$\quad Sn(OH)_4^{2-}\,(aq) + 2\,e^-$		
b. $MnO_4^-\,(aq) + 2\,H_2O\,(\ell) + 3\,e^- \rightarrow$	oxidizing agent	reduction
$\quad MnO_2\,(s) + 4\,OH^-\,(aq)$		
c. $ClO^-\,(aq) + H_2O\,(\ell) + 2\,e^- \rightarrow$	oxidizing agent	reduction
$\quad Cl^-\,(aq) + 2\,OH^-\,(aq)$		

18. Balancing redox equations in neutral or acidic solutions may be accomplished in several steps. They are:

1. Separating the equation into two equations which represent reduction and oxidation
2. Balancing mass of elements (other than H or O)
3. Balancing mass of O by adding H_2O
4. Balancing mass of H by adding H^+

5. Balancing charge by adding electrons

6. Balancing electron gain (in the reduction half-equation) with electron loss (in the oxidation half-equation)

7. Combining the two half equations

For the parts (a-c) of this problem, each step will be identified with a number corresponding to the list above. In addition, the physical states of all species will be omitted in all but the final step. While this omission is <u>not generally recommended</u>, it should increase the clarity of the steps involved. In addition when a step leaves a half equation unchanged from the previous step, we have omitted the half equation.

a. Cl_2 (aq) + Br^- (aq) $\rightarrow$ Br_2 (aq) + Cl^- (aq)

1.	$Cl_2 \rightarrow Cl^-$	$Br^- \rightarrow Br_2$
2.	$Cl_2 \rightarrow 2\,Cl^-$	$2\,Br^- \rightarrow Br_2$
5.	$Cl_2 + 2e^- \rightarrow 2\,Cl^-$	$2\,Br^- \rightarrow Br_2 + 2e^-$
7.	Cl_2 (aq) + 2 Br^- (aq) $\rightarrow$ 2 Cl^- (aq) + Br_2 (aq)	

b.   Sn (s) + H_3O^+ (aq) $\rightarrow$ Sn^{2+} (aq) + H_2 (g)

1.	$Sn \rightarrow Sn^{2+}$	$H_3O^+ \rightarrow H_2$
3.		$H_3O^+ \rightarrow H_2 + H_2O$
4.		$2\,H_3O^+ \rightarrow H_2 + 2\,H_2O$
5.	$Sn \rightarrow Sn^{2+} + 2e^-$	$2\,H_3O^+ + 2e^- \rightarrow H_2 + 2\,H_2O$
7.	Sn (s) + 2 H_3O^+ (aq) $\rightarrow Sn^{2+}$ (aq) + H_2 (g) + 2 H_2O (ℓ)	

c.   Zn (s) + VO^{2+} (aq) $\rightarrow$ Zn^{2+} (aq) + V^{3+} (aq)

1.	$Zn \rightarrow Zn^{2+}$	$VO^{2+} \rightarrow V^{3+}$
3.		$VO^{2+} \rightarrow V^{3+} + H_2O$
4.		$VO^{2+} + 2H^+ \rightarrow V^{3+} + H_2O$
5.	$Zn \rightarrow Zn^{2+} + 2\,e^-$	$VO^{2+} + 2\,H^+ + e^- \rightarrow V^{3+} + 2\,H_2O$
6.		$2\,VO^{2+} + 4\,H^+ + 2\,e^- \rightarrow 2\,V^{3+} + 2\,H_2O$
7.	Zn (s) + 2 VO^{2+} (aq) + 4 H^+ (aq) $\rightarrow Zn^{2+}$ (aq) + 2 V^{3+} (aq) + 2 H_2O (ℓ)	

20. See problem 18 for explanation of step numbers.

a. Ag^+ (aq) + HCHO (aq)→ Ag (s) + HCOOH (aq)

    1. $Ag^+ \rightarrow Ag$ HCHO $\rightarrow$ HCOOH
    3. H_2O + HCHO $\rightarrow$ HCOOH
    4. H_2O + HCHO $\rightarrow$ HCOOH + 2 H^+
    5. $Ag^+ + 1 e^- \rightarrow Ag$ H_2O + HCHO $\rightarrow$ HCOOH + 2 H^+
                                                                            + 2 e^-
    6. $2 Ag^+ + 2e^- \rightarrow 2 Ag$
    7. $2 Ag^+$ (aq) + H_2O (ℓ) + HCHO (aq) $\rightarrow$ 2 Ag (s) + HCOOH (aq) + 2 H^+ (aq)

b. H_2S (aq) + $Cr_2O_7^{2-}$ (aq) $\rightarrow$ S (s) + Cr^{3+} (aq)

    1. $H_2S \rightarrow S$ $Cr_2O_7^{2-} \rightarrow Cr^{3+}$
    2. $Cr_2O_7^{2-} \rightarrow 2 Cr^{3+}$
    3. $Cr_2O_7^{2-} \rightarrow 2 Cr^{3+} + 7 H_2O$
    4. $H_2S \rightarrow S + 2 H^+$ $14 H^+ + Cr_2O_7^{2-} \rightarrow 2 Cr^{3+} +$
                                                                            $7 H_2O$
    5. $H_2S \rightarrow S + 2 H^+ + 2e^-$ $14 H^+ + Cr_2O_7^{2-} + 6 e^- \rightarrow 2 Cr^{3+}$
                                                                            $+ 7 H_2O$
    6. $3 H_2S \rightarrow 3 S + 6 H^+ + 6e^-$
    7. $3 H_2S$ (aq) + 8 H^+ (aq) + $Cr_2O_7^{2-}$ (aq) $\rightarrow$ 3 S (s) + 2 Cr^{3+} (aq) + 7 H_2O (ℓ)
Note the removal of 6 H^+ (aq) from both sides of the equation.

c.   Zn (s) + VO_3^- (aq) $\rightarrow$ V^{2+} (aq) + Zn^{2+} (aq)

    1. $Zn \rightarrow Zn^{2+}$ $VO_3^- \rightarrow V^{2+}$
    3. $VO_3^- \rightarrow V^{2+} + 3 H_2O$
    4. $6 H^+ + VO_3^- \rightarrow V^{2+} + 3 H_2O$
    5. $Zn \rightarrow Zn^{2+} + 2 e^-$ $6 H^+ + VO_3^- + 3 e^- \rightarrow V^{2+}$
                                                                            $+ 3 H_2O$
    6. $3 Zn \rightarrow 3 Zn^{2+} + 6 e^-$ $12 H^+ + 2 VO_3^- + 6 e^- \rightarrow 2 V^{2+}$
                                                                            $+ 6 H_2O$
    7. $3 Zn$ (s) + 12 H^+ (aq) + 2 VO_3^- (aq) $\rightarrow$ 3 Zn^{2+} (aq) + 2 V^{2+} (aq) + 6 H_2O (ℓ)

22. One method for balancing redox reactions in basic solution is quite similar to the procedure in question 18 <u>with two exceptions:</u>

    3.   Balance mass of O by adding OH^-. Use twice as many OH^- as you need oxygens.

    4.   Balance H by adding H_2O.

Once again to clarify the steps, states of matter will be noted only in the final step. When no change is made from one step to the next, the equation will be written only for the first step.

a. $Zn\ (s) + ClO^-\ (aq) \rightarrow Zn(OH)_2\ (s) + Cl^-\ (aq)$

1.	$Zn \rightarrow Zn(OH)_2$	$ClO^- \rightarrow Cl^-$
3.	$Zn + 2\ OH^- \rightarrow Zn(OH)_2$	$ClO^- \rightarrow Cl^- + 2\ OH^-$
4.		$H_2O + ClO^- \rightarrow Cl^- + 2\ OH^-$
5.	$Zn + 2\ OH^- \rightarrow Zn(OH)_2 + 2\ e^-$	$2\ e^- + H_2O + ClO^- \rightarrow Cl^- + 2\ OH^-$
7.	$Zn\ (s) + H_2O\ (\ell) + ClO^-\ (aq) \rightarrow Zn(OH)_2\ (s) + Cl^-\ (aq)$	

b. $ClO^-\ (aq) + CrO_2^-\ (aq) \rightarrow Cl^-\ (aq) + CrO_4^{2-}\ (aq)$

1.	$ClO^- \rightarrow Cl^-$	$CrO_2^- \rightarrow CrO_4^{2-}$
3.	$ClO^- \rightarrow Cl^- + 2\ OH^-$	$4\ OH^- + CrO_2^- \rightarrow CrO_4^{2-}$
4.	$H_2O + ClO^- \rightarrow Cl^- + 2\ OH^-$	$4\ OH^- + CrO_2^- \rightarrow CrO_4^{2-} + 2\ H_2O$
5.	$H_2O + ClO^- + 2\ e^- \rightarrow Cl^- + 2\ OH^-$	$4\ OH^- + CrO_2^- \rightarrow CrO_4^{2-} + 2\ H_2O + 3\ e^-$
6.	$3\ H_2O + 3\ ClO^- + 6\ e^- \rightarrow 3\ Cl^- + 6\ OH^-$	$8\ OH^- + 2\ CrO_2^- \rightarrow 2\ CrO_4^{2-} + 4\ H_2O + 6\ e^-$
7.	$3\ ClO^-\ (aq) + 2\ OH^-\ (aq) + 2\ CrO_2^-\ (aq) \rightarrow 3\ Cl^-\ (aq) + 2\ CrO_4^{2-}\ (aq) + H_2O(\ell)$	

c. $Br_2\ (aq) \rightarrow Br^-\ (aq) + BrO_3^-\ (aq)$

1.	$Br_2 \rightarrow Br^-$	$Br_2 \rightarrow BrO_3^-$
2	$Br_2 \rightarrow 2\ Br^-$	$Br_2 \rightarrow 2\ BrO_3^-$
3.		$12\ OH^- + Br_2 \rightarrow 2\ BrO_3^-$
4.		$12\ OH^- + Br_2 \rightarrow 2\ BrO_3^- + 6\ H_2O$
5.	$2e^- + Br_2 \rightarrow 2\ Br^-$	$12\ OH^- + Br_2 \rightarrow 2\ BrO_3^- + 6\ H_2O + 10\ e^-$
6.	$10\ e^- + 5\ Br_2 \rightarrow 10\ Br^-$	

7. $6 Br_2 (aq) + 12 OH^- (aq) \rightarrow 10 Br^- (aq) + 2 BrO_3^- (aq) + 6 H_2O (\ell)$

Note that all coefficients in Step 7 are divisible by two. The overall balanced equation is
then

$$3 Br_2 (aq) + 6 OH^- (aq) \rightarrow 5 Br^- (aq) + BrO_3^- (aq) + 3 H_2O (\ell)$$

Electrochemical Cells and Cell Potentials

24.   a. Half equations   b. Processes   c. Compartment
       $Cr(s) \rightarrow Cr^{3+}(aq) + 3 e^-$ oxidation   anode
       $Fe^{2+}(aq) + 2 e^- \rightarrow Fe(s)$ reduction   cathode

26. The copper electrode is found to be the external anode (+) and the tin electrode the external
    cathode (-). The copper electrode is therefore the **internal cathode** and the tin electrode
    the **internal anode**.

    The half-reactions occurring in the half-cells are: Processes: Compartment
              $Cu^{2+} (aq) + 2 e^- \rightarrow Cu (s)$ reduction   cathode
              $Sn (s) \rightarrow Sn^{2+} (aq) + 2 e^-$ oxidation   anode

28. For a cell with $E° = +2.91$, the $\Delta G°$ is:
       $\Delta G° = - nFE°$

       $= - 2 \text{ mole} \cdot 9.65 \times 10^4 \dfrac{J}{volt \cdot mol} \cdot 2.91 \text{ volt} \cdot \dfrac{1.000 \text{ kJ}}{1000 \text{ J}}$

       $= - 562 \text{ kJ}$

30. Calculate $E°$ and decide if each reaction is product-favored:
    a. $2 I^- (aq) + Zn^{2+} (aq) \rightarrow I_2 (s) + Zn (s)$

                                                    process   potential
       $2 I^- (aq) \rightarrow I_2 (s) + 2 e^-$ oxidation   - 0.535 V
       $Zn^{2+} (aq) + 2 e^- \rightarrow Zn (s)$ reduction   - 0.763 V
                        Process is not product-favored. $E°_{net} = $ - 1.298 V

292

b. Zn^{2+} (aq) + Ni (s) → Zn (s) + Ni^{2+} (aq)

	process	potential
Ni (s) → Ni^{2+} (aq) + 2 e$^-$	oxidation	+ 0.25 V
Zn^{2+} (aq) + 2 e$^-$ → Zn (s)	reduction	- 0.763 V

Process is not product-favored. $E°_{net}$ =   - 0.51 V

c. 2 Cl$^-$ (aq) + Cu^{2+} (aq) → Cu (s) + Cl_2 (g)

	process	potential
2 Cl$^-$ (aq) → Cl_2 (g) + 2 e$^-$	oxidation	- 1.358 V
Cu^{2+} (aq) + 2 e$^-$ → Cu (s)	reduction	+ 0.337 V

Process is not product-favored. $E°_{net}$ =   - 1.021 V

32.   a. Sn^{2+} (aq) + 2 Ag (s) → Sn (s) + 2 Ag^+ (aq)

Sn^{2+} is reduced (- 0.14 V); Ag is oxidized (- 0.7994 V)

E° = (- 0.14 - 0.7994) = - 0.94 V   not product-favored

b. Zn (s) + Sn^{4+} (aq) → Sn^{2+} (aq) + Zn^{2+} (aq)

Sn^{4+} is reduced (+ 0.15 V); Zn is oxidized (+ 0.763 V)

E° = (+ 0.15 + 0.763) = + 0.91 V   product-favored

c. I_2 (aq) + 2 Br $^-$ (aq) → 2 I $^-$ (aq) + Br_2 (ℓ)

I_2 is reduced (+0.535 V) ; Br $^-$ is oxidized (- 1.066 V)

E° = (+ 0.535 - 1.066) = - 0.531 V   not product-favored

34. For the half-reactions listed:

a. Weakest oxidizing agent: V^{2+}

The more positive the E°, the better the oxidizing ability of the specie.

b. Strongest oxidizing agent: Cl_2

c. Strongest reducing agent: V

The more negative the E°, the better the reducing ability of a substance.

d. Weakest reducing agent: Cl$^-$

e. Pb (s) **cannot** reduce V^{2+} (aq).

Since the reduction potential of lead is less negative than that of V^{2+}, lead cannot reduce V^{2+}.

f. I_2 cannot oxidize Cl^- to Cl_2

>The greater the value of $E°$, the better the oxidizing ability of a substance.
>The $E°$ for Cl_2 is greater than that for I_2, hence I_2 cannot oxidize Cl^- to Cl_2.

g. $Pb(s)$ can reduce Cl_2, and I_2

>The comment from part c applies. The reduction potential of Pb is more negative than that for Cl_2 or I_2, making Pb capable of reducing either of these substances.

36. a. Maximum positive standard potential:

$$Cl_2\ (g)\ +\ V\ (s)\ \rightarrow\ 2\ Cl^-\ (aq)\ +\ V^{2+}\ (aq) \qquad E°\ =\ +2.54\ V$$

		potential
b.	$I_2\ (g)\ +\ 2\ e^-\ \rightarrow\ 2\ I^-\ (aq)$	$+\ 0.535\ V$
	$V\ (s)\ \rightarrow\ V^{2+}\ (aq)\ +\ 2\ e^-$	$\underline{+\ 1.18\ \ V}$
	$E°net\ =$	$+\ 1.72\ \ V$

The product-favored reaction is: $I_2\ (g)\ +\ V\ (s)\ \rightarrow\ 2\ I^-\ (aq)\ +\ V^{2+}\ (aq)$

38. The standard reduction potentials for the two species are:

$$Zn^{2+}\ (aq)\ +\ 2\ e^-\ \rightarrow\ Zn\ (s) \qquad E°\ =\ -\ 0.763\ V$$
$$Ag^+\ (aq)\ +\ 1\ e^-\ \rightarrow\ Ag\ (s) \qquad E°\ =\ +\ 0.7994\ V$$

a. The product-favored reaction would be:

$$2\ Ag^+\ (aq)\ +\ Zn\ (s)\ \rightarrow\ 2\ Ag\ (s)\ +\ Zn^{2+}\ (aq)$$

The potential for the cell would be: $\qquad E°\ =\ (+\ 0.7994\ +\ 0.763)\ =\ +\ 1.56\ V$

b. Silver is the cathode and zinc is the anode.

c. The diagram for the cell:

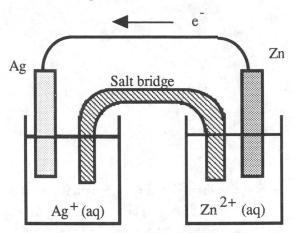

d. A strip of silver would serve as the cathode.

e. Electrons would flow from the zinc electrode to the silver electrode.

f. Nitrate ions would flow from the Ag compartment (site of increasing electrons) to the Zn compartment (as electrons leave the external Zn electrode, an increasing positive charge would develop).

Cells Under Nonstandard Conditions, E° and K

40. The Nernst equation for this reaction can be written:
$$E_{net} = E°_{net} - \frac{0.0592}{n} \log \frac{[Fe^{2+}]^2}{[Fe^{3+}]^2[I^-]^2}$$

Substituting appropriate values:
$$E_{net} = (0.771 - 0.535) - \frac{0.0592}{2} \log \frac{(0.1)^2}{(0.1)^2(0.1)^2}$$

$$= 0.236 - 0.0296 \cdot \log (100)$$

$$= 0.177 \text{ V}$$

Note that the potential of the cell, E_{net}, is decreased from that of the standard potential of the cell.

42. From Appendix J:
$$Ag^+ (aq) + 1 e^- \rightarrow Ag (s) \qquad E° = + 0.7994 \text{ V}$$
$$Fe^{3+} (aq) + 1 e^- \rightarrow Fe^{2+} (aq) \qquad E° = + 0.771 \text{ V}$$

a. The reaction for the cell operation:
$$Ag^+ (aq) + Fe^{2+} (aq) \rightarrow Ag (s) + Fe^{3+} (aq)$$

b. $E°_{net} = (+ 0.7994 - 0.771) = + 0.028 \text{ V}$

c. **Assume** that the reaction is the same under the non-standard conditions as under the standard conditions. Q can be calculated:

$$Q = \frac{[Fe^{3+}]}{[Fe^{2+}][Ag^+]} = \frac{(1.0 \text{ M})}{(1.0 \text{ M})(0.1 \text{ M})} = 10$$

Substitution into the Nernst equation yields:

$$E_{net} = E^{\circ}_{net} - \frac{0.0592}{n} \log Q$$

$$= +0.028 \text{ V} - \frac{0.0592}{1} \log (10) = -0.031 \text{ V}$$

The cell reaction--contrary to our assumption--is not the same as under standard conditions. The net reaction is now:

$$Ag \text{ (s)} + Fe^{3+} \text{ (aq)} \rightarrow Ag^+ \text{ (aq)} + Fe^{2+} \text{ (aq)}$$

44. To calculate the equilibrium constants for the reactions, we use the Nernst equation expressed as follows:

$$E_{net} = E^{\circ}_{net} - \frac{0.0592}{n} \log K$$

At equilibrium $E_{net} = 0$. E°_{net} must be calculated.

potential

a.
$$2 \text{ I}^- \text{ (aq)} \rightarrow \text{I}_2 \text{ (aq)} + 2 \text{ e}^- \qquad -0.535 \text{ V}$$
$$Fe^{3+} \text{ (aq)} + 1 \text{ e}^- \rightarrow Fe^{2+} \text{ (aq)} \qquad \underline{+0.771 \text{ V}}$$
$$E^{\circ}_{net} = +0.236 \text{ V}$$

The number of electrons in the balanced overall equation (n) is 2.
Substituting we get:

$$0 = 0.236 \text{ V} - \frac{0.0592}{2} \log K$$

$$\frac{(2)(0.236)}{(0.0592)} = \log K = 7.97 \text{ and } K = 9.5 \times 10^7$$

potential

b.
$$\text{I}_2 \text{ (aq)} + 2 \text{ e}^- \rightarrow 2 \text{ I}^- \text{ (aq)} \qquad +0.535 \text{ V}$$
$$2 \text{ Br}^- \text{ (aq)} \rightarrow \text{Br}_2 \text{ (aq)} + 2 \text{ e}^- \qquad \underline{-1.066 \text{ V}}$$
$$E^{\circ}_{net} = -0.531 \text{ V}$$

The number of electrons in the balanced overall equation (n) is 2.

Substituting we get:

$$0 = -0.531 \text{ V} - \frac{0.0592}{2} \log K$$

$$\frac{(2)(-0.531)}{(0.0592)} = \log K = -17.94 \text{ and } K = 1.1 \times 10^{-18}$$

Electrolysis, Electrical Energy, and Power

46. $1.00 \text{ g Au}^{3+} \cdot \dfrac{1 \text{ mol Au}^{3+}}{197.0 \text{ g Au}^{3+}} \cdot \dfrac{3 \text{ mol e}^-}{1 \text{ mol Au}^{3+}} \cdot \dfrac{96500 \text{ C}}{1 \text{ mol e}^-} \cdot \dfrac{1 \text{ A} \cdot \text{s}}{1 \text{ C}} \cdot \dfrac{1}{2.00 \text{ A}} =$

$$735 \text{ s} \quad (\text{or } 12.2 \text{ min})$$

48. Solutions to problems of this sort are best solved by beginning with a factor containing the desired units. Connecting this factor to data provided usually gives a direct path to the answer.

units desired
↓

$$\frac{63.55 \text{ g Ni}}{1 \text{ mol Ni}} \cdot \frac{1 \text{ mol Ni}}{2 \text{ mol e}^-} \cdot \frac{1 \text{ mol e}^-}{9.65 \times 10^4 \text{ C}} \cdot \frac{1 \text{ C}}{1 \text{ amp} \cdot \text{s}}$$

$$\cdot \frac{2.50 \text{ amps}}{1} \cdot \frac{3600 \text{ s}}{\text{hr}} \cdot \frac{2.00 \text{ hr}}{1} = 5.47 \text{ g Ni}$$

The second factor $\left(\dfrac{1 \text{ mol Ni}}{2 \text{ mol e}^-}\right)$ is arrived at by looking at the reduction half-reaction:

$$\text{Ni}^{2+} (\text{aq}) + 2 \text{ e}^- \rightarrow \text{Ni(s)}$$

All other factors are either data or common unity factors (e.g. $\dfrac{3600 \text{ s}}{1 \text{ hr}}$).

50. Current flowing if 0.052 g Ag are deposited in 450 s:

$$\frac{0.052 \text{ g Ag}}{1} \cdot \frac{1 \text{ mol Ag}}{108 \text{ g Ag}} \cdot \frac{1 \text{ mol e}^-}{1 \text{ mol Ag}} \cdot \frac{9.65 \times 10^4 \text{ C}}{1 \text{ mol e}^-} = 46 \text{ C}$$

and with this charge flowing in 450 s, the current is: $\dfrac{46 \text{ C}}{450 \text{ s}} = 0.10 \text{ amperes}$

52. The mass of aluminum produced in 8.0 hr by a current of 1.0×10^5 amp:

$$\frac{26.98 \text{ g Al}}{1 \text{ mol Al}} \cdot \frac{1 \text{mol Al}}{3 \text{ mol e}^-} \cdot \frac{1 \text{ mol e}^-}{9.65 \times 10^4 \text{ C}} \cdot \frac{1 \text{ C}}{1 \text{ amp} \cdot \text{s}}$$

$$\cdot \frac{1.0 \times 10^5 \text{ amp}}{1} \cdot \frac{3600 \text{ s}}{1 \text{ hr}} \cdot \frac{8.0 \text{ hr}}{1} = 2.7 \times 10^5 \text{ g Al}$$

54. The mass of lead consumed by a current of 1.0 amp for 50. hours:

During discharge of the lead storage battery, the anode reaction is
$$Pb \text{ (s)} \rightarrow Pb^{2+} \text{ (aq)} + 2e.$$

$$\frac{207 \text{ g Pb}}{1 \text{ mol Pb}} \cdot \frac{1 \text{ mol Pb}}{2 \text{ mol e}^-} \cdot \frac{1 \text{ mol e}^-}{9.65 \times 10^4 \text{ C}} \cdot \frac{1 \text{ C}}{1 \text{ amp} \cdot \text{s}}$$

$$\cdot \frac{1.0 \text{ amp}}{1} \cdot \frac{3600 \text{ s}}{1 \text{ hr}} \cdot \frac{50. \text{ hr}}{1} = 190 \text{ g Pb}$$

56. Calculate the number of Joules (V • C) associated with 1.0 amp of current for 100. hours.

$$\frac{1.0 \text{ amp}}{1} \cdot \frac{1 \text{ C}}{1 \text{ amp} \cdot \text{s}} = \frac{1.0 \text{ C}}{1 \text{ s}}$$ With a voltage of 12.0 volts, the number of joules is:

$$\frac{12.0 \text{ V}}{1} \cdot \frac{1.0 \text{ C}}{1 \text{ s}} = \frac{12.0 \text{ V} \cdot \text{C}}{1 \text{ s}}$$ and since $1 \text{ V} \cdot \text{C} = 1$ Joule we obtain

$$\frac{12 \text{ J}}{1 \text{ s}}$$ and since 1 watt $= \frac{1 \text{ J}}{\text{s}}$ then the power is 12 watts.

58. 10.9×10^7 kg $Cl_2 \cdot \dfrac{1 \times 10^8 \text{ g Cl}_2}{1.0 \text{ kg Cl}_2} \cdot \dfrac{1 \text{ mol Cl}_2}{70.906 \text{ g Cl}_2} \cdot \dfrac{2 \text{ mol e}^-}{1 \text{ mol Cl}_2} \cdot \dfrac{9.65 \times 10^4 \text{ C}}{1 \text{ mol e}^-}$

$$= 2.97 \times 10^{16} \text{ C}$$

The power is then:

$$\frac{2.97 \times 10^{16} \text{ C}}{1} \cdot \frac{4.6 \text{ V}}{1} \cdot \frac{1 \text{ J}}{1 \text{ V} \cdot \text{C}} \cdot \frac{1 \text{ kwh}}{3.6 \times 10^6 \text{ J}} = 3.8 \times 10^{10} \text{ kwh.}$$

60.   a. At the anode, oxidation of I^- occurs: $2 I^- \text{ (aq)} \rightarrow I_2 \text{ (aq)} + 2 e^-$ Sign: **positive**

b. At the cathode, reduction of water occurs:

$$2 H_2O \ (\ell) \ + \ 2 \ e^- \ \rightarrow \ H_2 \ (g) \ + 2 \ OH^- \ (aq) \qquad \text{Sign: \textbf{negative}}$$

c. $\dfrac{0.050 \text{ amp}}{1} \cdot \dfrac{5.0 \text{ hr}}{1} \cdot \dfrac{3600 \text{ s}}{1 \text{ hr}} \cdot \dfrac{1 \text{ C}}{1 \text{ amp} \bullet \text{s}} \cdot \dfrac{1 \text{ mol } e^-}{9.65 \times 10^4 \text{ C}} = 9.3 \times 10^{-3} \text{ mol } e^-$

Amount of products expected:

$9.3 \times 10^{-3} \text{ mol } e^- \cdot \dfrac{1 \text{ mol } I_2}{2 \text{ mol } e^-} \cdot \dfrac{253.8 \text{ g } I_2}{1 \text{ mol } I_2} = 1.2 \text{ g } I_2$

$9.3 \times 10^{-3} \text{ mol } e^- \cdot \dfrac{2 \text{ mol } OH^-}{2 \text{ mol } e^-} \cdot \dfrac{1 \text{ mol KOH}}{1 \text{ mol } OH^-} \cdot \dfrac{56.11 \text{ g KOH}}{1 \text{ mol KOH}} = 0.52 \text{ g KOH}$

$9.3 \times 10^{-3} \text{ mol } e^- \cdot \dfrac{1 \text{ mol } H_2}{2 \text{ mol } e^-} \cdot \dfrac{2.02 \text{ g } H_2}{1 \text{ mol } H_2} = 9.4 \times 10^{-3} \text{ g } H_2$

62. a. Electrolysis of KBr (aq):

Reduction of K^+	$E° = -2.925$ V
Reduction of H_2O	$= -0.83$ V

So H_2O will be reduced to H_2 at the cathode (in preference to elemental K).

Oxidation of Br^-	$E° = -1.066$ V
Oxidation of H_2O	$= -1.23$ V (+0.6 V overvoltage)

So Br^- will be oxidized to Br_2 at the anode.

b. Electrolysis of NaF (molten):

At the anode: $2 F^- \ (aq) \ \rightarrow \ F_2 \ (g) \ + \ 2 \ e^-$

At the cathode: $Na^+ \ (aq) + \ e^- \ \rightarrow \ Na \ (s)$

c. Electrolysis of NaF (aq):

Reduction of Na^+	$E° = -2.71$ V
Reduction of H_2O	$= -0.83$ V

So H_2O will be reduced to H_2 at the cathode.

Oxidation of F^-	$E° = -2.87$ V
Oxidation of H_2O	$= -1.23$ V (+0.6 V overvoltage)

So H_2O will be oxidized to O_2 at the anode.

a. $3 Br_2 (\ell) + 6 e^- \rightarrow 6 Br^-(aq)$

$\underline{I^-(aq) + 3 H_2O(\ell) \rightarrow IO_3^-(aq) + 6 H^+(aq) + 6 e^-}$

$I^-(aq) + 3 Br_2 (\ell) + 3 H_2O(\ell) \rightarrow IO_3^-(aq) + 6 Br^-(aq) + 6 H^+(aq)$

b. $5 U^{4+}(aq) + 10 H_2O(\ell) \rightarrow 5 UO_2^+(aq) + 20 H^+(aq) + 5 e^-$

$\underline{MnO_4^-(aq) + 8 H^+(aq) + 5 e^- \rightarrow Mn^{2+}(aq) + 4 H_2O(\ell)}$

$5 U^{4+}(aq) + 6 H_2O(\ell) + MnO_4^-(aq) \rightarrow 5 UO_2^+(aq) + Mn^{2+}(aq) + 12 H^+(aq)$

c. $MnO_2(s) + 4 H^+(aq) + 2 e^- \rightarrow Mn^{2+}(aq) + 2 H_2O(\ell)$

$\underline{2 I^-(aq) \rightarrow I_2(s) + 2 e^-}$

$MnO_2(s) + 2 I^-(aq) + 4 H^+(aq) \rightarrow Mn^{2+}(aq) + I_2(s) + 2 H_2O(\ell)$

66.   a. $Ni^{2+}(aq) + Cd(s) \rightarrow Ni(s) + Cd^{2+}(aq)$

b. Ni^{2+} is reduced (and the oxidizing agent).

Cd is oxidized (and the reducing agent).

c. **Cd** is the **anode**, and **Ni** is the **cathode**. Since Cd furnishes electrons to the external circuit, **Cd is negative**.

d. $E°_{net} = E°_{cathode} - E°_{anode}$

$= -0.25 - (-0.40) = +0.15 V$

e. Electrons flow from the Cd electrode to the Ni electrode.

f. As Cd is oxidized, there will be a shortage of anions in the anode compartment. Similarly as Ni^{2+} is reduced, there will be an excess of anions in the cathode compartment. The NO_3^- will migrate from the cathode compartment (nickel) to the anode compartment (cadmium).

g. $E_{net} = E°_{net} - \dfrac{0.0592}{n} \log K$

At equilibrium $E_{net} = 0$

$\dfrac{- nFE°}{2.303 \ RT} = \log K$

$$E_{net} = E°_{net} - \frac{0.0592}{n} \log K \text{ and since } E_{net} = 0$$

$$\frac{nE°}{0.0592} = \log K = \frac{(2)(+0.15 \text{ V})}{0.0592} = 5.07$$

$$K = 1.17 \times 10^5$$

h. For $[Cd^{2+}] = 0.010 \text{ M}$ and $[Ni^{2+}] = 1.0 \text{ M}$

$$E_{net} = E°_{net} - \frac{0.0592}{n} \log\frac{[Cd^{2+}]}{[Ni^{2+}]}$$

$$E_{net} = +0.15 - \frac{0.0592}{n} \log(\frac{1 \times 10^{-2}}{1.0})$$

$$= +0.15 + 0.0592$$

$$= +0.21 \text{ V}$$

The reaction is still the reaction given in part a.

i. Battery lifetime:

   Which reactant is consumed first?

$$50.0 \text{ g Ni} \cdot \frac{1 \text{ mol Ni}}{58.69 \text{ g Ni}} = 0.852 \text{ mol Ni}$$

$$50.0 \text{ g Cd} \cdot \frac{1 \text{ mol Cd}}{112.4 \text{g Cd}} = 0.445 \text{ mol Cd}$$

Cd is the limiting reagent.

$$0.445 \text{ mol Cd} \cdot \frac{2 \text{ mol e}^-}{1 \text{ mol Cd}} \cdot \frac{9.65 \times 10^4 \text{ C}}{1 \text{ mol e}^-} \cdot \frac{1 \text{ A} \cdot \text{s}}{1 \text{ C}} \cdot \frac{1}{0.050 \text{ A}} = 1.7 \times 10^6 \text{ s}$$

$$\text{or 480 hrs}$$

68. For the reaction:

$$V^{2+}(aq) + Md^{3+}(aq) \longrightarrow Md^{2+}(aq) + V^{3+}(aq) \qquad K = 15$$

We can calculate the E_{net}

$$E°_{net} = \frac{0.0592}{n} \log k$$

then $$E°_{net} = \frac{0.0592}{1} \log (15) = 0.0696 \text{ V}$$

$$E°_{net} = E°_{oxidation} + E°_{reduction}$$

$$0.0696 \text{ V} = +0.255 + E°_{reduction}$$

$$-0.185 \text{ V} = E°_{reduction}$$

he net process is

$$Pb(s) + PbO_2(s)\ 2\ H_2SO_4(aq) \longrightarrow 2\ PbSO_4(s) + 2\ H_2O(\ell)$$

$$
\begin{aligned}
E_{net} &= E°_{net} - \frac{0.0592}{n} \log \frac{1}{[H_2SO_4]^2} \\
&= 2.041\ V - \frac{0.0592}{n} \log \frac{1}{(6.00)^2} \\
&= 2.041\ V - \frac{0.0592}{n} \log(0.0278) \\
&= 2.041\ V + 0.047 = 2.087\ V
\end{aligned}
$$

72. For $Cu^{2+}(aq) + Sn(s) \longrightarrow Sn^{2+} + Cu(s)$, the net cell potential is:

$$E°_{net} = (+0.337 + 0.14) = +0.477\ \text{or}\ +0.48\ V$$

Calculate the E_{net} once 48.0 hrs have passed:

Determine the amount of Cu^{2+} and Sn^{2+} consumed:

$$0.400\ A \cdot \frac{48.0\ hr}{1} \cdot \frac{3600\ s}{1\ hr} \cdot \frac{1\ C}{1\ A \cdot s} \cdot \frac{1\ mol\ e^-}{9.65 \times 10^4\ C} \cdot \frac{1\ mol\ metal}{2\ mol\ e^-} =$$

$$0.358\ mol\ metal$$

Note that the last factor reflects the dipositive ions.

For one 1.00 L $Cu^{2+} = 1.00\ mol - 0.358\ mol = 0.64\ mol$

and $Sn^{2+} = 1.00\ mol + 0.358\ mol = 1.36\ mol$

$$
\begin{aligned}
E_{net} &= E°_{net} - \frac{0.0592}{n} \log \frac{[Sn^{2+}]}{[Cu^{2+}]} \\
&= +0.48\ V - \frac{0.0592}{2} \log \frac{1.36}{0.64} \\
&= +0.48\ V - 0.00969\ \text{or}\ +0.47\ V
\end{aligned}
$$

74.   a. Since the magnesium is more easily oxidized than the iron, the iron is "forced" to be reduced—a process that doesn't occur. This is sometimes referred to as cathodic protection.

b. Time to consume 5.0 kg of Mg

$$5.0 \times 10^3\ g\ Mg \cdot \frac{1\ mol\ Mg}{24.3\ g\ Mg} \cdot \frac{2\ mol\ e^-}{1\ mol\ Mg} \cdot \frac{9.65 \times 10^4\ C}{1\ mol\ e^-} \cdot \frac{1\ A \cdot s}{1\ C} \cdot \frac{1}{0.030\ A} \cdot$$

$$\frac{1\ hr}{3600\ s} \cdot \frac{1\ day}{24\ hr} \cdot \frac{1\ yr}{365.25\ day} = 42\ yr$$

76.   a. Balanced equation: $2\ Al(s)\ +\ 3\ Cl_2(g)\ \longrightarrow\ 2\ Al^{3+}(aq)\ +\ 6\ Cl^-(aq)$

       b. The **aluminum oxidation is the anode**; chlorine—the cathode.

       c. $E^\circ_{net}\ =\ E^\circ_{chlorine}\ -\ E^\circ_{aluminum}$
           $=\ 1.358\ -\ (-1.66)\ =\ 3.02\ V$

       d. $E_{net}\ =\ E^\circ_{net}\ -\ \dfrac{0.0592}{n}\ \log\dfrac{[Al^{3+}]^2[Cl^-]^2}{[Cl_2]^3}$

           $=\ 3.02\ V\ -\ \dfrac{0.0592}{6}\ \log\dfrac{1}{[0.50]^3}$

           $=\ 3.02\ -\ \dfrac{0.0592}{6}\ \log(\dfrac{1}{0.125})$

           $=\ 3.02\ -\ 0.0089\ =\ 3.01\ V$

       Note the voltage decreases if the pressure of chlorine is less than 1 atm.

       e. Time to consume 30.0 g Al:

       $30.0\ g\ Al\ \bullet\ \dfrac{1\ mol\ Al}{26.98\ g\ Al}\ \bullet\ \dfrac{3\ mol\ e^-}{1 mol\ Al}\ \bullet\ \dfrac{9.65\ x\ 10^4\ C}{1\ mol\ e^-}\ \bullet\ \dfrac{1\ A\ \bullet\ s}{1\ C}\ \bullet\ \dfrac{1}{0.75\ A}\ =$

                                                    $4.29\ x\ 10^5\ s$ or 119 hrs

78. a. The stoichiometry of the silver/zinc battery indicates a reaction of one mole each of silver
         oxide, zinc, and water. The mass of one mole of each of these three substances is:

                1 mol Ag_2O 231.7 g
                1 mol Zn   65.4 g
                1 mol H_2O 18.0 g
                                 315.1 g

         The energy associated with the battery is: $\dfrac{0.10\ C}{1\ s}\ \bullet\ \dfrac{1.59\ V}{1}\ =\ \dfrac{0.159\ V\ \bullet\ C}{s}$

         Since 1 V • C = 1 Joule, this energy corresponds to 0.159 J/s. [1 watt = 1 J/s]

         The energy/gram for the silver/zinc battery is: $\dfrac{0.159\ J/s}{315.1\ g}\ =\ 5.0\ x\ 10^{-4}$ watts/gram. (2 sf)

b. Performing the same calculations for the lead storage battery, using a stoichiometric amount for the overall battery reaction:

$$1 \text{ mol Pb} \qquad 207.2 \text{ g}$$
$$1 \text{ mol PbO}_2 \quad 239.2 \text{ g}$$
$$\underline{2 \text{ mol H}_2\text{SO}_4 \quad 196.2 \text{ g}}$$
$$642.6 \text{ g}$$

The energy associated with the battery is: $\dfrac{0.10 \text{ C}}{1 \text{ s}} \cdot \dfrac{2.0 \text{ V}}{1} = \dfrac{0.20 \text{ V} \cdot \text{C}}{\text{s}}$

The energy/gram for the lead storage battery is: $\dfrac{0.20 \text{ J/s}}{642.6 \text{ g}} = 3.1 \times 10^{-4} \text{ watts/g}$

c. The silver/zinc battery produces more energy (and more power)/gram.

Conceptual Questions

80.   a. Since A and C reduce H^+ to elemental hydrogen, they are both stronger reducing agnets than H_2.

Reducing agent strength: $H_2 < A, C$

b. C reduces A, B, and D—hence it is the strongest reducing agent.
A, B, D < C and from part a. above $B, D < H_2 < A < C$

c. D reduces B^{n+}, so D is a stronger reducing agent than B.
So $B < D < H_2 < A < C$

82.   a. Calculate $E^\circ{}_{net}$ then ΔH° and ΔG°

E° for $O_2(g) + 2 H_2O(\ell) + 4 e^- \longrightarrow 4 OH^-(aq)$ 0.040 V
   for $2 H_2O(\ell) + 2 e^- \longrightarrow H_2(g) + 2 OH^-(aq)$ -0.8277 V

reversing the 2nd equation

oxidation	$2 H_2(g) + 4 OH^-(aq) \longrightarrow 4 H_2O(\ell) + 4 e^-$	
reduction	$O_2(g) + 2 H_2O(\ell) + 4 e^- \longrightarrow 4 OH^-(aq)$	
net reaction	$2 H_2(g) + O_2(g) \longrightarrow 2 H_2O(\ell)$	$E^\circ{}_{net} = 1.23 \text{ V}$

$$\Delta G° = -nFE° = -(4 \text{ mol e}^-) \left(\frac{9.65 \times 10^4 \text{ C}}{1 \text{ mol e}^-}\right)(+1.23 \text{ V})$$

$$= -4.75 \times 10^5 \text{ V} \cdot \text{C} \quad \text{or} \quad -4.75 \times 10^5 \text{ J} \quad \text{or } -2.37 \times 10^2 \text{ kJ/mol } H_2O$$

Note that the ΔG for this process from the Appendix is -237.129 kJ/mol

$$\Delta H°rxn = \Delta H°_f H_2O(\ell) - [\Delta H°_f H_2(g) + 1/2 \Delta H°_f O_2(g)]$$

$$\Delta H°rxn = (-285.830 \text{ kJ/mol})(1 \text{ mol}) - 0 = -285.830 \text{ kJ/mol}$$

So efficiency $= \dfrac{-237.1 \text{ kJ/mol}}{-285.830 \text{ kJ/mol}} \times 100\% = 82.96\%$

b. For $H_2O(g)$:

Using $\Delta H°$ and $\Delta G°$ data from the Appendix

Efficiency $= \dfrac{-228.572 \text{ kJ/mol}}{-241.818 \text{ kJ/mol}} \times 100\% = 94.54\%$

c. The **increased efficiency of the cell when water vapor is produced** reflects the saving of "thermodynamic energy" of water vapor compared to water liquid.

84. Since the reaction depends on the oxidation of elemental hydrogen to water (2 mol e⁻ per mol H_2), we must detemine the amount of H_2 present:

$$n = \frac{(200. \text{ atm})(1.0 \text{ L})}{(0.0821 \frac{\text{L} \cdot \text{atm}}{\text{K} \cdot \text{mol}})(298 \text{ K})} = 8.2 \text{ mol } H_2$$

The amount of time this can produce current:

$$8.2 \text{ mol } H_2 \cdot \frac{2 \text{ mol e}^-}{1 \text{ mol } H_2} \cdot \frac{9.65 \times 10^4 \text{ C}}{1 \text{ mol e}^-} \cdot \frac{1 \text{ A} \cdot \text{s}}{1 \text{ C}} \cdot \frac{1}{1.5 \text{ A}} = 1.1 \times 10^6 \text{ s}$$

$$(290 \text{ hrs})$$

Chapter 22:
The Chemistry of The Main Group Elements

Review Questions

8. $[X]$ ns^2 np^2 where X represents the noble (or inert) gas for the $(n - 1)$ period and n represents the period in which the element is located.

    e.g. C(in period 2) has the configuration $[He]$ $2s^2$ $2p^2$

10. Four ions with an electron configuration that is the same as argon's

$$S^{2-}, \qquad Cl^-, \qquad K^+, \qquad Ca^{2+}$$

    sulfide ion, chloride ion, potassium ion, calcium ion

12. $$2\,M + X_2 \longrightarrow 2\,MX$$

    Given the stability of M+ cations and X- anions compared to the metal and nonmetal, the **reaction will likely be exothermic**, and an ionic product will result. Reactions between elements in "widely separated groups" tend to form ionic products (metal + nonmetal) while covalent products frequently are formed between elements in "closer groups" (e.g. two nonmetals).

14. 10 most **abundant elements in earth's crust** from lowest to greatest abundance:

Abundance	Element	Specie
0.6 %	Ti	$FeTiO_3$, ilmenite; TiO_2, rutile
0.9 %	H	$H2O$
1.9 %	Mg	oxides & chlorides; e.g. $MgCl_2 \cdot KCl \cdot 6\,H_2O$, carnallite
2.4 %	K	KCl
3.4 %	Ca	dolomite, $CaCO_3 \cdot MgCO_3$; limestone, $CaCO_3$
4.7 %	Fe	oxide ores: Fe_2O_3 , Fe_3O_4
7.4 %	Al	bauxite, $Al_2O_3 \cdot x\,H_2O$
25.7 %	Si	SiO_2 ; silicates, $SiO_4{}^{4-}$
49.5 %	O	oxides of all types; water; $CaCO_3$

16. The **ease of oxidation of calcium** would indicate that elemental calcium would not be likely in the earth's crust.

18. In general metal oxides (also commonly called base anhydrides) are more basic than nonmetal oxides (also known as acid anhydrides). Hence we would predict the order to be:
$$SO_3 \; < \; SiO_2, \; Al_2O_3 \; < \; Na_2O$$

20. Balanced equations:

    a. $2\,Na(s) + I_2(g) \longrightarrow 2\,NaI(s)$
    b. $8\,Ca(s) + S_8(g) \longrightarrow 8\,CaS(s)$
    c. $4\,Ca(s) + 3\,O_2(g) \longrightarrow 2\,Al_2O_3(s)$
    d. $Si(s) + 2\,Cl_2(g) \longrightarrow SiCl_4(\ell)$

Hydrogen

22. $2\,Na(s) + H_2(g) \longrightarrow 2\,NaH(s)$ (sodium hydride)

The compound will be ionic. One would anticipate the ionic bond to indicate a high melting point (i.e. **NaH is a solid**).
Chemically, the hydride ion is one of the best reducing agents known.

24. Volume of 1.0 kg of H_2 (g) at 25 °C and 1.0 atm:
The amount of H_2 corresponding to 1.0×10^3 g H_2 :

$$1.0 \times 10^3 \text{ g } H_2 \cdot \frac{1 \text{ mol } H_2}{2.02 \text{ g } H_2} = 5.0 \times 10^2 \text{ mol } H_2 \quad \text{(to 2 significant figures)}$$

This amount of H_2 would occupy a volume of :

$$V = \frac{nRT}{P} = \frac{(500 \text{ mol } H_2)(0.082057 \frac{L \cdot atm}{K \cdot mol})(298 \text{ K})}{1.0 \text{ atm}} = 1.2 \times 10^4 \text{ L}$$

26.

$CH_4(g)$	+	$H_2O(g)$	$\longrightarrow$	$3\,H_2(g)$	+	$CO(g)$	
186.264		188.825		130.684		197.674	$S°\,(\frac{J}{K \cdot mol})$
-74.81		-241.818		0		-110.525	$\Delta H°\,(\frac{kJ}{mol})$

$$\Delta H^\circ = [(3\ mol)(0) + (1\ mol)(-110.525\ \tfrac{kJ}{mol})] - [(1\ mol)(-74.81\ \tfrac{kJ}{mol})$$
$$+ (1\ mol)(-241.818\ \tfrac{kJ}{mol})]$$

$$= 206.10\ kJ$$

$$\Delta S^\circ = [(3\ mol)(130.684\ \tfrac{J}{K\bullet mol}) + (1\ mol)(197.674\ \tfrac{J}{K\bullet mol})] -$$
$$[(1\ mol)(186.264\ \tfrac{J}{K\bullet mol}) + (1\ mol)(188.825\ \tfrac{J}{K\bullet mol})]$$

$$= 214.637\ J/K$$

$$\Delta G^\circ = 206.10\ kJ - (298\ K)(214.637\ J/K)(\tfrac{1\ kJ}{1000\ J})$$

$$= 142.14\ kJ$$

Alkali Metals

28. $2\ Na(s) + F_2(g) \longrightarrow 2\ NaF(s)$

       $2\ Na(s) + Cl_2(g) \longrightarrow 2\ NaCl(s)$

       $2\ Na(s) + Br_2(\ell) \longrightarrow 2\ NaBr(s)$

       $2\ Na(s) + I_2(s) \longrightarrow 2\ NaI(s)$

We anticipate that these salts are ionic, are good conductors (in the liquid state or solution), have high melting points, and are readily soluble in water.

30. $4\ Li(s) + O_2(g) \longrightarrow 2\ Li_2O(s)$ oxides

       $2\ Na(s) + O_2(g) \longrightarrow Na_2O_2(s)$ peroxides

       $K(s) + O_2(g) \longrightarrow KO_2(s)$ superoxide

32.   a. $2\ NaCl(aq) + 2\ H_2O(\ell) \longrightarrow 2\ NaOH(aq) + Cl_2(g) + H_2(g)$

      b. Anticipated masses ratios:

$$\frac{1\ mol\ Cl_2}{2\ mol\ NaOH} = \frac{71\ g\ Cl_2}{80\ g\ NaOH} = 0.89\ g\ Cl_2/g\ NaOH$$

Actual: $\dfrac{24.06 \times 10^9\ pounds\ Cl_2}{25.71 \times 10^9\ pounds\ NaOH} = 0.94\ \dfrac{pounds\ Cl_2}{pounds\ NaOH}$

The difference in ratios means that alternative methods of producing chlorine are used. One of these is the Kel-Chlor process which uses HCl, $NOCl$, and O_2. Other products are NO and H_2O.

Alkaline Earths

34. $3 Mg(s) + N_2(g) \longrightarrow Mg_3N_2(s)$

        $2 Mg(s) + O_2(g) \longrightarrow 2 MgO(s)$

36. Uses of limestone:

        agricultural: to furnish Ca^{2+} to plants and neutralize acidic soils

        building: lime (CaO) is used in mortar and absorbs CO_2 to form $CaCO_3$

        steel-making: $CaCO_3$ furnishes lime (CaO) in the basic oxygen process. The lime reacts with gangue (SiO_2) to form calcium silicate.

        $CaCO_3(s) + H_2O(\ell) + CO_2(g) \rightleftharpoons Ca^{2+}(aq) + 2 HCO_3^-(aq)$

        This reaction is important in the formation of "hardwater" (not particularly a great happening for plumbing) and stalagmites and stalactites (aesthetically pleasing in caves).

38.   The amount of SO_2 that could be removed by 1000. kg of CaO by the reaction:

$$CaO\ (s) + SO_2\ (g) \longrightarrow CaSO_3\ (s)$$

$$1.000 \times 10^6 \text{ g CaO} \cdot \frac{1 \text{ mol CaO}}{56.079 \text{ g CaO}} \cdot \frac{1 \text{ mol } SO_2}{1 \text{ mol CaO}} \cdot \frac{64.059 \text{ g } SO_2}{1 \text{ mol } SO_2}$$

$$= 1.142 \times 10^6 \text{ g } SO_2$$

Aluminum

40. $2 Al(s) + 6 HCl(aq) \longrightarrow 2 Al^{3+}(aq) + 6 Cl^-(aq)$

        $2 Al(s) + 3 Cl_2(g) \longrightarrow 2 AlCl_3(s)$

        $4 Al(s) + 3 O_2(g) \longrightarrow 2 Al_2O_3(s)$

42. $2 Al (s) + 2 NaOH (aq) + 6 H_2O (\ell) \longrightarrow 2 NaAl(OH)_4 (aq) + 3 H_2 (g)$

Volume of H_2 (in mL) produced when 13.2 g of Al react:

$$13.2 \text{ g Al} \cdot \frac{1 \text{ mol Al}}{26.98 \text{ g Al}} \cdot \frac{3 \text{ mol } H_2}{2 \text{ mol Al}} = 0.734 \text{ mol } H_2$$

$$V = \frac{(0.734 \text{ mol } H_2)(0.082057 \frac{L \cdot atm}{K \cdot mol})(295.7 \text{ K})}{735 \text{ mm Hg} \cdot \frac{1 \text{ atm}}{760 \text{ mmHg}}} = 18.4 \text{ L}$$

or 1.84×10^4 mL

44. $Al_2O_3(s) + 3 H_2SO_4(aq) \longrightarrow Al_2(SO_4)_3 (aq) + 3 H_2O(\ell)$

$$1.00 \times 10^3 \text{ g } Al_2(SO_4)_3 \cdot \frac{1 \text{ mol } Al_2(SO_4)_3}{342.1 \text{ g } Al_2(SO_4)_3} \cdot \frac{1 \text{ mol } Al_2O_3}{1 \text{ mol } Al_2(SO_4)_3} \cdot$$

$$\frac{102.1 \text{ g } Al_2O_3}{1 \text{ mol } Al_2O_3} \cdot \frac{1 \text{ kg}}{1 \times 10^3 \text{ g}} = 0.298 \text{ kg } Al_2O_3$$

$$1.00 \times 10^3 \text{ g } Al_2(SO_4)_3 \cdot \frac{1 \text{ mol } Al_2(SO_4)_3}{342.1 \text{ g } Al_2(SO_4)_3} \cdot \frac{3 \text{ mol } H_2SO_4}{1 \text{ mol } Al_2(SO_4)_3} \cdot$$

$$\frac{98.07 \text{ g } H_2SO_4}{1 \text{ mol } H_2SO_4} \cdot \frac{1 \text{ kg}}{1 \times 10^3 \text{ g}} = 0.860 \text{ kg } H_2SO_4$$

46. The $AlCl_4^-$ will have a **tetrahedral geometry**. The aluminum atom would utilize **sp^3 hybridization**.

Silicon

48. The structure of SiO_2 is tetrahedral with the silicon atom being surrounded by four oxygen atoms.

a. The energies of four Si - O single bonds is greater than two Si=O double bonds. CO_2, on the other hand, forms discrete C=O double bonds.

b. The high melting point of SiO_2 is due to the **network structure** of the SiO_4 tetrahedron.

50.   a. $Si (s) + 2 CH_3Cl (g) \longrightarrow (CH_3)_2SiCl_2 (\ell)$

b. Stoichiometric amount of CH_3Cl :

$$2.65 \text{ g Si} \cdot \frac{1 \text{ mol Si}}{28.09 \text{ g Si}} \cdot \frac{2 \text{ mol } CH_3Cl}{1 \text{ mol Si}} = 0.189 \text{ mol } CH_3Cl$$

$$P = \frac{(0.189 \text{ mol } CH_3Cl) (0.082057 \frac{L \cdot atm}{K \cdot mol}) (297.7 \text{ K})}{5.60 \text{ L}} = 0.823 \text{ atm}$$

c. Mass of $(CH_3)_2SiCl_2$ produced:

$$0.0943 \text{ mol Si} \cdot \frac{1 \text{ mol } (CH_3)_2SiCl_2}{1 \text{ mol Si}} \cdot \frac{129.1 \text{ g } (CH_3)_2SiCl_2}{1 \text{ mol } (CH_3)_2SiCl_2} =$$

$$12.2 \text{ g } (CH_3)_2SiCl_2$$

Nitrogen and Phosphorus

52. $2 NO(g) \quad + \quad O_2(g) \longrightarrow \quad 2 NO_2(g)$

$\quad$ +90.25 $\qquad\qquad$ 0 $\qquad\qquad\qquad$ +33.18 $\qquad\qquad$ $\Delta H° (\frac{kJ}{mol})$

$$\Delta H°_{rxn} = [(2 \text{ mol})(+33.18 \frac{kJ}{mol})] - [(2 \text{ mol})(+90.25 \frac{kJ}{mol})]$$

$$= -114.14 \text{ kJ}$$

a. Note that neither of the nitrogen oxides is stable with respect to their elements.

b. The **reaction is exothermic**.

c. **Both** NO and NO_2 have an odd number of electrons, and **are paramagnetic**.

54.   a. The reaction of hydrazine with dissolved oxygen:

$$N_2H_4 (aq) + O_2 (aq) \longrightarrow N_2 (g) + 2 H_2O(\ell)$$

b. Mass of hydrazine to consume the oxygen in 3.00×10^4 L of water:

$$3.00 \times 10^4 \text{ L} \cdot \frac{3.08 \text{ cm}^3 O_2}{0.100 \text{ L}} \cdot \frac{1 \text{ mol } O_2}{22400 \text{ cm}^3} \cdot \frac{1 \text{ mol } N_2H_4}{1 \text{ mol } O_2} \cdot \frac{32.05 g N_2H_4}{1 \text{ mol } N_2H_4}$$

$$\uparrow \text{ at STP}$$

$$= 1320 \text{ g } N_2H_4$$

311

56. The half equations:

$$N_2H_5^+ \text{ (aq)} \longrightarrow N_2 \text{ (g)} + 5\,H^+ \text{ (aq)} + 4\,e^-$$

$$IO_3^- \text{ (aq)} \longrightarrow I_2 \text{ (s)}$$

are balanced according to the procedure in Chapter 5 to give:

$$5\,N_2H_5^+ \text{ (aq)} + 4\,IO_3^- \text{ (aq)} \longrightarrow 5\,N_2 \text{ (g)} + 2\,I_2 \text{ (s)} + 11\,H_2O(\ell) + H_3O^+ \text{ (aq)}$$

$E°$ for the reaction is: $E°$ for the hydrazine equation (oxidation) $= +\ 0.23$ V

$E°$ for the iodate equation (reduction) $= \underline{+\ 1.195\ \text{V}}$

$E°_{net}\ = +\ 1.42$ V

58. The dot structure for the azide ion:

$$\left[\ \ddot{N}\!=\!N\!=\!\ddot{N}\ \right]^-$$

Oxygen and Sulfur

60.   a. Allowable release of SO_2: (0.30 %)

$$1.80 \times 10^6 \text{ kg } H_2SO_4 \cdot \frac{1 \text{ mol } H_2SO_4}{98.07 \text{ g } H_2SO_4} \cdot \frac{1 \text{ mol } SO_2}{1 \text{ mol } H_2SO_4} \cdot \frac{64.06 \text{ g } SO_2}{1 \text{ mol } SO_2}$$

$$\cdot \frac{0.0030 \text{ kg } SO_2 \text{ released}}{1.00 \text{ kg } SO_2 \text{ produced}} = 3.53 \times 10^3 \text{ kg } SO_2$$

$$(3.5 \times 10^3 \text{ —to 2 sf})$$

b. Mass of $Ca(OH)_2$ to remove 3.92 T SO_2:

$$3.53 \times 10^3 \text{ kg } SO_2 \cdot \frac{1 \text{ mol } SO_2}{64.06 \text{ g } SO_2} \cdot \frac{1 \text{ mol } Ca(OH)_2}{1 \text{ mol } SO_2} \cdot \frac{74.09 \text{ g } Ca(OH)_2}{1 \text{ mol } Ca(OH)_2}$$

$$= 4.0 \times 10^3 \text{ kg } Ca(OH)_2$$

62. The polysulfide ion S_3^{2-} can be pictured as a tetrahedral structure:

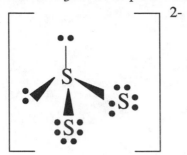

Chlorine

64. Calculate the equivalent net cell potential for the oxidation:

$$6[Mn^{2+}(aq) + 4 H_2O(\ell) \longrightarrow MnO_4^-(aq) + 8 H^+(aq) + 5 e^-] \qquad -1.51 \text{ V}$$

$$5[BrO_3^-(aq) + 6 H^+(aq) + 6 e^- \longrightarrow Br^-(aq) + 3 H_2O(\ell)] \qquad +1.44 \text{ V}$$

net $\; 6 Mn^{2+}(aq) + 9 H_2O(\ell) + 5 BrO_3^-(aq) \longrightarrow 6 MnO_4^-(aq) \qquad -0.07 \text{ V}$

$$+ 18 H^+(aq) + Br^-(aq)$$

The negative net potential indicates that this **process doesn't favor products** with 1.0 M bromate ion.

66. The molecule will be T-shaped.

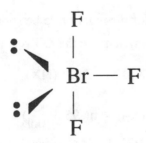

The lone pairs of electrons will **reduce the angles** from their ideal values (of 90°).

General Questions

68. $MCO_3(s) \longrightarrow MO(s) + CO_2(g)$ **M**

  $-1012.1 \dfrac{kJ}{mol}$ -569.43 -394.359 (Mg) $\Delta G°$ values

  -1128.79 -604.03 (Ca) $\Delta G°$ values

  -1137.6 -525.1 (Ba) $\Delta G°$ values

$\Delta G°_{rxn} = \Delta G°_f MO + \Delta G°_f CO_2 - \Delta G°_f MCO_3$

$MgCO_3 = (-569.43 \dfrac{kJ}{mol})(1 \text{ mol}) + (-394.359 \dfrac{kJ}{mol})(1 \text{ mol}) - (-1012.1 \dfrac{kJ}{mol})(1 \text{ mol})$

$\qquad = 48.3 \text{ kJ}$

$CaCO_3 = (-604.03 \dfrac{kJ}{mol})(1 \text{ mol}) + (-394.359 \dfrac{kJ}{mol})(1 \text{ mol}) - (-1128.79 \dfrac{kJ}{mol})(1 \text{ mol})$

$\qquad = 130.40 \text{ kJ}$

$BaCO_3 = (-525.1 \dfrac{kJ}{mol})(1 \text{ mol}) + (-394.359 \dfrac{kJ}{mol})(1 \text{ mol}) - (-1137.6 \dfrac{kJ}{mol})(1 \text{ mol})$

$\qquad = 218.1 \text{ kJ}$

The relative tendency for decomposition is then

$\qquad MgCO_3 > CaCO_3 > BaCO_3$

70.   a. Since $\Delta G°_{rxn} < 0$ for the reaction to be spontaneous (or product-favored), calculate the value for $\Delta G°_f MX$ that will make the $\Delta G°_{rxn}$ zero.

  $\Delta G°_{rxn} = \Delta G°_f (MX_n) - n \Delta G°_f (HX)$

  for HCl $= \Delta G°_f (MX_n) - n(-95.3 \dfrac{kJ}{mol})$

  so if $n(-95.3 \text{ kJ}) = \Delta G°_f(MX_n)$ then $\Delta G° = 0$
  and if $n (-95.3 \text{ kJ}) > \Delta G° (MX_n)$ then $\Delta G°_{rxn} < 0$.

  b. Examine $\Delta G°MX$ values for

metal:	Ba	Pb	Hg	Ti
$\Delta G°mx$:	-810.4	-314.10	-58.539	-737.2
n:	2	2	2	4
n(-93.3):	-190.6	-190.6	-190.6	-109.6

314

For Barium , Lead, and Titanium, n (-95.3) > $\Delta G°(MX)$ and we would expect these reactions to be spontaneous.

72.   a. The N-O bonds are the same length owing to the delocalization of a second pair of electrons between the two N-O bonds. A dot picture of this would show two reasonable structures (resonance hybrids). In essence there is a bond order of 1.5 for these two bonds; compared to a bond order of 1 for the N-OH bond.

b. The bond angle for the oxygens involved in the delocalized bond is only slightly larger than anticipated for the trigonal planar geometry (120°). The larger angle reflects the increased electron density (and repulsion) of this bond. The increased bond angle would result in a slightly smaller-than-ideal angle between the two "non-H" oxygens and the O-H bond. The bond angle N-O-H is only slightly less than anticipated for the tetrahedral orientation around the oxygen (two atoms and two lone pairs), a finding consistent with the two lone pairs of electrons on that oxygen.

c. The central **N atom has sp^2 hybridization**. The "unhybridized p" orbital on the N can participate in a π - type overlap with the orbitals on the two oxygen atoms, resulting in the pi bond.

Conceptual Questions

74. Reaction scheme:

Clue

1.   1.00 g A   + heat $\longrightarrow$ B   +   gas (P = 209 mm; V = 450 mL; T = 298 K)
        white solid   white solid

2.   Gas (from 1) + Ca(OH)$_2$ (aq) $\longrightarrow$ C (s)
                                                     white solid

3.   Aqueous solution of B is basic (turns red litmus paper blue)

4.   B (aq) + HCl (aq) + heat $\longrightarrow$ D
                                             white solid

5.   Flame test for B: green flame

6.   B (aq) + H$_2$SO$_4$ (aq) $\rightarrow$ E
                                        white solid

Clue 5 indicates that **B** is a barium salt.

Clue 2 suggests that the gas evolved in Clue 1 is CO_2, and that **C** would be $CaCO_3$.

Heating of carbonates liberates CO_2 (g).

Compound **B** is a metal oxide (Clue 3), and probably has the formula BaO.

Clues 3 and 5 suggest the oxide reacts with HCl and H_2SO_4 to form $BaCl_2$ (compound **D**) and $BaSO_4$ (compound **E**) respectively.

Since **B** is most likely BaO, compound **A** must be $BaCO_3$.

One gram of $BaCO_3$ (Molar mass 197) corresponds to 5.06×10^{-3} mol $BaCO_3$.

Compare this amount of substance to the amount of gas liberated when substance **A** is heated. Substitution of data from clue 1 yields:

$$n = \frac{(209 \text{ mmHg}) (0.450 \text{ L})}{(62.4 \frac{\text{L} \cdot \text{mmHg}}{\text{K} \cdot \text{mol}}) (298 \text{ K})} = 5.06 \times 10^{-3} \text{ mol gas}$$

This is the quantity of CO_2 anticipated from the thermal decomposition of $BaCO_3$.

Summary Questions

79.   a. Volume of seawater to obtain 1.00 kg Mg:

$$\frac{1.00 \times 10^3 \text{ g Mg}}{1} \cdot \frac{1 \text{ mol Mg}}{24.31 \text{ g Mg}} \cdot \frac{1 \text{ L seawater}}{0.050 \text{ mol Mg}} = 823 \text{ L seawater}$$

(820 L to 2 sf)

Mass of CaO to precipitate the magnesium:

The precipitation equation may be written as two steps:

1. $CaO(s) + H_2O(\ell) \longrightarrow Ca(OH)_2(s)$

2. $Ca(OH)_2(s) + Mg^{2+}(aq) \longrightarrow Ca^{2+}(aq) + Mg(OH)_2(s)$

The result of these processes is that 1 mole CaO precipitates 1 mole of Magnesium ions.

$$\frac{1.00 \times 10^3 \text{ g Mg}}{1} \cdot \frac{1 \text{ mol Mg}}{24.31 \text{ g Mg}} \cdot \frac{1 \text{ mol CaO}}{1 \text{ mol Mg}} \cdot \frac{56.08 \text{ g CaO}}{1 \text{ mol CaO}} = 2.31 \times 10^3 \text{ g CaO}$$

or 2.31 kg CaO

b. $MgCl_2 (\ell) \xrightarrow{\text{electricity}} Mg (s) + Cl_2 (g)$

Mass of Mg produced at the cathode:

$$1000. \text{ kg } MgCl_2 \cdot \frac{1 \text{ mol } MgCl_2}{95.211 \text{ g } MgCl_2} \cdot \frac{1 \text{ mol } Mg}{1 \text{ mol } MgCl_2} \cdot \frac{24.305 \text{ g } Mg}{1 \text{ mol } Mg} = 255.3 \text{ kg } Mg$$

Note the absence of a conversion of mass of $MgCl_2$ from kg to grams. Since the answer was requested in units of kg, any conversion to units of grams would have necessitated a conversion back to kg at the end of the calculation. The two conversion factors would cancel each other, and "leaving them out" causes no harm to the integrity of the reasoning (or the answer).

At the anode, chlorine is produced. $[2 \text{ Cl}^- \rightarrow \text{Cl}_2 + 2 \text{ e}^-]$
Mass of Cl_2 produced:

$$1000. \text{ kg } MgCl_2 \cdot \frac{1 \text{ mol } MgCl_2}{95.211 \text{ g } MgCl_2} \cdot \frac{1 \text{ mol } Cl_2}{1 \text{ mol } MgCl_2} \cdot \frac{70.906 \text{ g } Cl_2}{1 \text{ mol } Cl_2} = 744.8 \text{ kg } Cl_2$$

Faradays used in the process:
The reduction of magnesium requires 2 Faradays per mole: $Mg^{2+} + 2e^- \rightarrow Mg$

$$255.3 \text{ kg } Mg \cdot \frac{1.000 \times 10^3 \text{ g } Mg}{1.0 \text{ kg } Mg} \cdot \frac{1 \text{ mol } Mg}{24.305 \text{ g } Mg} \cdot \frac{2 \text{ F}}{1 \text{ mol } Mg} = 2.100 \times 10^4 \text{ F}$$

The oxidation of chlorine requires 2 Faradays per mole of chlorine: $2 \text{ Cl}^- \rightarrow \text{Cl}_2 + 2 \text{ e}^-$

$$744.7 \text{ kg } Cl_2 \cdot \frac{1.000 \times 10^3 \text{ g } Cl_2}{1.0 \text{ kg } Cl_2} \cdot \frac{1 \text{ mol } Cl_2}{70.906 \text{ g } Cl_2} \cdot \frac{2 \text{ F}}{1 \text{ mol } Cl_2} = 2.101 \times 10^4 \text{ F}$$

The total number of Faradays of electricity used in the process is 2.101×10^4 F.

c. Joules required per mole of magnesium:

$$\frac{8.4 \text{ kwh}}{1 \text{ lb } Mg} \cdot \frac{3.60 \times 10^6 \text{ J}}{1 \text{ kwh}} \cdot \frac{1 \text{ lb } Mg}{454 \text{ g } Mg} \cdot \frac{24.305 \text{ g } Mg}{1 \text{ mol } Mg} = 1.6 \times 10^6 \frac{\text{J}}{\text{mol } Mg}$$

The reaction, $MgCl_2 (s) \rightarrow Mg (s) + Cl_2 (g)$, represents the reverse of the formation of magnesium chloride from its elements (each in their standard states). From Appendix K, the ΔH for the process is $+641.32$ kJ/mol or 6.4×10^5 J/mol. The difference between this value and the value calculated above may be attributed to the energy required to melt the $MgCl_2$.

Chapter 23:
The Transition Elements

Configurations and Physical Properties

14. Cr^{3+} 3d [↑ | ↑ | ↑ | |]   4s [ ]   paramagnetic

Cr^{6+} 3d [| | | |]   4s [ ]

16. Transition metal ions with the electron configuration
   a. [Ar] $3d^6$ Fe^{2+}, Co^{3+}
   b. [Ar] $3d^{10}$ Cu^{1+}, Zn^{2+}
   c. [Ar] $3d^5$ Mn^{2+}, Fe^{3+}
   d. [Ar] $3d^8$ Ni^{2+}, Cu^{3+}

18. Figure 23.7 illustrates the variation of density with group number. In general the most dense elements are in the middle of any given group, with density also increasing from the nth period to the $n + 1^{st}$ period.

   a. Fe is more dense than Ti (smaller with greater mass)
   b. Os is more dense than Ti (smaller with greater mass)
   c. Zr is more dense than Ti (radius is larger, but mass is greater)
   d. Hf is more dense than Zr (radius is similar, but Hf has greater mass)

Metallurgy

19. Balance:
   a. $Cr_2O_3(s) + 2\ Al(s) \longrightarrow Al_2O_3(s) + 2\ Cr(s)$
   b. $TiCl_4(\ell) + 2\ Mg(s) \longrightarrow Ti(s) + 2\ MgCl_2(s)$
   c. $2\ [Ag(CN)_2]^-(aq) + Zn(s) \longrightarrow 2\ Ag(s) + [Zn(CN)_4]^{2-}(aq)$

21. Volume of 18.0 M H_2SO_4 required

$$1.00 \times 10^3 \text{ g FeTiO}_3 \cdot \frac{1 \text{ mol FeTiO}_3}{151.7 \text{ g FeTiO}_3} \cdot \frac{3 \text{ mol H}_2\text{SO}_4}{1 \text{ mol FeTiO}_3} \cdot \frac{1 \text{ L}}{18.0 \text{ mol H}_2\text{SO}_4} =$$

$$1.10 \text{ L H}_2\text{SO}_4$$

$$1.00 \times 10^3 \text{ g FeTiO}_3 \cdot \frac{1 \text{ mol FeTiO}_3}{151.7 \text{ g FeTiO}_3} \cdot \frac{1 \text{ mol TiO}_2}{1 \text{ mol FeTiO}_3} \cdot \frac{79.88 \text{ g TiO}_2}{1 \text{ mol TiO}_2} = 527 \text{ g TiO}_2$$

$$\text{or } 0.527 \text{ kg TiO}_2$$

23. Mass of NiS required:

$$908 \text{ kg Ni} \cdot \frac{90.76 \text{ kg NiS}}{58.69 \text{ kg Ni}} = 1.40 \times 10^3 \text{ kg NiS}$$

Mass of H_2 required:

$$908 \text{ kg Ni} \cdot \frac{2.016 \text{ kg H}_2}{58.69 \text{ kg Ni}} = 31.2 \text{ kg H}_2$$

Mass of CO required:

$$908 \text{ kg Ni} \cdot \frac{(4 \times 28.01) \text{ kg CO}}{58.69 \text{ kg Ni}} = 1.73 \times 10^3 \text{ kg CO}$$

Liquids and Formulas of Complexes

24. Classify each of the following as monodentate or multidentate:

    a. CH_3NH_2 monodentate (lone pair on N)

    b. $C_2O_4^{2-}$ bidentate (lone pairs on terminal O atoms)

    c. Br^- monodentate (lone pair on Br)

    d. $H_3C-C\equiv N$ monodentate (lone pair on N)

    e. ethylenediamine   bidentate (lone pairs on terminal N atoms)

    f. phenanthroline   bidentate (lone pairs on N atoms)

    g. N_3^- monodentate (lone pair on a N atom)

26.

	Compound	Metal	Oxidation Number
a.	$[Mn(NH_3)_6]SO_4$	Mn	+2
b.	$K_3[Co(CN)_6]$	Co	+3
c.	$[Co(NH_3)_4Cl_2]Cl$	Co	+3
d.	$Mn(en)_2Cl_2$	Mn	+2

28. $[Co(en)_2(H_2O)Cl]^{2+}$

Naming

29. Formulas for
    a. dichlorobis(ethylenediamine)nickel(II) $Ni(en)_2Cl_2$
    b. potassium tetrachloroplatinate(II) $K_2[PtCl_4]$
    c. potassium dicyanocuprate(I) $K[Cu(CN)_2]$
    d. diaquatetraammineiron(II) $[Fe(H_2O)_2(NH_3)_4]^{2+}$

31.
	Ligands	Names
a.	OH^-	hydroxo
b.	O^{2-}	oxo
c.	I^-	iodo
d.	$C_2O_4^{2-}$	oxalato

33.
	Formula	Name
a.	$[Ni(C_2O_4)_2(H_2O)_2]^{2-}$	diaquabis(oxalato)nickelate(II) ion
b.	$[Co(en)_2(NO_2)_2]^+$	bis(ethylenediamine)dinitritocobalt(III) ion
c.	$[Co(en)_2(NH_3)Cl]^{2+}$	amminechlorobis(ethylenediamine)cobalt(III) ion
d.	$Pt(NH_3)_2(C_2O_4)$	diammineoxalatoplatinum(II)

35.
	Formula	Name
a.	$[CrCl_2(H_2O)_4]Cl$	dichlorotetraaquachromium(III) chloride
b.	$[Cr(NH_3)_5SO_4]Cl$	pentaamminesulfatochromium(III) chloride
c.	$Na_2[CoCl_4]$	sodium tetrachlorocobaltate(II)
d.	$[Co(C_2O_4)_3]^{3-}$	tris(oxalato)cobaltate(III) ion

Isomerization

36. Geometric Isomers of

    a. Fe(NH3)2Cl2

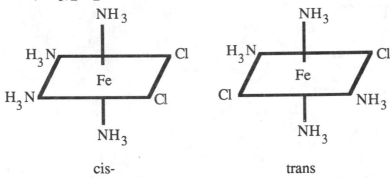

cis-   trans

    b.   Pt(NH3)2(NCS)(Br)

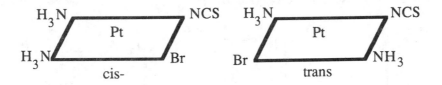

cis-   trans

    c.   Co(NH3)3(NO2)3

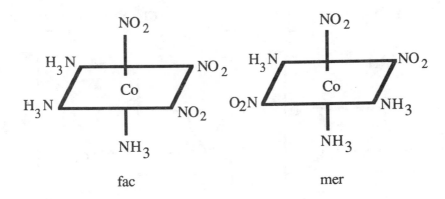

fac   mer

d.

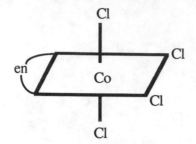

38. Molecules possess a chiral center:

    a. CH_2Cl_2 no

    b. $H_2NCH(CH_3)CO_2H$ yes

    c. $ClCH(OH)CH_2Cl$ yes

    d. $CH_3CH_2CH=CHC_6H_5$ no

40.   The four isomers of $[Co(en)(NH_3)_2(H_2O)Cl]^{2+}$:

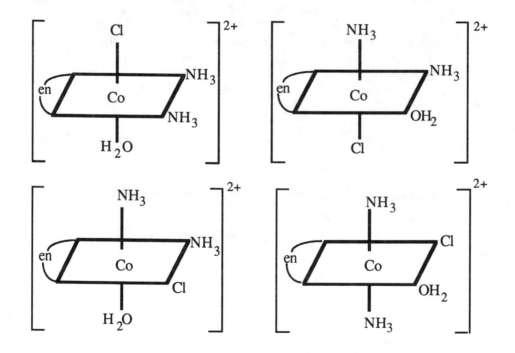

The ligand ethylenediamine is represented in
these drawings by the symbolism
shown to the right:

$$H_2C \overset{H_2}{\underset{H_2}{\overset{N:}{\underset{N:}{\begin{matrix} \\ \\ \end{matrix}}}}} = en$$

42. Chiral center
    a. [Fe(en)₃]²⁺   yes
    b. fac-[Co(en)(H₂O)Cl₃]⁺   no
    c. cis-[Co(en)₂Br₂]⁺   yes
    d. Pt(NH₃)(H₂O)Cl(NO₂)   no

a. $[Fe(en)_3]^{2+}$ yes

b. fac-$[Co(en)(H_2O)Cl_3]^+$ no

c. cis-$[Co(en)_2Br_2]^+$ yes

d. $Pt(NH_3)(H_2O)Cl(NO_2)$ no

Magnetism of Coordination Complexes

43. Only d^4 through d^7 metal ions can exhibit both high and low spin. With fewer than four
electrons, all the electrons would occupy the three "lower energy" orbitals. Metals with d^8
through d^{10} could not place any electrons in one of the lower energy orbitals—since they
would be full.

45.   a. $[Fe(CN)_6]^{4-}$ Fe has a d^6 configuration

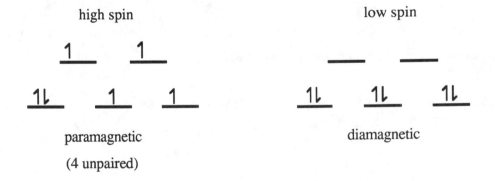

high spin   low spin

paramagnetic   diamagnetic
(4 unpaired)

    b. $[Co(NH_3)_6]^{3+}$ Co^{3+} has a d^6 configuration—like part a above.

c. $[Fe(H_2O)_6]^{3+}$ Fe^{3+} has a d^5 configuration

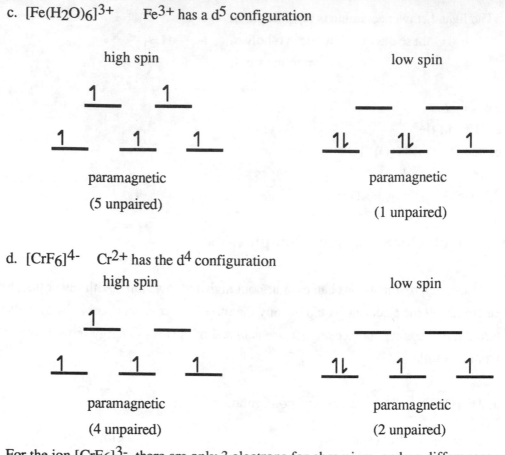

high spin	low spin
paramagnetic	paramagnetic
(5 unpaired)	(1 unpaired)

d. $[CrF_6]^{4-}$ Cr^{2+} has the d^4 configuration

high spin	low spin
paramagnetic	paramagnetic
(4 unpaired)	(2 unpaired)

For the ion $[CrF_6]^{3-}$, there are only 3 electrons for chromium, and no differences would exist between high and low spin complexes.

47. The configuration for Mn^{2+} is a d^5 configuration

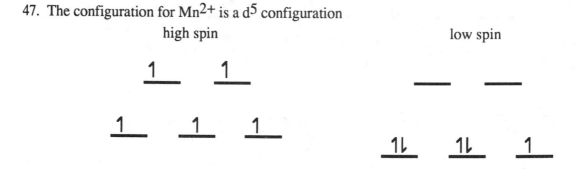

| high spin | low spin |

H_2O is a weak-field ligand, resulting in 5 unpaired electrons. CN^- however is a very strong field ligand, resulting in the pairing of four electrons leaving one electron unpaired. Note that CN^- ligand results in an increased value of Δ_o.

Color

48. Increasing crystal field splitting:

$$F^- < H_2O < NH_3 < CN^-$$

50. If light in the 700 nm wavelength (red) is absorbed, cyan is transmitted.

General Questions

51. For $[Fe(H_2O)_6]^{2+}$

    a. The coordination number of iron is 6.

       6 monodentate ligands are attached.

    b. The coordination geometry is octahedral.

    c. The oxidation state of iron is 2+.

    d. Fe^{2+} is a d^6 case. For a high spin core there are 4 unpaired electrons.

    e. The complex would be paramagnetic.

53.

complex	spin	character	unpaired electrons
a. $[Fe(CN)_6]^{4-}$	low	diamagnetic	
b. $[MnF_6]^{4-}$	high	paramagnetic	5
c. $[Cr(en)_3]^{3+}$	high	paramagnetic	3
d. $[Cu(phen)_3]^{2+}$	either	paramagnetic	1

Electron configuration:

a. Fe^{2+} _____ _____

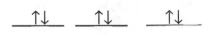

c. Cr^{3+} _____ _____

b. Mn^{2+}

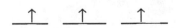

d. Cu^{2+}

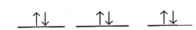

55.   A + $BaCl_2$ ppt ($BaSO_4$) $\Rightarrow$ A = $[Co(NH_3)_5Br]SO_4$

       B + $AgNO_3$ ppt (AgBr) $\Rightarrow$ B = $[Co(NH_3)_5SO_4]Br$

57. $Pt(NO_2)Cl(NH_3)_2$ diamminechloronitritoplatinum(II)

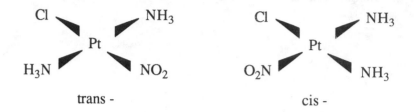

trans -   cis -

These are geometric isomers.

59. Aqueous cobalt(III) sulfate is diamagnetic. H_2O provides a large enough Δ to force $[Co(H_2O)_6]^{3+}$ to have no unpaired electrons. An excess of F^- results in the conversion to the $[CoF_6]^{3-}$ ion. The weaker F^- ligands doesn't separate the d orbital energies as much as H_2O, resulting in four unpaired electrons.

61. Substituting 10^8 and 10^{18} into the expression (-RTlnK), produces ΔG values of -45.6 kJ (ammine) and -102.7 (en) respectively. Since the differences in ΔH values are much less than this [~ 8 kJ], entropy must play a role. While there are fewer molecules in the second reaction, the change in entropy—as the much larger en ligands (compared to the ammine ligands) forms the complex—is greater.

63.   3.8249 g Al_2O_3 $\cdot$ $\dfrac{53.96 \text{ g Al}}{101.96 \text{ g } Al_2O_3}$ = 2.0243 g Al

       Percentage of Al = $\dfrac{2.0243 \text{ g Al}}{2.1309 \text{ g alloy}}$ x 100 = 95.00 %

       Percentage of Cu = 5.00 %

Conceptual Questions

65. Co^{2+} < Ni^{2+} < Cu^{2+} > Zn^{2+} ion

        7.7×10^4 < 5.6×10^8 < 6.8×10^{12} > 2.9×10^9 K_f (ammine)

67. Two geometric isomers of $[Cr(dmen)_3]^{3+}$, a fac and a mer, can exist.

69. The bond angles and lengths are such that the nitrogen atoms (with their lone pairs of electrons) can not span the diagonal of the Pt complex and accomplish reasonable overlap with the Pt orbitals to form the bonds.

Chapter 24:
Nuclear Chemistry

Nuclear Reactions

11. Balance the following nuclear equations, supplying the missing particle.
[The missing particle is emboldened.]

a. $^{54}_{26}\text{Fe} + ^{4}_{2}\text{He} \rightarrow 2\,^{1}_{1}\text{H} + \mathbf{^{56}_{26}\text{Fe}}$

b. $^{27}_{13}\text{Al} + ^{4}_{2}\text{He} \rightarrow ^{30}_{15}\text{P} + \mathbf{^{1}_{0}\text{n}}$

c. $^{32}_{16}\text{S} + ^{1}_{0}\text{n} \rightarrow ^{1}_{1}\text{H} + \mathbf{^{32}_{15}\text{P}}$

d. $^{96}_{42}\text{Mo} + ^{2}_{1}\text{H} \rightarrow ^{1}_{0}\text{n} + \mathbf{^{97}_{43}\text{Tc}}$

e. $^{98}_{42}\text{Mo} + ^{1}_{0}\text{n} \rightarrow ^{99}_{43}\text{Tc} + \mathbf{^{0}_{-1}\text{e}}$

13. Balance the following nuclear equations, supplying the missing particle.

[The missing particle is emboldened.]

a. $^{104}_{47}\text{Ag} \rightarrow ^{104}_{48}\text{Cd} + \mathbf{^{0}_{-1}\text{e}}$

b. $^{87}_{36}\text{Kr} \rightarrow ^{0}_{-1}\text{e} + \mathbf{^{87}_{37}\text{Rb}}$

c. $^{231}_{91}\text{Pa} \rightarrow ^{227}_{89}\text{Ac} + \mathbf{^{4}_{2}\text{He}}$

d. $^{230}_{90}\text{Th} \rightarrow ^{4}_{2}\text{He} + \mathbf{^{226}_{88}\text{Ra}}$

e. $^{82}_{35}\text{Br} \rightarrow ^{82}_{36}\text{Kr} + \mathbf{^{0}_{-1}\text{e}}$

f. $\mathbf{^{24}_{11}\text{Na}} \rightarrow ^{24}_{12}\text{Mg} + ^{0}_{-1}\text{e}$

15. $^{235}_{92}\text{U} \rightarrow ^{231}_{90}\text{Th} \rightarrow ^{231}_{91}\text{Pa} \rightarrow ^{227}_{89}\text{Ac} \rightarrow ^{227}_{90}\text{Th} \rightarrow ^{223}_{88}\text{Ra}$
$$\qquad\quad + \qquad\quad + \qquad\quad + \qquad\quad + \qquad\quad +$$
$$\qquad\quad ^{4}_{2}\text{He} \qquad ^{0}_{-1}\text{e} \qquad ^{4}_{2}\text{He} \qquad ^{0}_{-1}\text{e} \qquad ^{4}_{2}\text{He}$$

17. The change in mass (Δ m) for ^{10}B is:

$$\Delta m = 10.01294 - [5(1.00783) + 5(1.00867)]$$
$$= 10.01294 - 10.0825$$
$$= -0.06956 \text{ g/mol}$$

while that for ^{11}B is:

$$\Delta m = 11.00931 - [5(1.00783) + 6(1.00867)]$$
$$= 11.00931 - 11.09117$$
$$= -0.08186 \text{ g/mol}$$

The binding energies for the two isotopes are:

^{10}B : $\Delta E = (-6.956 \times 10^{-5} \text{ kg/mol})(2.998 \times 10^8 \text{ m/s})^2 \left(\dfrac{1 \text{ J}}{1 \text{ kg} \cdot \text{m}^2 \cdot \text{s}^{-2}} \right)$

$$= -6.252 \times 10^{12} \text{ J/mol}$$

^{11}B : $\Delta E = (-8.186 \times 10^{-5} \text{ kg/mol})(2.998 \times 10^8 \text{ m/s})^2 \left(\dfrac{1 \text{ J}}{1 \text{ kg} \cdot \text{m}^2 \cdot \text{s}^{-2}} \right)$

$$= -7.358 \times 10^{12} \text{ J/mol}$$

The **binding energy per nucleon** is:

For ^{10}B : $\dfrac{6.252 \times 10^9 \text{ kJ}}{10 \text{ mol nucleons}} \cdot \dfrac{1 \text{ mol nucleons}}{6.0223 \times 10^{23} \text{ nucleons}} = 1.038 \times 10^{-12} \dfrac{\text{J}}{\text{nucleon}}$

and

For ^{11}B : $\dfrac{-7.358 \times 10^9 \text{ kJ}}{11 \text{ mol nucleon}} \cdot \dfrac{1 \times 10^3 \text{ J}}{1 \text{ kJ}} = 1.111 \times 10^{-12} \dfrac{\text{J}}{\text{nucleon}}$

Rates of Disintegration Reactions:

19. For ^{64}Cu, $t_{1/2} = 128$ hr

The fraction remaining as ^{64}Cu following n half-lives is equal to $(\frac{1}{2})^n$. Note that 2 days and 16 hrs (64 hrs) corresponds to exactly **five** half-lives.

The <u>fraction</u> remaining as ^{64}Cu is the $(\frac{1}{2})^5$ or $\frac{1}{32}$ or 0.03125.

The mass remaining is:

$$(0.03125)(15.0 \text{ mg}) = 0.469 \text{ mg}.$$

21. a. The equation for β–decay of ^{131}I is:

$$^{131}_{53}\text{I} \rightarrow \ ^{0}_{-1}\text{e} + \ ^{131}_{54}\text{Xe}$$

b. The amount of ^{131}I remaining after 32.2 days:

For ^{131}I, $t_{1/2}$ is 8.05 days--so 32.2 days is exactly **four** half-lives:
The fraction of ^{131}I remaining is $\left(\frac{1}{2}\right)^4$ or $\frac{1}{16}$ or 0.0625.

The amount of the original 25.0 mg remaining will be :

$$(0.0625)(25.0 \text{ mg}) = 1.56 \text{ mg}$$

23. Mass of Cobalt-60 ($t_{1/2}$ = 5.3 yrs) remaining after 21.2 yr:

$$k = \frac{0.693}{t_{1/2}} = \frac{0.693}{5.3 \text{ yr}} = 0.131 \text{ yr}^{-1}$$

$$\ln\left(\frac{N}{N_o}\right) = -kt \qquad \text{where} \quad N = \text{amt. remaining after time t and}$$
$$N_o = \text{initial amount}$$

$$\ln\left(\frac{N}{N_o}\right) = -(0.131 \text{ yr}^{-1})(21.2 \text{ yr})$$

$$\ln\left(\frac{N}{10.0 \text{ mg}}\right) = -2.77$$

$$\frac{N}{10.0 \text{ mg}} = 0.0625$$

$$N = 0.625 \text{ mg}$$

Mass remaining after 1 century:

$$\ln\left(\frac{N}{10.0 \text{ mg}}\right) = -(0.131 \text{ yr}^{-1})(100 \text{ yr})$$

and $N = 2 \times 10^{-5}$ mg

25. For a sample of NaI containing ^{131}I, the time required for activity to fall to 5.0 % of original activity:

This isotope has a half-life of 8.05 days. We begin by calculating the rate constant, k.

$$k = \frac{0.693}{t_{1/2}} = \frac{0.693}{8.05 \text{ days}} = 0.0861 \text{ days}^{-1}$$

The time required for the sample to have only 5% of the original activity is then:

$$\ln\left(\frac{N}{N_o}\right) = -(0.0861 \text{ days}^{-1})\, t$$

$$\ln\left(\frac{5.00}{100.}\right) = -(0.0861 \text{ days}^{-1})\, t$$

$$35 \text{ days} = t$$

27. Using the activity in disintegrations per minute per gram compared to the activity of ^{14}C in living material, we can calculate the elapsed time since the death of the tree, if we first calculate the rate constant, k:

$$k = \frac{0.693}{t_{1/2}} = \frac{0.693}{5.73 \times 10^3 \text{ yr}} = 1.21 \times 10^{-4} \text{ yr}^{-1}$$

$$\ln\left(\frac{11.2 \text{ dis/min} \cdot g}{14.0 \text{ dis/min} \cdot g}\right) = -1.21 \times 10^{-4} \text{ yr}^{-1} \cdot t$$

Solving for t yields approximately **1850 years**.

Subtracting the current year from 1850 gives an approximate date of 145 AD.

Nuclear Transmutations

29. The formation of ^{241}Am from ^{239}Pu may be written:

$$^{239}_{94}Pu + 2\,^{1}_{0}n \rightarrow\ ^{241}_{95}Am + \,^{0}_{-1}e$$

31. A proposed method for producing ^{246}Cf:

$$^{238}_{92}U + \,^{12}_{6}C \rightarrow\ ^{246}_{98}Cf + 4\,^{1}_{0}n$$

Nuclear Fission and Power

33. The energy liberated by one pound of ^{235}U:

$$\frac{2.1 \times 10^{10} \text{ kJ}}{1 \text{ mol } ^{235}U} \cdot \frac{1 \text{ mol } ^{235}U}{235 \text{ g U}} \cdot \frac{454 \text{ g U}}{1 \text{ pound U}} = \frac{4.1 \times 10^{10} \text{ kJ}}{1 \text{ pound U}}$$

comparing this amount of energy to the energy per ton of coal yields:

$$\frac{4.1 \times 10^{10} \text{ kJ}}{1 \text{ pound U}} \cdot \frac{1 \text{ ton coal}}{2.6 \times 10^7 \text{ kJ}} = \frac{1600 \text{ tons coal}}{1 \text{ pound U}}$$

Uses of Radioisotopes

35. This "dilution" problem may be solved using the equation which is useful for solutions:

$$M_c \times V_c = M_d \times V_d$$

where c and d represent the concentrated and diluted states, respectively. We'll use the number of disintegrations/second as our "molarity."

$$2.0 \times 10^6 \text{ dps} \cdot 1.0 \text{ mL} = 1.5 \times 10^4 \text{ dps} \cdot V_d$$

$$130 \text{ mL} = V_d$$

The approximate volume of the circulatory system is 130 mL.

General Questions

37. Balance the following nuclear reactions, supplying the missing particle.

[The missing particle is emboldened.]

a. $^{13}_{6}C + \mathbf{^{1}_{0}n} \rightarrow ^{14}_{6}C$

e. $^{212}_{84}Po \rightarrow ^{208}_{82}Pb + \mathbf{^{4}_{2}He}$

b. $^{40}_{18}Ar + \mathbf{^{4}_{2}He} \rightarrow ^{43}_{19}K + ^{1}_{1}H$

f. $^{122}_{53}I \rightarrow \mathbf{^{122}_{52}Te} + ^{0}_{1}e$

c. $^{250}_{98}Cf + ^{11}_{5}B \rightarrow 4\ ^{1}_{0}n + \mathbf{^{257}_{103}Lr}$

g. $\mathbf{^{23}_{10}Ne} \rightarrow ^{23}_{11}Na + ^{0}_{-1}e$

d. $^{53}_{24}Cr + ^{4}_{2}He \rightarrow \mathbf{^{1}_{0}n} + ^{56}_{26}Fe$

h. $^{137}_{53}I \rightarrow ^{1}_{0}n + \mathbf{^{136}_{53}I}$

39. The fossil cells age can be calculated using the first order kinetics expressions:

Since $t_{1/2} = 4.9 \times 10^{10}$ yr, $k = \dfrac{0.693}{4.9 \times 10^{10} \text{ yr}}$ and $\ln\left(\dfrac{C_t}{C_o}\right) = -kt$

then $\ln(0.951) = -\dfrac{0.693}{4.9 \times 10^{10} \text{ yr}}$ t and

$\dfrac{4.9 \times 10^{10} \text{ yr}}{0.693} \cdot \ln(0.951) = t$ and 3.6×10^9 yr $= t$

The fossil cells are about 3.6 billion years old.

41. The time when natural uranium contained 3.0% ^{235}U is:

Since $t_{1/2} = 7.04 \times 10^8$ yr, $k = \dfrac{0.693}{7.04 \times 10^8 \text{ yr}}$ and $\ln\left(\dfrac{C_t}{C_o}\right) = -kt$

then $\ln\left(\dfrac{0.72}{3.0}\right) = -\dfrac{0.693}{7.04 \times 10^8 \text{ yr}} \cdot t$ and

$\dfrac{7.04 \times 10^8 \text{ yr}}{0.693} \cdot \ln\left(\dfrac{0.72}{3.0}\right) = t$ and $t = 1.5 \times 10^9$ years